Fractals in Science

Airways of the lung. The bronchi and bronchioles of the lung form a "tree" that has multiple generations of branchings. The small-scale branching of the airways look like branching at larger scales. Courtesy of Christopher Burke, Quesada/Burke Studios, New York.

Armin Bunde   Shlomo Havlin (Eds.)

Fractals in Science

With a Macintosh Program Diskette,
120 Figures and 10 Color Plates

Springer-Verlag
Berlin Heidelberg New York
London Paris Tokyo
Hong Kong Barcelona
Budapest

Professor Dr. Armin Bunde

Institut für Theoretische Physik
Universität Giessen
Heinrich-Buff-Ring 16
D-35392 Giessen
Germany

Professor Dr. Shlomo Havlin

Department of Physics
Bar-Ilan University
Ramat Gan
Israel

First Edition 1994
Second Printing 1995

ISBN 3-540-56221-4 Springer-Verlag Berlin Heidelberg New York
ISBN 0-387-56221-4 Springer-Verlag New York Berlin Heidelberg

CIP data applied for.

Camera-ready copy from the authors/editors using a Springer TeX macro package
Production Editor: P. Treiber
SPIN: 10489206   56/3144-543210 - Printed on acid-free paper

Preface

Applying fractal geometry to science is bringing about a breakthrough in our understanding of complex systems in nature that show self-similar or self-affine features. Self-similar and self-affine processes appear everywhere in nature, in galaxies and landscapes, in earthquakes and geological cracks, in aggregates and colloids, in rough surfaces and interfaces, in glassy materials and polymers, in proteins as well as in other large molecules. Fractal structures appear also in the human body; well known examples include the lung and the vascular system. Furthermore, fractal geometry is an important tool in the analysis of phenomena as diverse as rhythms in music melodies and in the human heartbeat and DNA sequences.

Since the pioneering work of B.B. Mandelbrot, this interdisciplinary field has expanded very rapidly. The scientific community applying fractal concepts is very broad and ranges from astronomers, geoscientists, physicists, chemists and engineers to biologists and those engaging in medical research.

The purpose of this book is to provide easy access to fractals in science and to bridge the gap between the different disciplines. Similar in style to the previous book *Fractals and Disordered Systems* in which the main emphasis was on fractals in materials science, all chapters are written in a uniform notation, and cross-references in each chapter to related subjects in other chapters are provided. In each chapter emphasis is placed on the various connections between theory and experiment. A special chapter (Chap. 9) entitled "Computer Exploration of Fractals, Chaos, and Cooperativity" presents computer demonstrations of fractal models. A diskette of these interactive programs, for either Macintosh or PC-compatible computers, is enclosed.

The first chapter, by A. Bunde and S. Havlin, is for beginners in the field and serves to introduce the basic ideas and concepts of fractal geometry. The second chapter, by P. Bak and M. Creutz, deals with self-organized criticality, a process that may explain why fractals occur so widely in nature. In the third chapter, S.V. Buldyrev, A.L. Goldberger, S. Havlin, C.-K. Peng, and H.E. Stanley describe fractal processes in biology and medicine with particular emphasis on novel applications of fractal landscape analysis to DNA sequences and cardiac rhythms. In Chap. 4, J. Kertész and T. Vicsek present an introduction to the new and fascinating field of self-affine fractal surfaces generated by natural processes like fractures, erosion, imbibition, and burning. In Chap. 5, G. H. Weiss introduces the reader to the theory of diffusion and random walks, which represent the basic mechanisms for disorder in nature, and describes several applications to disordered media, semiconductors, and ecology. M. Daoud reviews, in Chap. 6, fractal applications to polymer science, with emphasis on single polymer chains, polymer solutions, melts, branched polymers, and gels. Chapter 7, by S. Redner and F. Leyvraz, introduces the reader to the recent developments in chemical reactions controlled by diffusion, a study relevant to a wide range of processes including electron-hole recombination in semiconductors, and catalytic reactions. In Chap. 8, D. Avnir, R. Gutfraind and D. Farin discuss the use of fractal analysis in heterogeneous chemistry and demonstrate the importance of fractal geometry to relevant chemical processes, including the fundamental pharmacological problem of controlled drug release. In Chap. 9, D. Rapaport and M. Meyer present interactive computer demonstrations of basic fractal models.

We wish to thank first and foremost the authors, and also our colleagues H. Bolterauer, H. Brender, L. Lam, R. Nossal, S. Rabinovich, H. E. Roman, and H. Taitelbaum for useful discussions. We kindly acknowledge the help of S.V. Buldyrev, S. Glotzer, S. Harrington, M. Meyer, M. Sernetz, and P. Trunfio, who contributed the color figures. We also wish to thank H.J. Kölsch and P. Treiber from Springer-Verlag Heidelberg for their continuous help during the preparation of this book. We hope that *Fractals in Science* can be used as a textbook for graduate students, for teachers at universities preparing courses or seminars and for researchers in a variety of fields who are about to encounter fractals in their own work.

Armin Bunde   Giessen,
Shlomo Havlin   Ramat-Gan,
                                                              February 1994

Contents

1 A Brief Introduction to Fractal Geometry

By A. Bunde and S. Havlin (With 22 Figures)

2 Fractals and Self-Organized Criticality

By P. Bak and M. Creutz (With 10 Figures)

3   Fractals in Biology and Medicine:
From DNA to the Heartbeat

By S.V. Buldyrev, A.L. Goldberger, S. Havlin, C.-K. Peng,
and H.E. Stanley (With 17 Figures)

4 Self-Affine Interfaces

By J. Kertész and T. Vicsek (With 10 Figures)

5 A Primer of Random Walkology

By G.H. Weiss (With 11 Figures)

6   Polymers

By M. Daoud (With 14 Figures)

7   Kinetics and Spatial Organization
of Competitive Reactions

By S. Redner and F. Leyvraz (With 6 Figures)

8   Fractal Analysis in Heterogeneous Chemistry

By D. Avnir, R. Gutfraind, and D. Farin (With 12 Figures)

9   Computer Exploration of Fractals, Chaos, and Cooperativity

By Dennis C. Rapaport and Martin Meyer (With 10 Figures)

XII   Contents

List of Contributors

David Avnir

Department of Organic Chemistry, The Hebrew University of Jerusalem
Jerusalem 91904, Israel

Per Bak

Department of Physics, Brookhaven National Laboratory
Upton, NY 11973, USA

Sergey V. Buldyrev

Center for Polymer Studies, Boston University
Boston, MA 02215, USA

Armin Bunde

Institut für Theoretische Physik, Justus-Liebig-Universität
D-35392 Giessen, Germany

Michael Creutz

Department of Physics, Brookhaven National Laboratory
Upton, NY 11973, USA

Mohamed Daoud

Service de Physique Théorique de Saclay
F-91191 Gif-Sur-Yvette Cedex, France

Dina Farin

Department of Organic Chemistry, The Hebrew University of Jerusalem
Jerusalem 91904, Israel

Ary L. Goldberger

Cardiovascular Division, Harvard Medical School
Beth Israel Hospital, Boston, MA 02215, USA

Ricardo Gutfraind

Department of Organic Chemistry, The Hebrew University of Jerusalem
Jerusalem 91904, Israel

Shlomo Havlin

Department of Physics, Bar-Ilan University
Ramat-Gan 52100, Israel

János Kertész

Technical University of Budapest, Budafoki út 8
H-1521 Budapest, Hungary

Francois Leyvraz

Instituto de Fisica, Laboratorio de Cuernavaca
UNAM, Mexico

Martin Meyer

Institut für Theoretische Physik, Justus-Liebig-Universität
D-35392 Giessen, Germany

C.-K. Peng

Center for Polymer Studies, Boston University
Boston, MA 02215, USA

Dennis C. Rapaport

Department of Physics, Bar-Ilan University
Ramat-Gan 52100, Israel

Sidney Redner

Center for Polymer Studies, Boston University
Boston, MA 02215, USA

H. Eugene Stanley

Center for Polymer Studies, Boston University
Boston, MA 02215, USA

Tamás Vicsek

Eötvös University, Department of Atomic Physics
Puskin u. 5–7, 1088 Budapest, Hungary

George H. Weiss

Physical Sciences Laboratory, Division of Computer Research and Technology
National Institutes of Health, Bethesda, MD 20205, USA

1 A Brief Introduction to Fractal Geometry

Armin Bunde and Shlomo Havlin

1.1 Introduction

In this chapter we present some definitions related to the fractal concept as well as several methods for calculating the fractal dimension. The purpose is to introduce the reader to the basic properties of fractals so that this book will be self contained. Because of space constraints, we do not give references to most of the original works. We refer mostly to books and reviews on fractal geometry where the original references can be found.

Fractal geometry is a mathematical tool for dealing with complex systems that have no characteristic length scale. A well known example is the shape of a coastline. When we see two pictures of a coastline on two different scales, with 1 cm corresponding for example to 0.1 km or 10 km, we cannot tell which scale belongs to which picture: both look the same. This means that the coastline is scale invariant or, equivalently, has no characteristic length scale. Other examples in nature are rivers, cracks, mountains, and clouds.

Scale-invariant systems are usually characterized by noninteger ("fractal") dimensions. The notion of noninteger dimensions and several basic properties of fractal objects were studied as long ago as the last century by Georg Cantor, Giuseppe Peano, and David Hilbert, and in the beginning of this century by Helge von Koch, Waclaw Sierpinski, Gaston Julia, and Felix Hausdorff. Even earlier traces of this concept can be found in the study of arithmetic - geometric averages by Carl Friedrich Gauss about 200 years ago and in the artwork of Albrecht Dürer (see Fig. 1.0) about 500 years ago. (For literature on the history of fractals, see for example [1.1–3]). Benoit Mandelbrot [1.1] showed the relevance of fractal geometry to many systems in nature and presented many important features of fractals. For recent books and reviews on fractals see [1.4–15].

◄ **Fig. 1.0.** The Dürer pentagon after five iterations. For the generating rule, see Fig. 1.7. The Dürer pentagon is in *blue*, its external perimeter is in *red*. Courtesy of M. Meyer

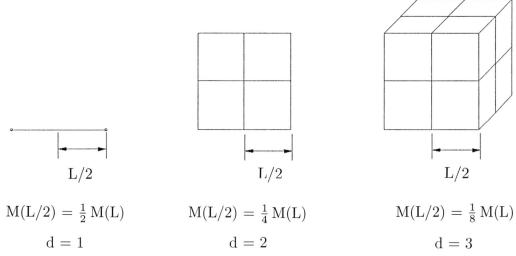

$$M(L/2) = \tfrac{1}{2}M(L) \qquad M(L/2) = \tfrac{1}{4}M(L) \qquad M(L/2) = \tfrac{1}{8}M(L)$$

$$d = 1 \qquad\qquad\qquad d = 2 \qquad\qquad\qquad d = 3$$

Fig. 1.1. Examples of regular systems with dimensions $d = 1$, $d = 2$, and $d = 3$

Before introducing the concept of fractal dimension, we should like to remind the reader of the concept of dimension in regular systems. It is well known that in regular systems (with uniform density) such as long wires, large thin plates, or large filled cubes, the dimension d characterizes how the mass $M(L)$ changes with the linear size L of the system. If we consider a smaller part of the system of linear size bL ($b < 1$), then $M(bL)$ is decreased by a factor of b^d, i.e.,

$$M(bL) = b^d M(L). \tag{1.1}$$

The solution of the functional equation (1.1) is simply $M(L) = AL^d$. For the long wire the mass changes linearly with b, i.e., $d = 1$. For the thin plates we obtain $d = 2$, and for the cubes $d = 3$; see Fig. 1.1.

Next we consider fractal objects. Here we distinguish between deterministic and random fractals. Deterministic fractals are generated iteratively in a deterministic way, while random fractals are generated using a stochastic process. Although fractal structures in nature are random, it is useful to study deterministic fractals where the fractal properties can be determined exactly. By studying deterministic fractals one can gain also insight into the fractal properties of random fractals, which usually cannot be treated rigorously.

1.2 Deterministic Fractals

In this section, we describe several examples of deterministic fractals and use them to introduce useful fractal concepts such as fractal and chemical dimension, self similarity, ramification, and fractal substructures (minimum path, external perimeter, backbone, and red bonds).

1.2.1 The Koch Curve

One of the most common deterministic fractals is the Koch curve. Figure 1.2 shows the first $n = 3$ iterations of this fractal curve. By each iteration the length of the curve is increased by a factor of 4/3. The mathematical fractal is defined in the limit of infinite iterations, $n \to \infty$, where the total length of the curve approaches infinity.

   The dimension of the curve can be obtained as for regular objects. From Fig. 1.2 we notice that, if we decrease the linear size by a factor of $b = 1/3$, the total length (mass) of the curve decreases by a factor of $1/4$, i.e.,

$$M(\frac{1}{3}L) = \frac{1}{4}M(L). \tag{1.2}$$

This feature is very different from regular curves, where the length of the object decreases proportional to the linear scale. In order to satisfy (1.1) and (1.2) we are led to introduce a *noninteger* dimension, satisfying $1/4 = (1/3)^d$, i.e., $d = \log 4/\log 3$. For such non integer dimensions Mandelbrot coined the name "fractal dimension" and those objects described by a fractal dimension are called fractals. Thus, to include fractal structures, (1.1) is generalized by

$$M(bL) = b^{d_f}M(L), \tag{1.3}$$

and

$$M(L) = AL^{d_f}, \tag{1.4}$$

where d_f is the fractal dimension.

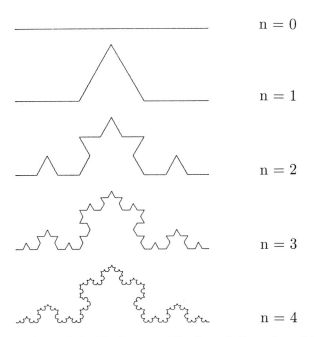

Fig. 1.2. The first iterations of the Koch curve. The fractal dimension of the Koch curve is $d_f = \log 4/\log 3$

When generating the Koch curve and calculating d_f, we observe the striking property of fractals – the property of *self-similarity*. If we examine the Koch curve, we notice that there is a central object in the figure that is reminiscent of a snowman. To the right and left of this central snowman there are two other snowmen, each being an exact reproduction, only smaller by a factor of $1/3$. Each of the smaller snowmen has again still smaller copies (by $1/3$) of itself to the right and to the left, etc. Now, if we take any such triplet of snowmen (consisting of $1/3^m$ of the curve), for any m, and magnify it by 3^m, we will obtain exactly the original Koch curve. This property of self-similarity or scale invariance is the basic feature of all deterministic and random fractals: if we take a part of a fractal and magnify it by the same magnification factor in all directions, the magnified picture cannot be distinguished from the original. The self-similarity of a random fractal is demonstrated in Fig. 8.0.

For the Koch curve as well as for all deterministic fractals generated iteratively, (1.3) and (1.4) are of course valid only for length scales L below the linear size L_0 of the curve (see Fig. (1.2)). If the number of iterations n is finite, then (1.3) and (1.4) are valid only above a lower cut off length $L_{\min}$, $L_{\min} = L_0/3^n$ for the Koch curve. Hence, for a finite number of iterations, there exist two cut-off length scales in the system, an upper cut-off $L_{\max} = L_0$ representing the total linear size of the fractal, and a lower cut-off $L_{\min}$. This feature of having two characteristic cut-off lengths is shared by all fractals in nature.

An interesting modification of the Koch curve is shown in Fig. 1.3, which demonstrates that the *chemical distance* is an important concept for describing structural properties of fractals (for a review see, for example, [1.15] and Chap. 2 in [1.13]. The chemical distance ℓ is defined as shortest path on the fractal between two sites of the fractal. In analogy to the fractal dimension d_f that

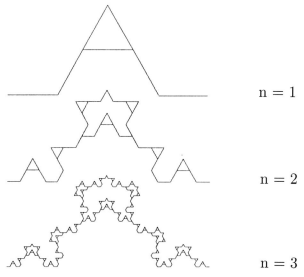

$n = 1$

$n = 2$

$n = 3$

Fig. 1.3. The first iterations of a modified Koch curve, which has a nontrivial chemical distance metric

characterizes how the mass of a fractal scales with (air) distance L, we introduce the *chemical dimension* d_ℓ in order to characterize how the mass scales with the chemical distance ℓ,

$$M(b\ell) = b^{d_\ell} M(\ell), \qquad \text{or} \qquad M(\ell) = B\ell^{d_\ell}. \qquad (1.5)$$

From Fig. 1.3 we see that if we reduce ℓ by a factor of 5, the mass of the fractal within the reduced chemical distance is reduced by a factor of 7, i.e., $M(\frac{1}{5}\ell) = \frac{1}{7}M(\ell)$, yielding $d_\ell = \log 7 / \log 5 \cong 1.209$. Note that the chemical dimension is smaller than the fractal dimension $d_f = \log 7 / \log 4 \cong 1.404$, which follows from $M(\frac{1}{4}L) = \frac{1}{7}M(L)$.

The structure of the shortest path between two sites represents an interesting fractal by itself. By definition, the length of the path is the chemical distance ℓ, and the *fractal dimension of the shortest path*, $d_{\min}$, characterizes how ℓ scales with (air) distance L. Using (1.4) and (1.5) we obtain

$$\ell \sim L^{d_f / d_\ell} \equiv L^{d_{\min}}, \qquad (1.6)$$

from which follows $d_{\min} = d_f / d_\ell$. For our example we find that $d_{\min} = \log 5 / \log 4 \cong 1.161$. For the Koch curve, as well as for any linear fractal, one simply has $d_\ell = 1$ and hence $d_{\min} = d_f$. Since, by definition, $d_{\min} \geq 1$, it follows that $d_\ell \leq d_f$ for all fractals.

1.2.2 The Sierpinski Gasket, Carpet, and Sponge

Next we discuss the Sierpinski fractal family: the "gasket", the "carpet", and the "sponge".

a) **The Sierpinski gasket.** The Sierpinski gasket is generated by dividing a full triangle into four smaller triangles and removing the central triangle (see Fig. 1.4). In the following iterations, this procedure is repeated by dividing each of the remaining triangles into four smaller triangles and removing the central triangles.

To obtain the fractal dimension, we consider the mass of the gasket within a linear size L and compare it with the mass within $\frac{1}{2}L$. Since $M(\frac{1}{2}L) = \frac{1}{3}M(L)$, we have $d_f = \log 3 / \log 2 \cong 1.585$. It is easy to see that $d_\ell = d_f$ and $d_{\min} = 1$.

b) **The Sierpinski carpet.** The Sierpinski carpet is generated in close analogy to the Sierpinski gasket.

Instead of starting with a full triangle, we start with a full square, which we divide into n^2 equal squares. Out of these squares we choose k squares and remove them. In the next iteration, we repeat this procedure by dividing each of the small squares left into n^2 smaller squares and removing those k squares that are located at the same positions as in the first iteration. This procedure is repeated again and again.

Figure 1.5 shows the Sierpinski carpet for $n = 5$ and the specific choice of $k = 9$. It is clear that the k squares can be chosen in many different ways,

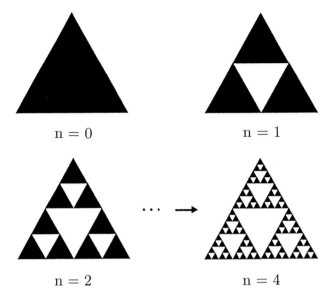

$n = 0$ $n = 1$

$n = 2$ $\cdots \rightarrow$ $n = 4$

Fig. 1.4. The Sierpinski gasket. The fractal dimension of the Sierpinski gasket is $d_f = \log 3/\log 2$

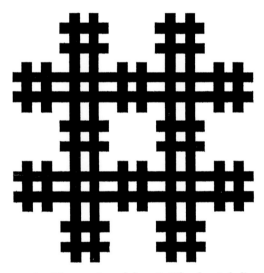

Fig. 1.5. A Sierpinski carpet with $n = 5$ and $k = 9$. The fractal dimension of this structure is $d_f = \log 16/\log 5$

and the fractal structures will all look very different. However, since $M(\frac{1}{n}L) = \frac{1}{n^2-k}M(L)$ it follows that $d_f = \log(n^2 - k)/\log n$, irrespective of the way the k squares are chosen. Similarly to the gasket, we have $d_\ell = d_f$ and hence $d_{\min} = 1$.

In contrast, the *external perimeter* ("hull", see also Fig. 1.0) of the carpet and its fractal dimension d_h depend strongly on the way the squares are chosen. The hull consists of those sites of the cluster, which are adjacent to empty sites and are connected with infinity via empty sites. In our example, Fig. 1.5, the hull is a fractal with the fractal dimension $d_h = \log 9/\log 5 \cong 1.365$. On the

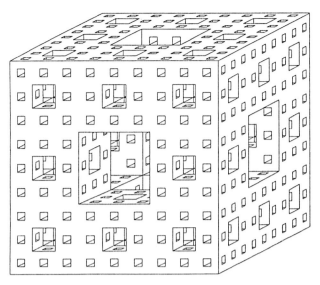

Fig. 1.6. The Sierpinski sponge (third iteration). The fractal dimension of the Sierpinski sponge is $d_f = \log 20/\log 3$

other hand, if a Sierpinski gasket is constructed with the $k = 9$ squares chosen from the center, the external perimeter stays smooth and $d_h = 1$.

Although the rules for generating the Sierpinski gasket and carpet are quite similar, the resulting fractal structures belong to two different classes, to *finitely ramified* and *infinitely ramified* fractals. A fractal is called finitely ramified if any bounded subset of the fractal can be isolated by cutting a *finite* number of bonds or sites. The Sierpinski gasket and the Koch curve are finitely ramified, while the Sierpinski carpet is infinitely ramified. For finitely ramified fractals like the Sierpinski gasket many physical properties, such as conductivity and vibrational excitations, can be calculated exactly. These exact solutions help to provide insight onto the anomalous behavior of physical properties on fractals, as was shown in Chap. 3 in [1.13].

c) The Sierpinski sponge. The Sierpinski sponge shown in Fig. 1.6 is constructed by starting from a cube, subdividing it into $3 \times 3 \times 3 = 27$ smaller cubes, and taking out the central small cube and its six nearest neighbor cubes. Each of the remaining 20 small cubes is processed in the same way, and the whole procedure is iterated ad infinitum. After each iteration, the volume of the sponge is reduced by a factor of $\frac{20}{27}$, while the total surface area increases. In the limit of infinite iterations, the surface area is infinite, while the volume vanishes. Since $M(\frac{1}{3}L) = \frac{1}{20}M(L)$, the fractal dimension is $d_f = \log 20/\log 3 \cong 2.727$. We leave it to the reader to prove that both the fractal dimension d_h of the external surface and the chemical dimension d_ℓ is the same as the fractal dimension d_f.

Modification of the Sierpinski sponge, in analogy to the modifications of the carpet can lead to fractals, where the fractal dimension of the hull, d_h, differs from d_f.

1.2.3 The Dürer Pentagon

Five-hundred years ago the artist Albrecht Dürer designed a fractal based on regular pentagons, where in each iteration each pentagon is divided into six smaller pentagons and five isosceles triangles, and the triangles are removed (see Fig. 1.7). In each triangle, the ratio of the larger side to the smaller side is the famous *proportio divina* or golden ratio, $g \equiv 1/(2\cos 72^o) \equiv (1+\sqrt{5})/2$. Hence, in each iteration the sides of the pentagons are reduced by $1+g$. Since $M(\frac{L}{1+g}) = \frac{1}{6}M(L)$, the fractal dimension of the Dürer pentagon is $d_f = \log 6/\log(1+g) \cong 1.862$. The external perimeter of the fractal (see Fig. 1.0) forms a fractal curve with $d_h = \log 4/\log(1+g)$.

A nice modification of the Dürer pentagon is a fractal based on regular hexagons, where in each iteration one hexagon is divided into six smaller hexagons, six equilateral triangles, and a David-star in the center, and the triangles and the David-star are removed (see Fig. 1.8). We leave it as an exercise to the reader to show that $d_f = \log 6/\log 3$ and $d_h = \log 4/\log 3$.

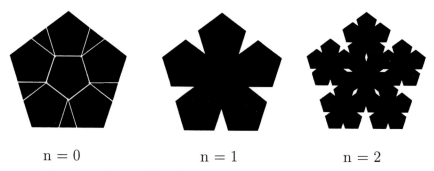

n = 0   n = 1   n = 2

Fig. 1.7. The first iterations of the Dürer pentagon. The fractal dimension of the Dürer pentagon is $d_f = \log 6/\log(1+g)$

1.2.4 The Cantor Set

Cantor sets are examples of disconnected fractals (*fractal dust*). The simplest set is the triadic Cantor set (see Fig. 1.9). We divide a unit interval [0,1] into three equal intervals and remove the central one. In each following iteration, each of the remaining intervals is treated in this way. In the limit of $n = \infty$ iterations one obtains a set of points. Since $M(\frac{1}{3}L) = \frac{1}{2}M(L)$, we have $d_f = \log 2/\log 3 \cong 0.631$, which is smaller than one.

In chaotic systems, strange fractal attractors occur. The simplest strange attractor is the Cantor set. It occurs, for example, when considering the one-dimensional *logistic map*

$$x_{t+1} = \lambda x_t(1 - x_t). \tag{1.7}$$

The index $t = 0, 1, 2, \ldots$ represents a discrete time. For $0 \leq \lambda \leq 4$ and x_0 between 0 and 1, the trajectories x_t are bounded between 0 and 1. The dynamical behavior of x_t for $t \to \infty$ depends on the parameter λ. Below $\lambda_1 = 3$, only one

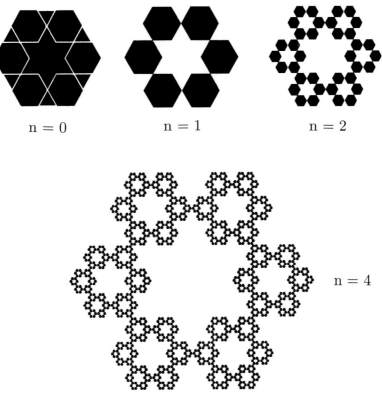

n = 0   n = 1   n = 2

n = 4

Fig. 1.8. The first iterations of the David fractal. The fractal dimension of the David fractal is $d_f = \log 6 / \log 3$

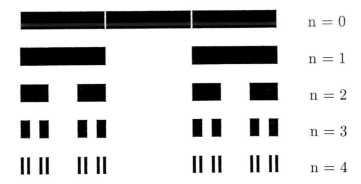

n = 0

n = 1

n = 2

n = 3

n = 4

Fig. 1.9. The first iterations of the triadic Cantor set. The fractal dimension of this Cantor set is $d_f = \log 2 / \log 3$

stable fixed-point exists to which x_t is attracted. At λ_1, this fixed-point becomes unstable and bifurcates into two new stable fixed-points. At large times, the trajectories move alternately between both fixed-points, and the motion is periodic with period 2. At $\lambda_2 = 1 + \sqrt{6} \cong 3.449$ each of the two fixed-points bifurcates into two new stable fix points and the motion becomes periodic with period 4. As λ is increased, further bifurcation points λ_n occur, with periods of 2^n between λ_n and λ_{n+1} (see Fig. 9.6).

For large n, the differences between λ_{n+1} and λ_n become smaller and smaller, according to the law $\lambda_{n+1} - \lambda_n = (\lambda_n - \lambda_{n-1})/\delta$, where $\delta \cong 4.6692$ is the so-called Feigenbaum constant. The Feigenbaum constant is "universal", since it applies to all nonlinear "single-hump" maps with a quadratic maximum [1.16].

At $\lambda_\infty \cong 3.569\,945\,6$, an infinite period occurs, where the trajectories x_t move in a "chaotic" way between the infinite attractor points. These attractor points define the strange attractor, which forms a Cantor set with a fractal dimension $d_f \cong 0.538$ [1.17]. For a further discussion of strange attractors and chaotic dynamics we refer to [1.3,8,9].

1.2.5 The Mandelbrot-Given Fractal

This fractal was suggested as a model for percolation clusters and its substructures (see Sect. 1.3.4 and Chap. 2 in [1.13]). Figure 1.10 shows the first three generations of the Mandelbrot-Given fractal [1.18] At each generation, each segment of length a is replaced by 8 segments of length $a/3$. Accordingly, the fractal dimension is $d_f = \log 8/\log 3 \cong 1.893$, which is very close to $d_f = 91/46 \cong 1.896$ for percolation in two dimensions. It is easy to verify that $d_\ell = d_f$, and therefore $d_{\min} = 1$. The structure contains loops, branches, and dangling ends of all length scales.

Fig. 1.10. Three generations of the Mandelbrot-Given fractal. The fractal dimension of the Mandelbrot-Given fractal is $d_f = \log 8/\log 3$

Imagine applying a voltage difference between two sites at opposite edges of a metallic Mandelbrot-Given fractal: the *backbone* of the fractal consists of those bonds which carry the electric current. The *dangling ends* are those parts of the cluster which carry no current and are connected to the backbone by a single bond only. The *red bonds* (or singly connected bonds) are those bonds that carry the total current; when they are cut the current flow stops. The

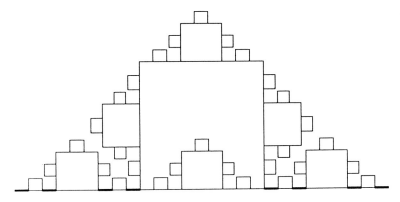

Fig. 1.11. The backbone of the Mandelbrot-Given fractal, with the red bonds shown in *bold*

blobs, finally, are those parts of the backbone that remain after the red bonds have been removed.

The backbone of this fractal can be obtained easily by eliminating the dangling ends when generating the fractal (see Fig 1.11). It is easy to see that the fractal dimension of the backbone is $d_B = \log 6/\log 3 \cong 1.63$. The red bonds are all located along the x axis of the figure and form a Cantor set with the fractal dimension $d_{\text{red}} = \log 2/\log 3 \cong 0.63$.

1.2.6 Julia Sets and the Mandelbrot Set

A complex version of the logistic map (1.7) is

$$z_{t+1} = z_t^2 + c, \tag{1.8}$$

where both the trajectories z_t and the constant c are complex numbers. The question is: if a certain c-value is given, for example $c = -1.5652 - i1.03225$, for which initial values z_0 are the trajectories z_t bounded? The set of those values forms the *filled-in Julia set*, and the boundary points of them form the Julia set.

To clarify these definitions, consider the simple case $c = 0$. For $|z_0| > 1$, z_t tends to infinity, while for $|z_0| < 1$, z_t tends to zero. Accordingly, the filled-in Julia set is the set of all points $|z_0| \leq 1$, the Julia set is the set of all points $|z_0| = 1$.

In general, points on the Julia set form a chaotic motion on the set, while points outside the Julia set move away from the set. Accordingly, the Julia set can be regarded as a "repeller" with respect to (1.8). To generate the Julia set, it is thus practical to use the inverted transformation

$$z_t = \pm\sqrt{z_{t+1} - c}, \tag{1.9}$$

start with an arbirarily large value for $t+1$, and go backward in time. By going backward in time, even points far away from the Julia set are attracted by the Julia set.

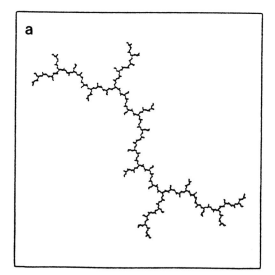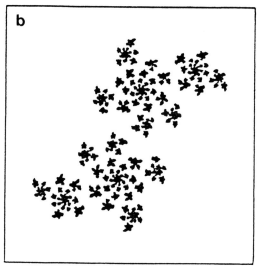

Fig. 1.12. Julia sets for (a) $c = i$ and (b) $c = 0.11031 - i0.67037$. After [1.10]

For obtaining the Julia set for a given value of c, one starts with some arbitrary value for z_{t+1}, for example, $z_{t+1} = 2$. To obtain z_t, we use (1.9), and determine the sign randomly. This procedure is continued to obtain z_{t-1}, z_{t-2}, etc. By disregarding the initial points, e.g., the first 1000 points, one obtains a good approximation of the Julia set.

The Julia sets can be connected (Fig 1.12a) or disconnected (Fig. 1.12b) like the Cantor sets. The self-similarity of the pictures is easy to see. The set of c values that yield connected Julia sets forms the famous Mandelbrot set (Fig. 9.0). It has been shown by Douady and Hubbard [1.19] that the Mandelbrot set is identical to that set of c values for which z_t converges starting from the initial point $z_0 = 0$. For a detailed discussion with beautiful pictures see [1.10] and Chaps. 13 and 14 in [1.3].

1.3 Random Fractal Models

In this section we present several random fractal models that are widely used to mimic fractal systems in nature. We begin with perhaps the simplest fractal model, the random walk.

1.3.1 Random Walks

Imagine a random walker on a square lattice or a simple cubic lattice. In one unit of time, the random walker advances one step of length a to a randomly chosen nearest neighbor site. Let us assume that the walker is unwinding a wire, which he connects to each site along his way. The length (mass) M of the wire that connects the random walker with his starting point is proportional to the number of steps n (Fig. 1.13) performed by the walker.

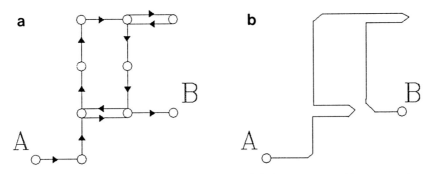

Fig. 1.13. (a) A normal random walk with loops. (b) A random walk without loops

Since for a random walk in any d-dimensional space the mean end-to-end distance R is proportional to $n^{1/2}$ (for a simple derivation see e.g., Chap. 3 in [1.13]), it follows that $M \sim R^2$. Thus (1.4) implies that the fractal dimension of the structure formed by this wire is $d_f = 2$, for all lattices.

The resulting structure has loops, since the walker can return to the same site. We expect the chemical dimension d_ℓ to be 2 in $d = 2$ and to decrease with increasing d, since loops become less relevant. For $d \geq 4$ we have $d_\ell = 1$. If we assume, however, that there is no contact between sections of the wire connected to the same site (Fig. 1.13b), the structure is by definition linear, i.e., $d_\ell = 1$ for all d. For more details on random walks and its relation to Brownian motion, see Chap. 5 and [1.20].

1.3.2 Self-Avoiding Walks

Self-avoiding walks (SAWs) are defined as the subset of all nonintersecting random walk configurations. An example is shown in Fig. 1.14a. As was found by Flory in 1944 [1.21], the end-to-end distance of SAWs scales with the number of steps n as

$$R \sim n^\nu, \tag{1.10}$$

with $\nu = 3/(d+2)$ for $d \leq 4$ and $\nu = 1/2$ for $d > 4$. Since n is proportional to the mass of the chain, it follows from (1.4) that $d_f = 1/\nu$. Self-avoiding walks serve as models for polymers in solution, see [1.22] and Chap. 6.

Subsets of SAWs do not necessarily have the same fractal dimension. Examples are the kinetic growth walk (KGW) [1.23] and the smart growth walk (SGW) [1.24], sometimes also called the "true" or "intelligent" self-avoiding walk. In the KGW, a random walker can only step on those sites that have not been visited before. Asymptotically, after many steps n, the KGW has the same fractal dimension as SAWs. In $d = 2$, however, the asymptotic regime is difficult to reach numerically, since the random walker can be trapped with high probability (see Fig. 1.14b). A related structure is the hull of a random walk in $d = 2$. It has been conjectured by Mandelbrot [1.1] that the fractal dimension of the hull is $d_h = 4/3$.

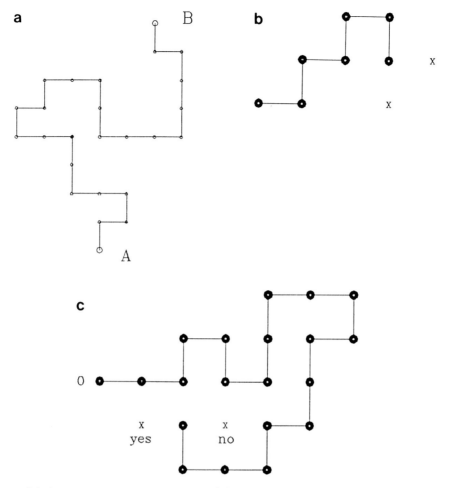

Fig. 1.14. (a) A typical self-avoiding walk. (b) A kinetic growth walk after 8 steps. The available sites are marked by *crosses*. (c) A smart growth walk after 19 steps. The only available site is marked by "yes"

In the SGW, the random walker avoids traps by stepping only at those sites from which he can reach infinity. The structure formed by the SGW is more compact and characterized by $d_f = 7/4$ in $d = 2$ [1.24]. Related structures with the same fractal dimension are the hull of percolation clusters (see also Sect. 1.3.4) and diffusion fronts (for a detailed discussion of both systems see also Chaps. 2 and 7 (by J.F. Gouyet, M. Rosso, and B. Sapoval) in [1.13]).

1.3.3 Kinetic Aggregation

The simplest model of a fractal generated by diffusion of particles is the diffusion-limited aggregation (DLA) model, which was introduced by Witten and Sander in 1981 [1.25]. In the lattice version of the model, a seed particle is fixed at the origin of a given lattice and a second particle is released from a circle around the origin. This particle performs a random walk on the lattice. When it comes to a nearest neighbor site of the seed, it sticks and a cluster (aggregate) of 2 particles is formed. Next, a third particle is released from the

a

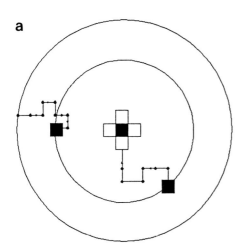

b

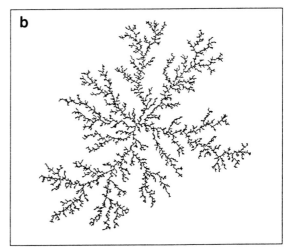

Fig. 1.15. (a) Generation of a DLA cluster. The inner release radius is usually a little larger than the maximum distance of a cluster site from the center, the outer absorbing radius is typically 10 times this distance. (b) A typical off-lattice DLA cluster of 10 000 particles

circle and performs a random walk. When it reaches a neighboring site of the aggregate, it sticks and becomes part of the cluster. This procedure is repeated many times until a cluster of the desired number of sites is generated. For saving computational time it is convenient to eliminate particles that have diffused too far away from the cluster (see Fig. 1.15).

In the continuum (off-lattice) version of the model, the particles have a certain radius a and are not restricted to diffusing on lattice sites. At each time step, the length ($\leq a$) and the direction of the step are chosen randomly. The diffusing particle sticks to the cluster, when its center comes within a distance a of the cluster perimeter. It was found numerically that for off-lattice DLA, $d_f = 1.71 \pm 0.01$ in $d = 2$ and $d_f = 2.5 \pm 0.1$ in $d = 3$ [1.26,27]. These results may be compared with the mean field result $d_f = (d^2 + 1)/(d + 1)$ [1.28]. For a renormalization group approach, see [1.29] and references therein. The chemical dimension d_ℓ is found to be equal to d_f [1.30].

Diffusion-limited aggregation serves as an archetype for a large number of fractal realizations in nature, including viscous fingering, dielectric breakdown, chemical dissolution, electrodeposition, dendritic and snowflake growth, and the growth of bacterial colonies. For a detailed discussion of the applications of DLA we refer to [1.5] and [1.13]. Models for the complex structure of DLA have been developed by Mandelbrot [1.32] and Schwarzer et al. [1.33].

A somehow related model for aggregation is cluster-cluster aggregation (CCA) [1.34]. In CCA, one starts from a very low concentration of particles diffusing on a lattice. When two particles meet, they form a cluster of 2, which can also diffuse. When the cluster meets another particle or another cluster, a larger cluster is formed. In this way, larger and larger aggregates are formed. The structures are less compact than DLA, with $d_f \cong 1.4$ in $d = 2$ and $d_f \cong 1.8$ in $d = 3$. CCA seems to be a good model for smoke aggregates in air and for gold colloids. For a discussion see Chap. 8 (by J.K. Kjems) in [1.13].

1.3.4 Percolation

Consider a square lattice, where each site is occupied randomly with probability p or empty with probability $1-p$. At low concentration p, the occupied sites are either isolated or form small clusters (Fig. 1.16a). Two occupied sites belong to the same cluster if they are connected by a path of nearest neighbor occupied sites. When p is increased, the average size of the clusters increases. At a critical concentration p_c (also called the percolation threshold) a large cluster appears which connects opposite edges of the lattice (Fig. 1.16b). This cluster is called the *infinite* cluster, since its size diverges when the size of the lattice is increased to infinity. When p is increased further, the density of the infinite cluster increases, since more and more sites become part of the infinite cluster, and the average size of the *finite* clusters decreases (Fig. 1.16c).

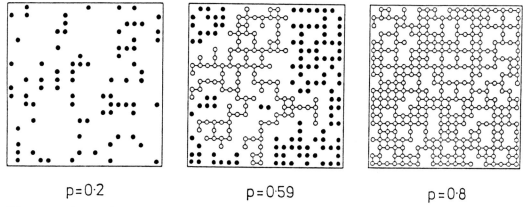

p=0·2   p=0·59   p=0·8

Fig. 1.16. Square lattice of size 20×20. Sites have been randomly occupied with probability p ($p = 0.20, 0.59, 0.80$). Sites belonging to finite clusters are marked by full circles, while sites on the infinite cluster are marked by open circles

The percolation transition is characterized by the geometrical properties of the clusters near p_c. The probability P_∞ that a site belongs to the infinite cluster is zero below p_c and increases above p_c as

$$P_\infty \sim (p - p_c)^\beta. \tag{1.11}$$

The linear size of the *finite* clusters, below and above p_c, is characterized by the *correlation length* ξ. The correlation length is defined as the mean distance between two sites on the same finite cluster and represents the characteristic length scale in percolation. When p approaches p_c, ξ increases as

$$\xi \sim \mid p - p_c \mid^{-\nu}, \tag{1.12}$$

with the same exponent ν below and above the threshold. While p_c depends explicitly on the type of the lattice (e.g., $p_c \cong 0.593$ for the square lattice and $1/2$ for the triangular lattice), the *critical exponents* β and ν are universal and depend only on the dimension d of the lattice, but not on the type of the lattice.

Near p_c, on length scales smaller than ξ, both the infinite cluster and the finite clusters are self-similar. Above p_c, on length scales larger than ξ, the infinite cluster can be regarded as an homogeneous system which is composed of many unit cells of size ξ. Mathematically, this can be summarized as

$$M(r) \sim \begin{cases} r^{d_f}, & r \ll \xi, \\ r^d, & r \gg \xi. \end{cases} \tag{1.13}$$

The fractal dimension d_f can be related to β and ν:

$$d_f = d - \frac{\beta}{\nu}. \tag{1.14}$$

Since β and ν are universal exponents, d_f is also universal. One obtains $d_f = 91/48$ in $d = 2$ and $d_f \cong 2.5$ in $d = 3$. The chemical dimension d_ℓ is smaller than d_f, $d_\ell \cong 1.15$ in $d = 2$ and $d_\ell \cong 1.33$ in $d = 3$. A large percolation cluster in $d = 3$ is shown in Fig. 1.17.

Interestingly, a percolation cluster is composed of several fractal substructures such as the backbone, dangling ends, blobs, external perimeter, and the red bonds, which are all described by different fractal dimensions.

The percolation model has found applications in physics, chemistry, and biology, where occupied and empty sites may represent very different physical, chemical, or biological properties. Examples are the physics of two component systems (the random resistor, magnetic or superconducting networks), the polymerization process in chemistry, and the spreading of epidemics and forest fires. For reviews with a comprehensive list of references, see Chaps. 2 and 3 of [1.13] and [1.35–37].

1.4 How to Measure the Fractal Dimension

One of the most important "practical" problems is to determine the fractal dimension d_f of either a computer generated fractal or a digitized fractal picture. Here we sketch the two most useful methods: the "sandbox" method and the "box counting" method.

1.4.1 The Sandbox Method

To determine d_f, we first choose one site (or one pixel) of the fractal as the origin for n circles of radii $R_1 < R_2 < ... < R_n$, where R_n is smaller than the radius R of the fractal, and count the number of points (pixels) $M_1(R_i)$ within each circle i. (Sometimes, it is more convenient to choose n squares of side length $L_1...L_n$ instead of the circles.) We repeat this procedure by choosing many other (altogether m) pixels as origins for the n circles and determine the corresponding number of points $M_j(R_i)$, $j = 2, 3, ..., m$ within each circle (see Fig. 1.18a). We obtain the mean number of points $M(R_i)$ within each circle by averaging, $M(R_i) = \frac{1}{m} \sum_{j=1}^{m} M_j(R_i)$, and plot $M(R_i)$ versus R_i in a double

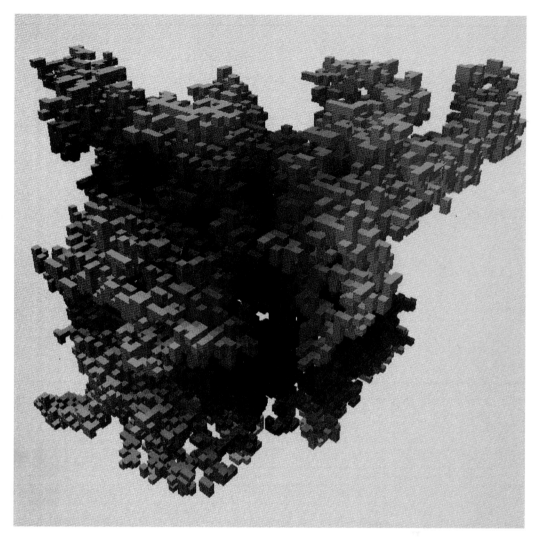

Fig. 1.17. A large percolation cluster in $d = 3$. The colors mark the topological distance from an arbitrary center of the cluster in the middle of the page. Courtesy of M. Meyer

logarithmic plot. The slope of the curve, for large values of R_i, determines the fractal dimension.

In order to avoid boundary effects, the radii must be smaller than the radius of the fractal, and the centers of the circles must be chosen well inside the fractal, so that the largest circles will be well within the fractal. In order to obtain good statistics, one has either to take a very large fractal cluster with many centers of circles or many realizations of the same fractal.

1.4.2 The Box Counting Method

We draw a grid on the fractal that consists of N_1^2 squares, and determine the number of squares $S(N_1)$ that are needed to cover the fractal (see Fig. 1.18b). Next we choose finer and finer grids with $N_1^2 < N_2^2 < N_3^2 < ... < N_m^2$ squares

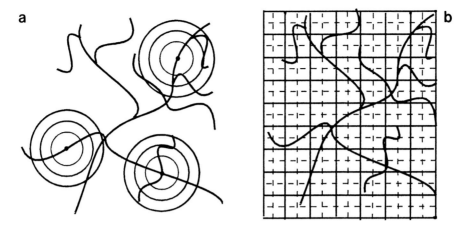

Fig. 1.18. Illustrations for determining the fractal dimension: (a) the sandbox method, (b) the box counting method

and calculate the corresponding numbers of squares $S(N_1)$ ... $S(N_m)$ needed to cover the fractal. Since $S(N)$ scales as

$$S(N) \sim N^{-d_f}, \tag{1.15}$$

we obtain the fractal dimension by plotting $S(N)$ versus $1/N$ in a double logarithmic plot. The asymptotic slope, for large N, gives d_f.

Of course, the finest grid size must be larger than the pixel size, so that many pixels can fall into the smallest square. To improve statistics, one should average $S(N)$ over many realizations of the fractal.

1.5 Self-Affine Fractals

The fractal structures we have discussed in the previous sections are self-similar: if we cut a small piece out of a fractal and magnify it isotropically to the size of the original, both the original and the magnification look the same. By magnifying isotropically, we have rescaled the x, y, and z axis by the same factor.

There exist, however, systems that are invariant only under *anisotropic* magnifications. These systems are called *self-affine* [1.1] (see Chap. 4). A simple model for a self-affine fractal is shown in Fig. 1.19. The structure is invariant under the anisotropic magnification $x \rightarrow 4x$, $y \rightarrow 2y$. If we cut a small piece out of the original picture (in the limit of $n \rightarrow \infty$ iterations), and rescale the x axis by a factor of 4 and the y axis by a factor of 2, we will obtain exactly the original structure. In other words, if we describe the form of the curve in Fig. 1.19 by the function $F(x)$, this function satisfies the equation $F(4x) = 2F(x) = 4^{1/2}F(x)$.

In general, if a self-affine curve is scale invariant under the transformation $x \rightarrow bx$, $y \rightarrow ay$, we have

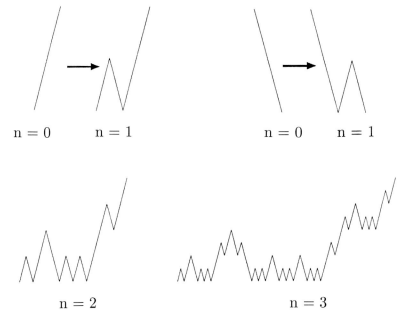

Fig. 1.19. A simple deterministic model of a self-affine fractal

$$F(bx) = aF(x) \equiv b^H F(x), \qquad (1.16)$$

where the exponent $H = \log a / \log b$ is called the Hurst exponent [1.1]. The solution of the functional equation (1.16) is simply $F(x) = Ax^H$. In the example of Fig. 1.19, $H = 1/2$.

Next we consider random self-affine structures, which are used as models for random surfaces. The simplest structure is generated by a one-dimensional random walk, where the abcissa is the time axis and the ordinate is the displacement $Z(t) = \sum_{i=1}^{t} e_i$ of the walker from its starting point. Here, $e_i = \pm 1$ is the unit step made by the random walker at time t. Since different steps of the random walker are uncorrelated, $\langle e_i e_j \rangle = \delta_{ij}$, it follows that $F(t) \equiv \langle Z^2(t) \rangle^{1/2} = t^{1/2}$, and the Hurst exponent of the structure is $H = 1/2$.

Next we assume that different steps i and j are correlated in such a way that $\langle e_i e_j \rangle = b|i-j|^{-\gamma}$, $1 > \gamma \geq 0$. To see how the Hurst exponent depends on γ, we have to evaluate the mean square displacement $\langle Z^2(t) \rangle = \sum_{i,j}^{t} \langle e_i e_j \rangle$. For calculating the double sum it is convenient to introduce the Fourier transform of e_i, $e_\omega = (1/\Omega)^{1/2} \sum_{l=1}^{\Omega} e_l \exp(-i\omega l)$, where Ω is the number of sites in the system. It is easy to verify that $\langle Z^2(t) \rangle$ can be expressed in terms of the power spectrum $\langle e_\omega e_{-\omega} \rangle$ [1.38]:

$$\langle Z^2(t) \rangle = \frac{1}{\Omega} \sum_\omega \langle e_\omega e_{-\omega} \rangle |f(\omega, t)|^2, \qquad (1.17a)$$

where

$$f(\omega, t) \equiv (e^{-i\omega(t+1)} - 1)/(e^{-i\omega} - 1).$$

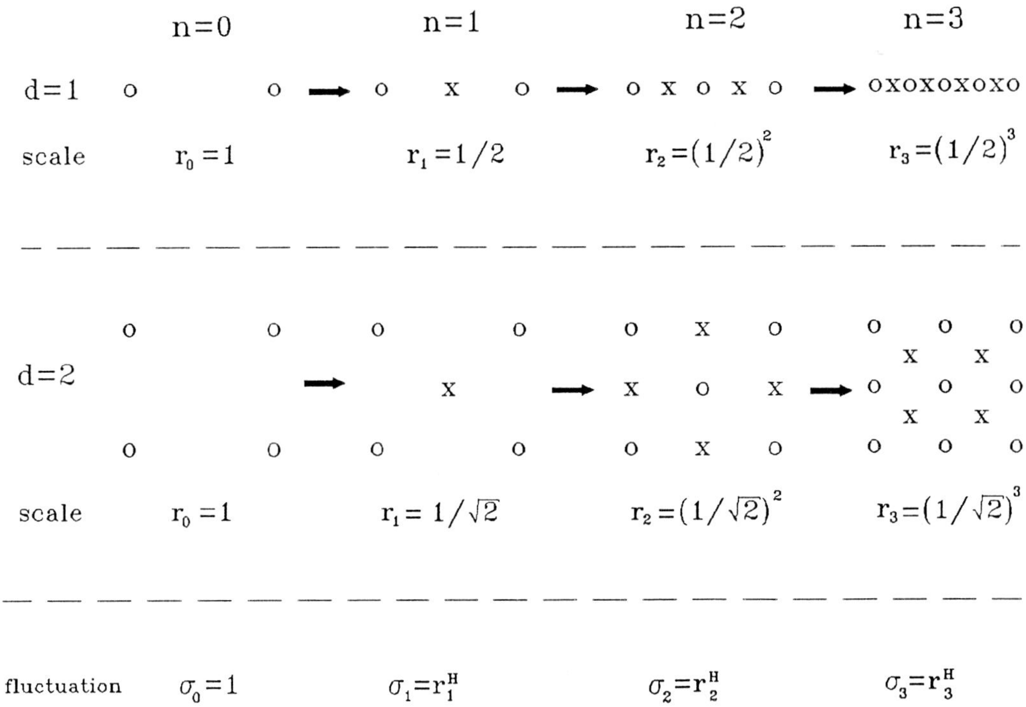

	n=0	n=1	n=2	n=3

d=1   o   o ➝ o x   o ➝ o x o x o   ➝ oxoxoxoxo

scale $r_0 = 1$ $r_1 = 1/2$ $r_2 = (1/2)^2$ $r_3 = (1/2)^3$

- -

d=2

```
    o        o        o        o        o   x   o        o      o      o
                                                               x      x
          ➝         x      ➝    x   o   x   ➝   o      o      o
                                                               x      x
    o        o        o        o        o   x   o        o      o      o
```

scale $r_0 = 1$ $r_1 = 1/\sqrt{2}$ $r_2 = (1/\sqrt{2})^2$ $r_3 = (1/\sqrt{2})^3$

- -

fluctuation $\sigma_0 = 1$ $\sigma_1 = r_1^H$ $\sigma_2 = r_2^H$ $\sigma_3 = r_3^H$

Fig. 1.20. Illustration of the successive random addition method in $d = 1$ and $d = 2$. The *circles* mark those points that have been considered already in the earlier iterations, the *crosses* mark the new midpoints added at the present iteration. At each iteration n, first the Z values of the midpoints are determined by linear interpolation from the neighboring points, and then random displacements of variance σ_n are added to all Z values

Since the power spectrum scales as

$$\langle e_\omega e_{-\omega} \rangle \sim \omega^{-(1-\gamma)}, \tag{1.17b}$$

the integration of (1.17a) yields, for large t,

$$\langle Z^2(t) \rangle \sim t^{2-\gamma}. \tag{1.17c}$$

Therefore, the Hurst exponent is $H = (2 - \gamma)/2$. According to (1.17b), for $0 < \gamma < 1$, $\langle Z^2(t) \rangle$ increases faster in time than the uncorrelated random walk, while for $1 < \gamma < 2$, $\langle Z^2(t) \rangle$ increases slower. The long range correlated random walks were called *fractional Brownian motion* by Mandelbrot [1.1].

There exist several methods to generate correlated random surfaces. We shall describe the *successive random additions* method [1.31], which iteratively generates the self affine function $Z(x)$ in the unit interval $0 \leq x \leq 1$.

In the $n = 0$ iteration, we start at the edges $x = 0$ and $x = 1$ of the unit interval and choose the values of $Z(0)$ and $Z(1)$ from a distribution with zero mean and variance $\sigma_0^2 = 1$ (see Fig. 1.20). In the $n = 1$ iteration, we choose the midpoint $x = 1/2$ and determine $Z(1/2)$ by linear interpolation, i.e., $Z(1/2) = (Z(0) + Z(1))/2$, and add to all so-far calculated Z values ($Z(0)$, $Z(1/2)$, and $Z(1)$) random displacements from the same distribution as before, but with a variance $\sigma_1 = (1/2)^H$ (see Fig. 1.20). In the $n = 2$ iteration we again first choose the midpoints ($x = 1/4$ and $x = 3/4$), determine their Z values by linear interpolation, and add to all so-far calculated Z values random displacements from the same distribution as before, but with a variance $\sigma_2 = (1/2)^{2H}$. In general, in the nth iteration one first interpolates the Z values of the midpoints and then adds random displacements to all existing Z values, with variance $\sigma_n = (1/2)^{nH}$. The procedure is repeated until the required resolution of the surface is obtained. Figure 1.21 shows the graphs of three random surfaces generated this way, with $H = 0.2$, $H = 0.5$, and $H = 0.8$. For an alternative method, see [1.39].

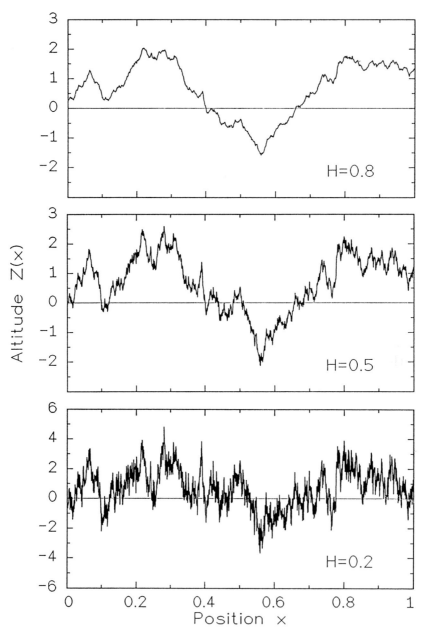

Fig. 1.21. Correlated random walks with $H = 0.2$, 0.5, and 0.8, generated by the successive random addition method in $d = 1$

The generalization of the successive random addition method to two dimensions is straightforward (see Fig. 1.20). We consider a function $Z(x,y)$ on the unit square $0 \leq x, y \leq 1$. In the $n = 0$ iteration, we start with the four corners $(x,y) = (0,0), (1,0), (1,1), (0,1)$ of the unit square and choose their Z values from a distribution with zero mean and variance $\sigma_0^2 = 1$ (see Fig. 1.20). In the $n = 1$ iteration, we choose the midpoint at $(x,y) = (1/2, 1/2)$ and determine $Z(1/2, 1/2)$ by linear interpolation, i.e., $Z(1/2, 1/2) = (Z(0,0) + Z(0,1) + Z(1,1) + Z(1,0))/4$. Then we add to all so far calculated Z-values ($Z(0,0)$, $Z(0,1)$, $Z(1,0)$, $Z(1,1)$ and $Z(1/2, 1/2)$) random displacements from the same distribution as before, but with a variance $\sigma_1 = (1/\sqrt{2})^H$ (see Fig. 1.20). In the $n = 2$ iteration we again choose the midpoints of the five sites $(0,1/2)$, $(1/2,0)$, $(1/2,1)$ and $(1,1/2)$, determine their Z value by linear interpolation, and add to all so far calculated Z values random displacements from the same distribution as before, but with a variance $\sigma_2 = (1/\sqrt{2})^{2H}$. This procedure is repeated again and again, until the required resolution of the surface is obtained. Fig. 9.1 shows the graphs of two random surfaces generated this way.

1.6 Fractals in Nature

Self-similar or self-affine fractal structures with features similar to those fractal models discussed above can be found in nature on all, astronomic as well as microscopic, length scales. Examples include clusters of galaxies (the fractal dimension of the mass distribution is about 1.2 [1.29]), the crater landscape of the moon, the distribution of earthquakes (see Chap. 2), and the structure of coastlines, rivers [1.40], mountains, and clouds. Fractal cracks (see, for example, Chap. 5 (by H.J. Herrmann) in [1.13]) occur on length scales ranging from 10^3 km (such as the San Andreas fault) to micrometers (like fractures in solid materials).

Many naturally growing plants show fractal structures: examples range from trees and the roots of trees to cauliflower and broccoli. The patterns of blood vessels in the human body, the kidney, the lung, and some types of nerve cells have fractal features (see Chap. 3). In materials science, fractals appear in polymers, gels, ionic glasses, aggregates, electrodeposition, rough interfaces and surfaces (see [1.13] and Chaps. 4 and 6 of this book), as well as in fineparticle systems [1.41]. In all these structures there is no characteristic length scale in the system besides the physical upper and lower cut-offs.

The occurrence of self-similar or self-affine fractals is not limited to structures in real space. In Sect. 1.5 we considered the displacement $Z(t)$ of a one-dimensional random walk as a function of time t. Although the abscissa is time, the *graph* $Z(t)$ of the random walk can be regarded as a self-affine structure. In this case, the length scale in the fractal definition is replaced by a time scale. Examples from biology and medicine include heartbeat time series with different Hurst exponents for healthy and sick individuals, and the variation of the electric current with time that passes through a single ion channel in a cell membrane (see Chap. 3 and [1.42]). Other examples which will be treated in this book are from ecology (Chap. 5), chemical reactions (Chap. 7), and heterogeneous chemistry (Chap. 8).

Acknowledgements. We should like to thank D. Ben-Avraham, H. Brender, A. Levinson, P. Maass, M. Meyer, U.A. Neumann, R. Nossal, S. Rabinowicz, H.E. Roman, A. Shehter, S. Schwarzer, H.E. Stanley, H. Taitelbaum, B.L. Trus, G.H. Weiss and B. Vilensky for valuable discussions and helpful remarks. We are grateful to the Alexander-von-Humboldt Foundation and the Deutsche Forschungsgemeinschaft for financial support.

References

1.1   B.B. Mandelbrot: *Fractals: Form, Chance and Dimension* (Freeman, San Francisco 1977);
      B.B. Mandelbrot: *The Fractal Geometry of Nature* (Freeman, San Francisco 1982)
1.2   H. Jones, in: *Fractals and Chaos*, ed. by T. Crilly, R.A. Earschaw, H. Jones (Springer, New York 1991)
1.3   H.-O. Peitgen, H. Jürgens, D. Saupe: *Chaos and Fractals* (Springer Verlag, New York 1992)
1.4   J. Feder: *Fractals* (Plenum, New York 1988)
1.5   T. Vicsek: *Fractal Growth Phenomena* (World Scientific, Singapore 1989)
1.6   D. Avnir, ed.: *The Fractal Approach to Heterogeneous Chemistry* (John Wiley, New York 1992)
1.7   M. Barnsley: *Fractals Everywhere* (Academic Press, San Diego 1988)
1.8   H. Takayasu: *Fractals in the Physical Sciences* (Manchester University Press, Manchester 1990)
1.9   H.G. Schuster: *Deterministic Chaos - An Introduction* (Physik Verlag, Weinheim 1984)
1.10 H.-O. Peitgen, P.H. Richter: *The Beauty of Fractals* (Springer Verlag, Heidelberg 1986)
1.11 H.E. Stanley, N. Ostrowsky, eds.: *Correlations and Connectivity: Geometric Aspects of Physics, Chemistry and Biology* (Kluwer, Dordrecht 1990)
1.12 H.-O. Peitgen, H. Jürgens, D. Saupe: *Chaos and Fractals* (Springer Verlag, Heidelberg 1991)
1.13 A. Bunde and S. Havlin, eds.: *Fractals and Disordered Systems* (Springer Verlag, Heidelberg 1991)
1.14 J.-F. Gouyet: *Physique et Structures Fractales* (Masson, Paris 1992)
1.15 S. Havlin, D. ben-Avraham: Adv. in Phys. **36**, 695 (1987)
1.16 M. Feigenbaum: J. Stat. Phys. **19** , 25 (1978)
1.17 P. Grassberger: J. Stat. Phys. **26**, 173 (1981)
1.18 B.B. Mandelbrot, J. Given: Phys. Rev. Lett. **52**, 1853 (1984)
1.19 A. Douady and J. H. Hubbard: CRAS Paris, **294**, 123 (1982); see also [1.10]
1.20 G.H. Weiss: *Random Walks* (North Holland, Amsterdam 1994)
1.21 P.J. Flory: *Principles of Polymer Chemistry* (Cornell University Press, New York 1971)
1.22 P.G. de Gennes: *Scaling Concepts in Polymer Physics* (Cornell University Press, Ithaca 1979)
1.23 I. Majid, N. Jan, A. Coniglio, H.E. Stanley: Phys. Rev. Lett. **52**, 1257 (1984);
      S. Havlin, B. Trus, H.E. Stanley: Phys. Rev. Lett. **53**, 1288 (1984);
      K. Kremer, J.W. Lyklema: Phys. Rev. Lett. **55**, 2091 (1985)
1.24 R.M. Ziff, P.T. Cummings, G. Stell: J. Phys. A **17**, 3009 (1984);
      A. Bunde, J.F. Gouyet: J. Phys. A **18**, L285 (1984);
      A. Weinrib, S. Trugman: Phys. Rev. B **31**, 2993 (1985);
      K. Kremer, J.W. Lyklema: J. Phys. A **18**, 1515 (1985);
      H. Saleur, B. Duplantier: Phys. Rev. Lett. **58**, 2325 (1987)

1.25   T. A. Witten, L.M. Sander: Phys. Rev. Lett. **47**, 1400 (1981)
1.26   P. Meakin: Phys. Rev. A **27**, 604, 1495 (1983)
1.27   P. Meakin, in: *Phase Transitions and Critical Phenomena,* Vol.12, ed. by C. Domb and J. Lebowitz (Academic Press, New York 1988) p. 335
1.28   M. Muthukumar: Phys. Rev. Lett. **50**, 839 (1983); M. Tokuyama, K. Kawasaki: Phys. Lett. A **100**, 337 (1984)
1.29   L. Pietronero: Physica A **191**, 85 (1992)
1.30   P. Meakin, I. Majid, S. Havlin, H.E. Stanley: Physica A **17**, L975 (1984)
1.31   R.F. Voss, in: *Fundamental Algorithms in Computer Graphics,* ed. by R.A. Earshaw (Springer, Berlin 1985) p. 805
1.32   B.B. Mandelbrot: Physica A **191**, 95 (1992); see also B.B. Mandelbrot, T. Vicsek: J. Phys. A **20**, L377 (1989)
1.33   S. Schwarzer, J. Lee, A. Bunde, S. Havlin, H.E. Roman, H.E. Stanley: Phys. Rev. Lett. **65**, 603 (1990)
1.34   P. Meakin: Phys. Rev. Lett. **51**, 1119 (1983); M. Kolb: Phys. Rev. Lett. **53**, 1653 (1984)
1.35   D. Stauffer, A. Aharony: *Introduction to Percolation Theory* (Taylor and Francis, London 1992)
1.36   H. Kesten: *Percolation Theory for Mathematicians* (Birkhauser, Boston 1982)
1.37   G.R. Grimmet: *Percolation* (Springer Verlag, New York 1989)
1.38   S. Havlin, R. Blumberg-Selinger, M. Schwartz, H.E. Stanley, A. Bunde: Phys. Rev. Lett. **61**, 1438 (1988)
1.39   C.-K. Peng, S. Havlin, M. Schwartz, H.E. Stanley: Phys. Rev. A **44**, R2239 (1991)
1.40   H. Inaoka, H. Takayasu: Phys. Rev. E **47**, 899 (1993)
1.41   B.H. Kaye: *A Random Walk Through Fractal Dimensions* (Verlag Chemie, Weinheim 1989)
1.42   L.S. Liebovitch, I.M. Sullivan: Biophys. J. **52**, 979 (1987)

2 Fractals and Self-Organized Criticality

Per Bak and Michael Creutz

2.1 Introduction

Many objects in nature are best described geometrically as fractals, with self-similar features on all length scales. The universe consists of clusters of galaxies, organized in clusters of clusters of galaxies and so on [2.1]. Mountain landscapes have peaks of all sizes, from kilometers down to millimeters. River networks consist of streams of all sizes. Turbulent fluids have vortices over a wide range of sizes. Earthquakes occur on structures of faults ranging from thousands of kilometers to centimeters. Fractals are scale-free in the sense that in viewing a picture of a part of a fractal one cannot deduce its actual size if a yardstick is not shown in the same picture.

If fractals are indeed the geometry of nature, one must still understand how nature produces them. A good deal of effort has been put into the geometrical characterization of these objects, but there has been practically no progress in understanding their dynamical origin. We have a tendency to think of the universe and the crust of the earth as static structures because the dynamics that forms these structures have much longer time scale than the observation period, which can be a human lifetime. The earthquakes that we observe last at most a few seconds, whereas the fault formations appear static and are built up over millions of years.

The origin of fractals is a dynamical, not a geometrical, problem. The laws of physics are local, but fractals are nevertheless organized over the greatest distances. The mystery is enhanced by the fact that large equilibrium systems, operating near their ground state, tend to be only locally correlated. Only at a critical point where a continuous phase transition takes place are those systems fractal.

◄   **Fig. 2.0.** The sandpile model with 49 152 grains of sand added to a single point. The heights z_i from 0 through 3 are represented by *white, red, blue,* and *green,* respectively

But real systems are dissipative, that is, they have friction, and rarely go to their ground state, unlike the ideal models discussed in freshman physics. Consider, for example, a pendulum. The ideal motion is periodic and for small amplitudes is well approximated by a sine wave. To make it more realistic one can put in a drag term, giving rise to a damped oscillatory behavior with a decreasing amplitude theoretically continuing forever. However, in the real world the motion will be impeded by some imperfections, perhaps in the form of dust. Once the amplitude gets small enough, the pendulum will suddenly stop, and this will generally occur at the end of a swing where the velocity is smallest. This is not the state of smallest energy, and indeed the probability is a minimum for stopping at exactly the bottom of the potential. In a sense, the system is most likely to settle near a "minimally stable" state, far from any "thermal equilibrium".

If we generalize to a multidimensional system of many coupled pendulums, a new issue arises. A minimally stable state will be particularly sensitive to small perturbations which can "avalanche" through the system. Thus small disturbances could grow and propagate through the system with little resistance despite the damping and other impediments. Since energy is dissipated through this process, the energy must be replenished for avalanches to continue. The systems that we shall study are ones where energy is constantly supplied and eventually dissipated in the form of avalanches.

The canonical metaphorical example is a simple pile of sand. Adding sand slowly to a flat pile will result only in some local rearrangement of particles. The individual grains, or degrees of freedom, do not interact over large distances. Continuing the process will result in the slope increasing to a critical value where an additional grain of sand gives rise to avalanches of any size, from a single grain falling up to the full size of the sand pile. The pile can no longer be described in terms of many local degrees of freedom, but only a holistic description in terms of one sandpile will do. The distribution of avalanches follows a power law.

"Self-organized criticality" (SOC) refers to this tendency of large dissipative systems to drive themselves to a critical state with a wide range of length and time scales [2.2-4]. The idea provides a unifying concept for large-scale behavior in systems with many degrees of freedom. It has been looked for in such diverse areas as earthquake structure, economics, and biological evolution.

The critical state is an attractor for the dynamics. If a slope were too steep one would obtain a large avalanche and a collapse to a flatter and more stable configuration. On the other hand, if it were too shallow the new sand would just accumulate to make the pile steeper. If the process is modified, for instance by using wet sand instead of dry sand, the pile will modify its slope during a transient period and return to a new critical state. It is this resiliency which suggests that self-organized criticality might be a quite universal phenomenon. If one builds snow screens locally to prevent avalanches, the pile will again respond by locally building up to steeper states, and large avalanches will resume. The large fluctuations associated with the avalanches are unavoidable;

this might provide some food for thought when applied to vast economic or political systems.

Self-organized criticality complements the concept of "chaos" wherein simple systems with a small number of degrees of freedom can display quite complex behavior. Chaos is associated with fractal "strange" attractors in the phase space spanned by nonlinear systems with only a few degrees of freedom. These self-similar structures need have little to do with fractals in real spatially extended physical systems. Specifically, chaotic systems exhibit white noise with short temporal correlations, whereas fractal systems are expected to have long-range temporal correlations. In contrast, self-organized criticality emphasizes unifying features in the coherent evolution of systems with many degrees of freedom.

2.2 Simulations of Sandpile Models

The convergence to the self-organized critical state can be demonstrated by computer simulations on toy sandpile models (see also Chap. 9). The simplest example is a cellular automaton formulated on a two-dimensional regular lattice of N sites. Integer variables z_i on each site i are used to represent the local sandpile height. Here we consider a two-dimensional lattice with open boundaries. Addition of a sand particle to a site i is represented by increasing the value of z_i at that site by unity. When the height somewhere exceeds a critical value z_{cr}, here taken to be 3, there is a toppling event wherein 1 grain of sand is transferred from the unstable site to each of the 4 neighboring sites; i.e, the value of z_i is reduced by 4 and the values of z at the 4 neighboring sites are increased by 1. The updating is done concurrently, with all sites updated simultaneously. The initial toppling may initiate a chain reaction, where the total number of topplings is a measure of the size of an avalanche. As an example of a state in this model, in Fig. 2.0 we show the pile of sand resulting from dropping 49 152 grains on a single site. Already we see signs of a fractal structure emerging.

To explore self-organized criticality in this model, one can randomly add sand and have the system relax. The result of such an addition becomes unpredictable, with one only being able to find the outcome by actually simulating the resulting avalanche.

In Fig. 2.1 we show a typical state of the sandpile after a large amount of sand has been dropped pseudo-randomly. The lattice here is 286 by 184 sites and the boundaries are open. At first glance, the system appears quite random. There are, however, some subtle correlations. For example, never do two black cells lie adjacent to each other, nor does any site have four black neighbors. This follows from the fact that in tumbling a site to height 0, a grain of sand is dumped onto each neighbor.

To such a configuration a small amount of sand was added to a site near the center, triggering an avalanche. We follow this avalanche in Fig. 2.2. To trace

Fig. 2.1. A typical critical state for the sandpile on a 286 by 184 lattice with open boundaries. The heights z_i from 0 through 3 are represented by *black, red, blue,* and *green*, respectively

the avalanche, we give the cells which have collapsed a cyan color. Figures 2.2a and 2.2b show intermediate active stages in the collapse. Yellow sites inside or outside the old avalanche area are still active. Figure 2.2c displays the final stable configuration.

This was a particularly large avalanche, selected for illustrative purposes. The initial state was not exactly that of Fig. 2.1 because in making this figure we first produced several uninteresting small avalanches. Indeed the ultimate size of the disturbance is unpredictable without actually running the simulation. Some cascades may involve a single tumbling, and others collapse most of the system.

Note in Fig. 2.2b the appearance of islands which have not yet collapsed. These have all disappeared in Fig. 2.2c, the final relaxed state. This is an exact result, special to this model; once in the critical ensemble, it is impossible to trigger a set of avalanches which will leave an isolated island of unaltered cells surrounded by disturbed ground.

Figure 2.3 shows a log - log plot of the distribution of the avalanche sizes s and durations t. The linearity indicates a power law,

$$P(s) \sim s^{1-\tau}, \quad \tau \simeq 2.1, \tag{2.1}$$

Fig. 2.2 a-c. The progress of an avalanche obtained by increasing one of the z_i in a critical ▶ state. The undisturbed stable sites are colored as in Fig. 2.1 while the still active sites in parts **a** and **b** are various shades of *yellow*, and the tumbled region is *cyan*

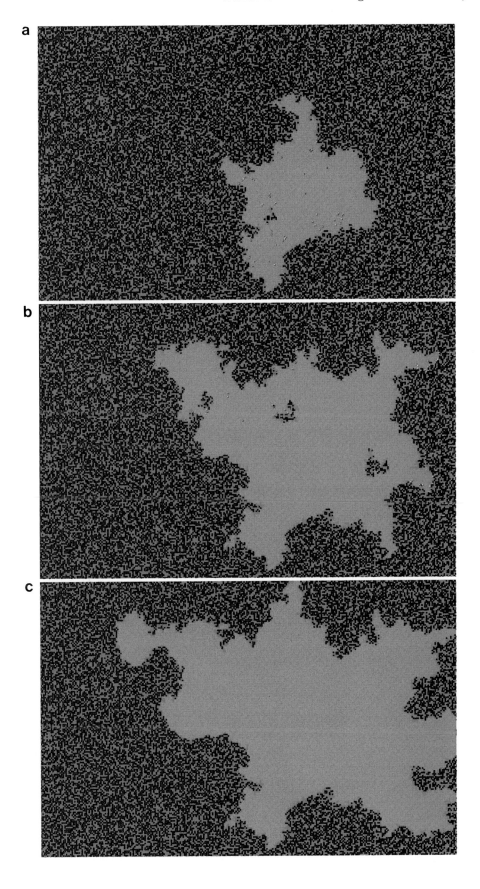

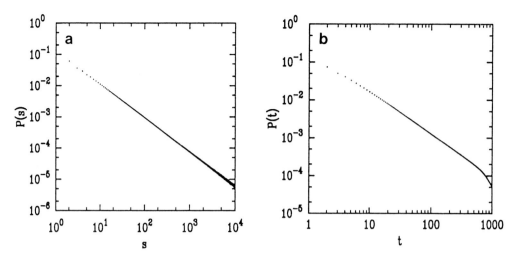

Fig. 2.3 a,b. A log - log plot of the distribution of (**a**) the avalanche sizes s and (**b**) durations t for the sandpile model

where s is the number of tumblings in an avalanche and P is the probability distribution for avalanches of a given size. Thus the state in Fig. 2.1, which at first appears featureless, is actually remarkably correlated. For a random distribution of z's one would expect the chain reaction generating the avalanche to be either subcritical, in which case the avalanche would die after a few steps and large avalanches would be exponentially unlikely, or supercritical, in which case the avalanche would explode with a collapse of the entire system. The power law indicates that the reaction is precisely critical, i.e., the probability that activity at some site branches into more than one active site is balanced by the probability that the activity dies. Thus, by evolving through avalanche after avalanche, the matrix has "learned" to respond critically to the next perturbation.

Several other quantities which obey fractal scaling laws can be defined for the sandpile. For instance, the duration t, that is, the number of updatings for an avalanche to complete, has a distribution

$$P(t) \sim t^{1-\tau_t}, \qquad \tau_t \simeq 2.14 \tag{2.2}$$

Also, the number of distinct tumbled sites, s_d, which is different from the total number of topplings since some sites topple more than once, goes as

$$P(s_d) \sim s_d^{1-\tau_d}, \qquad \tau_d \simeq 2.07. \tag{2.3}$$

For some applications it is useful to define conditional expectation values. For instance, one can define an exponent for the expectation value of the duration, t, for an avalanche of size s:

$$t \sim s^{\gamma_{ts}}, \qquad \gamma_{ts} \simeq 0.61. \tag{2.4}$$

Similar relations connect mutually all the quantities s, t, s_d, and the linear size r of the clusters [2.6]. The exponent relating s_d to r, $\gamma_{s_d r}$, appears to be 2, indicating that the clusters are compact. The existence of relations between the different stochastic variables suggests scaling relations between the exponents. In general

$$\tau_x = 2 + (\tau_y - 2)/\gamma_{xy}. \tag{2.5}$$

As long as the variables x and y are reasonably correlated, this is a mathematical identity and does not imply new physics. The compactness of clusters thus suggests that $(\tau_t - 2) = (\tau_r - 2)/2$.

In order to calculate the power spectrum it turns out that one should calculate the expectation value of s^2 vs duration t, that is,

$$\Lambda(t) = \sum s^2 P(s,t). \tag{2.6}$$

This quantity is given by the power law

$$\Lambda(t) \sim t^\mu, \quad \mu \simeq 2.1. \tag{2.7}$$

The values of the exponents quoted here were calculated by Kim Christensen [2.7]. The model can be defined in $d = 3, 4$, etc. dimensions. For instance, $\tau \simeq 2.31$ in three dimensions; thus, the values of the exponents depend on d. For an excellent discussion of these exponents and their relation to each other, see the paper by Christensen et al. [2.6].

2.3 Abelian Sandpile Models

It would be highly desirable to have an analytical theory, such as the renormalization group theory for equilibrium critical phenomena, by which one could estimate the exponents and at the same time gain insights into the mechanisms of self-organized criticality. We are not yet at that point. However, in a series of papers Deepak Dhar and coworkers have shown that the sand model has some rather remarkable mathematical properties [2.8-11]. In particular, the critical attractor of the system is characterized in terms of an Abelian group. The properties of the group can be used to calculate the number of states belonging to the critical attractor, and the rate of convergence to the attractor. Further consequences of the Abelian algebra have been explored by one of us [2.12-13], and in the following we shall generally follow the discussion given in [2.13].

2.3.1 The Abelian Group

Dhar introduced the useful toppling matrix $\Delta_{i,j}$ with integer elements representing the change in height, z, at site i resulting from a toppling at site j [2.8]. Under a toppling at site j, the height at site i becomes $z_i - \Delta_{i,j}$. For the simple two-dimensional sand model the toppling matrix is given as

$$\Delta_{i,j} = 4 \qquad i = j,$$
$$\Delta_{i,j} = -1 \qquad i, j \text{ nearest neighbors}, \qquad (2.8)$$
$$\Delta_{i,j} = 0 \qquad \text{otherwise}.$$

For this discussion there is little special about the specific lattice geometry; indeed, the following results easily generalize to other lattices and dimensions. In fact, on a Cayley tree the model can be solved exactly. The analysis requires only that under a toppling of a single site i, that site has its slope decreased ($\Delta_{i,i} > 0$), the slope at any other site is either increased or unchanged ($\Delta_{i,j} \leq 0, j \neq i$), the total amount of sand in the system does not increase ($\sum_j \Delta_{i,j} > 0$), and, finally, each site can be connected through topplings to some location where sand can be lost, such as at a boundary.

For the specific case in (2.8), the sum of slopes over all sites is conserved whenever a site away from the lattice edge undergoes a toppling. Only at the lattice boundaries can sand be lost. Thus the details of this model depend crucially on the boundaries, which we take to be open. A toppling at an edge loses one grain of sand and at a corner loses two.

The actual value of the threshold z_T is unimportant to the dynamics. This can be changed by simply adding constants to all the z_i. Thus without loss of generality we consider $z_T = 3$. With this convention, if all z_i are initially non-negative they will remain so, and we restrict ourselves to states C belonging to that set. The states where all z_i are positive and less than 4 are called stable; a state that has any z_i larger than or equal to 4 is called unstable. One conceptually important configuration is the minimally stable state C^* which has all the heights at the critical value z_T. By construction, any addition of sand to C^* will give an unstable state.

We now formally define various operators acting on the states C. First, the "adding sand" operator α_i acting on any C yields the state $\alpha_i C$ where $z_i = z_i + 1$ and all other z are unchanged. Next, the toppling operator t_i transforms C into the state with heights z'_j where $z'_j = z_j - \Delta_{i,j}$. The operator U which updates the lattice one time step is now simply the product of t_i over all sites where the slope is unstable,

$$UC = \prod_i t_i^{p_i} C, \qquad (2.9)$$

where $p_i = 1$ if $z_i \geq 4$ and 0 otherwise. Using U repeatedly we can define the relaxation operator R. Applied to any state C this corresponds to repeating U until no more z_i change. Neither U nor R have any effect on stable states. Finally, we define the avalanche operators a_i, which describe the action of adding a grain of sand followed by relaxation

$$a_i C = R \alpha_i C. \qquad (2.10)$$

At this point it is not entirely clear that the operator R exists; that is, it might be that the updating procedure enters a nontrivial cycle. We now prove that this is impossible. First note that a toppling in the interior of the lattice does not change the total amount of sand. A toppling on the boundary, however, decreases this sum due to sand falling off the edge. Thus the total sand in the system is a nonincreasing quantity. No cycle can have toppling at the boundary since this will decrease the sum. Next, the sand on the boundary will monotonically increase if there is any toppling one site away. This cannot happen in a cycle; thus there can be no topplings one site away from the edges. By induction there can be no toppling arbitrary distances from the boundary; thus there can be no cycle, and the relaxation operator exists. Note that for a general geometry this result requires that every site be eventually connected to an edge where sand can be lost. With periodic boundaries no sand would be lost and thus cycles are expected and observed. We call these unphysical systems "Escher models" after the artist who constructed drawings of water flowing perpetually downhill and yet circulating in the system.

It is useful to introduce the concept of recursive states. This set, denoted $\mathcal{R}$, includes those stable states which can be reached from any stable state by some addition of sand followed by relaxation. As the minimally stable state C^* can be obtained from any other state by adding just enough sand to each site to make z_i equal to three, it belongs to $\mathcal{R}$. Thus, one might conveniently define $\mathcal{R}$ as the set of states which can be obtained from C^* by acting with some product of the operators a_i.

It is easily shown that there exist nonrecursive, transient states; for instance, no recursive state can have two adjacent heights that are both zero. One can also show that the self-organized critical ensemble, reached under random addition of sand to the system, has equal probability for each state in the recursive set.

The crucial result of [2.8-11] is that the operators a_i acting on stable states commute, and that they generate an Abelian group when restricted to recursive states. We begin by showing that the operators commute, that is, that $a_i a_j C = a_j a_i C$ for all C. First we express the a's in terms of toppling and adding operators,

$$a_i a_j C = \left(\prod_{k=1}^{n_1} t_{l_k} \right) \alpha_i \left(\prod_{k=n_1+1}^{n} t_{l_k} \right) \alpha_j C, \qquad (2.11)$$

where the specific number of topplings n_1 and n depend on i, j, and C. Acting on general states, the operators t and α all commute because they merely linearly add or subtract heights. We can therefore shift α_i to the right in this expression:

$$a_i a_j C = \left(\prod_{k=1}^{n} t_{l_k} \right) \alpha_i \alpha_j C. \qquad (2.12)$$

Now we rearrange the product of topplings. In the nontrivial case that the α-operators render either i or j (or both) unstable, the product must contain toppling operators corresponding to those unstable sites. We shift those operators to the right. Those operators constitute by definition the update operator, U, so we can write

$$a_i a_j C = \left(\prod t_{l_k} \right) U \alpha_i \alpha_j C. \qquad (2.13)$$

The factors within the bracket are the remaining t's. Now, the update operator may leave some sites still unstable, and then the product must include further toppling operators; working on those sites, we can pull out another factor of the update operator. This procedure can be repeated until we have used all the toppling factors and the state is stable. Thus, we can identify the operator within the brackets in (2.12) as the relaxation operator R. But $\alpha_i \alpha_j C$ is the same state as $\alpha_j \alpha_i C$, so $a_i a_j C = a_j a_i C$.

A trivial consequence of this argument is that the total number of tumblings occurring in the operations $a_i a_j C$ and $a_j a_i C$ are the same. Of course, if a particular site k tumbles it can be caused by either addition; the orders of the tumblings may or may not be altered.

We now prove that the avalanche operators have unique inverses when restricted to recursive states; that is, there exists a unique operator a_i^{-1} such that $a_i(a_i^{-1}C) = C$ for all C in $\mathcal{R}$. This implies that the operators a_i acting on the recursive set generate an Abelian group. For any recursive state C we first find another recursive state such that a_i acting on it gives C, and we then show that this construction is unique.

We begin by adding a grain of sand to the state C and allowing it to relax. This generates a new recursive state $a_i C$. Now, since the state C is by assumption recursive, there is some way to add sand to regenerate C from any given state. In particular, there is some product P of addition operators a_j such that

$$C = P a_i C. \tag{2.14}$$

But the a's commute, so we have

$$C = a_i P C, \tag{2.15}$$

and thus PC is a recursive state on which a_i gives C.

We must still show that this state is unique. To do this consider repeating the above process to find a series of states C_n satisfying

$$(a_i)^n C_n = C. \tag{2.16}$$

Because on our finite system the total number of stable states is finite, this sequence must eventually enter a loop. As the original state C can be generated by running around the loop in the opposite direction, C must itself belong to the loop. Calling the length of the loop m, we have $(a_i)^m C = C$. We now uniquely define $a_i^{-1}C = a_i^{m-1}C$.

We now have sufficient machinery to count the number of recursive states. As all recursive states can be obtained by adding sand to C^*, we can write any state $C \in \mathcal{R}$ in the form

$$C = \left(\prod_i a_i^{n_i} \right) C^*. \tag{2.17}$$

Here the integers n_i represent the amount of sand to be added at the respective sites. However, in general there are several different ways to reach any given state. In particular, adding four grains of sand to any one site must force a toppling and is equivalent to adding a single grain to each of its neighbors. This can be expressed as the operator statement

$$a_i^4 = \prod_{j \in nn} a_j, \tag{2.18}$$

where the product is over the nearest neighbors of site i. We can rewrite this equation by multiplying it by the product of inverse avalanche operators on the nearest neighbors on both sides, thus obtaining

$$\prod_j a_j^{\Delta_{ij}} = E, \tag{2.19}$$

where E is the identity operator. This allows us to change the powers appearing in (2.17). If we now label states by the vector $\mathbf{n} = (n_1, n_2, n_3, \ldots, n_N)$ we see that two such states are equivalent if the difference of these vectors is of the form $\sum_j \beta_j \Delta_{ij}$, where the coefficients β_j are integers. These are the only constraints; if two states cannot be related by toppling they are independent. Thus any vector $\mathbf{n}$ can be translated repeatedly until it lies in an N- dimensional hyper-parallelopiped whose base edges are the vectors Δ_{ji}, $j = 1, \ldots \ldots, N$. The vertices of this object have integer coordinates and its volume is the number of integer coordinate points inside. This volume is just the absolute value of the determinant of Δ. *Thus the number of recursive states equals the absolute value of the determinant of the toppling matrix Δ.*

For large lattices this determinant can be found easily by Fourier transform. In particular, whereas there are 4^N stable states, there are only

$$\exp\left(N \int_{(-\pi,-\pi)}^{(\pi,\pi)} \frac{d^2q}{(2\pi)^2} \ln(4 - 2q_x - 2q_y)\right) \simeq (3.2102\ldots)^N \tag{2.20}$$

recursive states. Thus starting from an arbitrary state and adding sand, we see that the system "self-organizes" into an exponentially small subset of states forming the attractor of the dynamics.

2.3.2 An Isomorphism

Following [2.12], we now look into the consequences of stacking sandpiles on top of one another. Given stable configurations C and C' with configurations z_i and z_i', we define the state $C \oplus C'$ to be the state obtained by relaxing the configuration with heights $z_i + z_i'$. Clearly, if either C or C' are recursive states, so is $C \oplus C'$.

One can show that under the operation $\oplus$ the recursive states form an Abelian group isomorphic to the algebra generated by the a_i. First, the addition of a state C with heights z_i is equivalent to operating with a product of a_i raised to z_i, that is,

$$B \oplus C = \left(\prod_i a_i^{z_i}\right) B. \tag{2.21}$$

The operation $\oplus$ is associative and Abelian because the operators a_i are.

Since any element of a group raised to the order of the group gives the identity, it follows that $a_i^{|\Delta|} = E$. This implies the simple formula $a_i^{-1} = a_i^{|\Delta|-1}$. The analog of this for the states is the existence of an inverse state, $-C$,

Fig. 2.4. The identity state on a 286 by 184 lattice. The heights are color coded as in Fig. 2.1

$$-C = (|\Delta| - 1) \otimes C. \tag{2.22}$$

Here, $n \otimes C$ means adding n copies of C and relaxing. The state $-C$ has the property that for any state $B \oplus C \oplus (-C) = B$.

The state $I = C \oplus (-C)$ represents the identity and has the property $I \oplus B = B$ for every recursive state B. The state which is isomorphic to the operator a_i is simply $a_i I$. The identity state provides a simple way to check if a state, obtained for instance by a computer simulation, has reached the attractor, i.e., if a given state is a recursive state: A stable state is in $\mathcal{R}$ if and only if $C \oplus I = C$. The proof is simple. By construction, a recursive state has this property. On the other hand, since I is recursive, so is $C \oplus I$.

The identity state can be constructed by taking any recursive state, say C^*, and repeatedly adding it to itself to use $|\Delta| \otimes C = I$. However, on any but the smallest lattices, $|\Delta|$ is a very large integer. A simpler scheme is given in [2.12]. Figure 2.4 shows the identity state on a 286 by 184 lattice. Note the fractal structure, with features of many length scales.

Here we present another scheme to construct this state. Figure 2.5a,b shows a sequence of configurations obtained by pouring sand in from the boundaries:

Fig. 2.5 a - d. A procedure for constructing the identity. In parts **a** and **b** an empty table is having sand poured on from the edges. In part **c** the entire table is supercritical. In part **d** the boundaries have been opened, and the sand is running back off. The system finally relaxes the state shown in Fig. 2.4. In this sequence, heights are color coded as in Fig. 2.1, with active sites in various shades of *yellow*

the heights at the edges are kept very high, and the inside is initially empty. A variety of fractal structures emerge as the interior slowly fills up. Figure 2.5c shows the final stationary state where the sand falling in from the boundaries matches that falling off from the updating. Then we reverse the boundary conditions and allow the sand to fall back off. Figure 2.5d shows an intermediate state of this procedure. When the sand finally stops falling off, we obtain the identity state as shown in Fig. 2.4.

2.3.3 A Burning Algorithm and the $q = 0$ Potts Model

Majumdar and Dhar [2.11] have constructed a simple "burning" algorithm to check and enumerate the configurations belonging to the recursive set. The boundary is included as a single site, labeled 0. To decide whether a given configuration belongs to $\mathcal{R}$, consider first, at time $t = 0$, all the sites unburnt, except 0 which is burnt. An unburnt site i is defined to burn at time $t = 1$ if z_i exceeds the number of edges connecting i to other unburnt sites. Those sites burn at $t = 2$, and so on. The fires start from the edge and burn inward. Once some sites have burned, other sites may become burnable since the number of edges is reduced. The path of the fire may be indicated by bonds connecting burnt sites. If and only if all sites eventually burn under this algorithm does the state belong to the recursive set.

The burning algorithm can also be described in the following way. For a given configuration, add one particle to each of the edge sites, two particles to the corners, and update according to the usual rules. If the original state is recursive, this will generate an avalanche under which each site of the system will tumble exactly once. If the state is not recursive, some untumbled sites will remain. Figure 2.6 shows such a process underway on the lattice from Fig. 2.1. Here sites which have already burned are shown in cyan, while the remaining sites in the center have not yet burned. The small number of sites shown in light tan are the active burning sites. Note the fractal shape of the interface.

It is unclear whether the dynamics of the burning algorithm, intended to identify and count allowed configurations, has anything to do with the dynamics of avalanche formation. Majumdar and Dhar studied the scaling of the extension r of the burning cluster vs time and found

$$t \simeq r^{z'}, \quad z' \equiv \gamma'_{tr} = 5/4. \tag{2.23}$$

The spanning-tree problem corresponds to a problem in equilibrium statistical mechanics, namely the $q = 0$-state Potts model. The Potts model is defined in terms of the Hamiltonian

$$H = q^{1/2} \sum_{ij} \delta(\sigma_i, \sigma_j), \tag{2.24}$$

where the σ are q-state spins and the summation is over nearest neighbors on a two-dimensional lattice. Several results about the Potts model are known from an equivalence with the two-dimensional Coulomb gas model and from con-

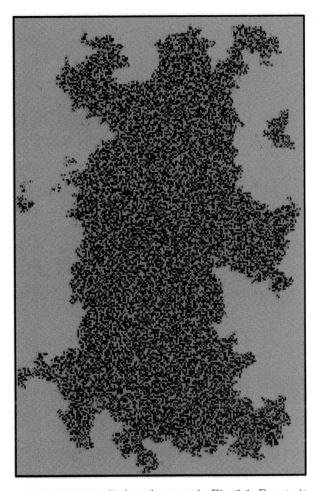

Fig. 2.6. The burning algorithm applied to the state in Fig. 2.1. Burnt sites are *cyan*, burning sites are *yellow*, and the remaining sites are colored as in Fig. 2.1

formal field theory. From this analogy Majumdar and Dhar [2.10] argue that the height - height correlation $\langle z(r')z(r'-r)\rangle - \langle z\rangle^2$ varies as r^{-2x_t} for large separations r, where $x_t = 2$ is the exponent for the energy - energy correlation function for the Potts model. The height variable in the sandpile model corresponds to the energy density for the Potts model. This again indicates that there are strong correlations built into the seemingly random configurations obtained numerically (Fig. 2.1) for the sandpile.

2.4 Real Sandpiles and Earthquakes

2.4.1 The Dynamics of Sand

"Sand" represents a "state of matter" to which not much attention has been paid. Sand can be shaped into many different forms; that is, it can exist in very many stable states, almost all of which are out of the flat equilibrium

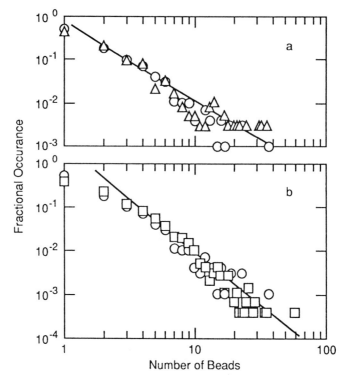

Fig. 2.7a,b. The avalanche distribution from recent experiments by Grumbacher et al. [2.16]

state. Thus, sand contains memory: one can write letters in sand. If a heap of sand is perturbed, for instance by adding more sand, by tilting the pile, or by shaking it, the system goes from one metastable state to another. In some sense this happens by a diffusion process, but this process is very different from the process which relaxes a glass of water to equilibrium after being shaken. The diffusion of sand can stop at any of many stable states, and the process is a threshold process, since nothing happens before the perturbation reaches a certain magnitude. It might therefore be reasonable to expect real sandpiles to exhibit self-organized critical behavior.

After a couple of false starts with inconclusive results, there are now several experiments reporting power law distributions of avalanches [2.14-16]. Figure 2.7 shows the results of recent experiments by Grumbacher et al. [2.16]. They built small heaps on a scale, and monitored the distribution of avalanches of particles falling off the edges. The figure shows log - log plots of the normalized distribution function of avalanches. The experiments were performed using iron spheres (*triangles*) and glass spheres (*circles*) of the same size. In all cases a power law distribution function was found. In a remarkable experiment Bretz et al. [2.15] were even able to image the flow of small avalanches not reaching the edge and measure their flow and size.

The threshold dynamics of sand is a paradigm of many processes in nature. Earthquakes occur only when the stress somewhere on the crust of the earth exceeds a critical value, and the earthquake takes the crust from one stable state to another. Economic systems are driven by threshold processes; the individual agents may change their behavior only when conditions reach a certain level.

Biological species emerge or die when specific conditions in the ecology are fulfilled. Neurons in a neural network fire when the input from other neurons reaches a threshold level, etc. Indeed, the ideas of SOC have been applied to these and many other natural phenomena, including volcanic activity [2.17] and solar flares [2.18].

2.4.2 Earthquakes and SOC

The idea of self-organized criticality as applied to earthquakes [2.19-23], may be visualized as follows. Think of the lithosphere of the earth as a collection of tectonic plates, being squeezed very, very slowly into each other. At some time in our geological history the stresses may have been small, without large ruptures or earthquakes. During millions of years, however, the system evolved into a kind of stationary state where the build-up of stress is balanced on average by its release during earthquakes. Because of the long evolutionary process, the crust has "learned," by suitably arranging the building blocks at hand into a very balanced network of faults, valleys, mountains, oceans, and other geological structures, to respond critically to any initial rupture. The earthquake can be thought of as a critical chain reaction where the process is just barely able to continue.

For a simple model, consider a two-dimensional lattice of interacting blocks. The initial block structure represents a discretization of the space in much the same way as the lattice in lattice gauge theories of particle physics. The block size does not represent an intrinsic length scale in the problem. On each block, at site (i,j) a force $F_{i,j}$ acts in an unspecified general direction of motion in some fault region. In the beginning, $F_{i,j}$ may assume small random values. The initial state is not important for the long term dynamics. Let the force increase uniformly by a infinitesimally small amount per unit time; this simulates the slow driving by the tectonic plate motion. Eventually, the force at some site (i,j) will exceed a critical threshold value F_C for rupture. The value chosen for F_C may be either uniform or random. The initial rupture is simulated by updating the forces at the critical site and the sites of the neighbors at $(i, j\pm 1)$ and $(i\pm 1, j)$:

$$
\begin{aligned}
F_{i,j} &\longrightarrow 0 , \\
F_{nn} &\longrightarrow F_{nn} + \alpha F_{i,j}
\end{aligned}
\tag{2.25}
$$

where nn denotes the nearest neighbors and α is an adjustable parameter.

These equations represent the transfer of force to the neighbors. This may cause the neighbors to be unstable and a chain reaction to take place. This chain reaction is the earthquake. The equations are completely deterministic, with no external noise. We are not dealing with a noise-driven phenomenon; on the contrary, the physics turns out to be stable with respect to noise, i.e., noise is irrelevant. When the earthquake stops, the system is quiescent until the force at some other location exceeds the critical value and a new event is initiated. The process continues again and again. One observes that for some time the earthquakes become bigger and bigger. When one is convinced that the system has self-organized into a stationary state, one can start measuring the energies of subsequent earthquakes as defined by the total number of rupture events following a single initial rupture.

This model was suggested by Olami, Christensen and Feder [2.22], who realized that the picture could be directly related to earlier spring-block models. The value of α is directly related to the elastic parameters of the crust of the earth. For $\alpha = 1/4$ the force is conserved, i.e., the amount lost on the unstable site equals the total amount gained by the 4 neighboring sites. The criticality in this case has been observed to prevail for values of α down to 0.05, with only 20% conservation. This came as a surprise since there was then a widespread belief that the lack of conservation would spontaneously generate a length scale, i.e., a "characteristic earthquake size." In fact, it seems that criticality occurs generically almost independently of the details of the toppling rule. Another model without conservation which appears to exhibit self-organized criticality is the "game of life" [2.23] (Sect. 9.4), a cellular automaton representing a society of living and dying individuals. Perhaps intermittent fluctuations in evolution, like the extinction of the dinosaurs, can be seen as manifestations of SOC.

Figure 2.8 shows the distribution of earthquakes for $\alpha = 0.20$. The straight line yields a power-law distribution with an exponent $b = 1 - \tau = 0.8$. The line indicates that the system has self-organized into the critical state. Indeed, real earthquakes exhibit a power law distribution, known as the Gutenberg-Richter law. The slope depends on the degree of dissipation, $(1/4 - \alpha)$, so there is no universality of the exponent b in the nonconservative case. One should not look for unique b-values in nature. Indeed different values have been observed in different geographical areas.

The power-law distribution of earthquakes stems from the fractal nature of the SOC state, with correlated regions ranging over all length scales; these correlated regions, generated by the long-term dynamics, are the equivalent of the active faults or fault segments in real earthquakes. The fault structure changes on large geological time scales. More realistic long-range SOC models produce faults which topologically look much more like a real fractal arrangement of two-dimensional faults in a three-dimensional matrix [2.21].

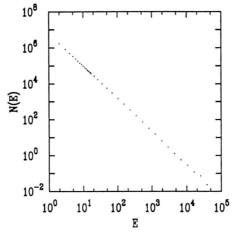

Fig. 2.8. The distribution of earthquakes for $\alpha = 0.20$ for the model described in the text. The straight line yields a power-law distribution with an exponent $b = 1 - \tau = 0.8$

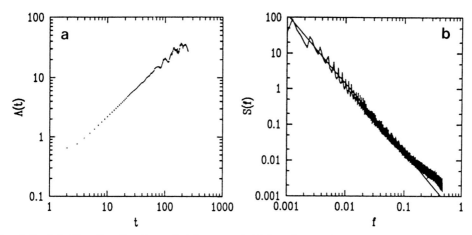

Fig. 2.9. (a) The distribution of Λ for $\alpha = 0.20$ for the earthquake model described in the text. The slope of the straight line gives $\mu = 0.8$. (b) A direct measurement of the power spectrum in the earthquake model

2.5 1/f Noise

One-over-f, $1/f$, noise can thought of as a signal arising from the superposition of avalanches occurring in a self-organized critical state [2.2]. In order to illustrate how this works, consider the weighted lifetime distribution of avalanches $\Lambda(t)$ as defined in Sect. 2.2. Christensen et al. [2.6] showed that if Λ has a scaling behavior, $\Lambda(t) \sim t^{\mu}$, then the power spectrum $S(f)$ becomes

$$S(f) \sim f^{-1-\mu}, \quad \text{for } -1 < \mu < 1. \tag{2.26}$$

Figure 2.9a shows the distribution of Λ for $\alpha = 0.20$ for the above earthquake model [2.24]. The slope of the straight line gives $b + 1 = 1.8$.

Figure 2.9b shows a direct measurement of the power spectrum; an exponent $\phi = 1.75$ was found from the slope of the log - log plot, in reasonable agreement with the value 1.8 expected from the lifetime distribution function defined above.

In nature, values of the exponent of the $1/f$ noise in the interval 0.6–2.0 have been reported [2.25]. A value of 1 corresponds to $\mu = 0$; this particular value is obtained for $\alpha \simeq 0.11$ in the model. The exponent does depend on the parameters of the model and thus is not universal, unlike the conventional exponents for equilibrium critical phenomena.

2.6 On Forest Fires and Turbulence

A liquid driven by imposing a velocity difference v over a length scale L undergoes a transition to a turbulent state with vortices exhibiting a large range of length scales. The temporal variations of the velocity at a given spot are intermittent, with large and small bursts of activity. The energy is dissipated locally within a short length scale known as the Kolmogorov length.

Mandelbrot [2.1] has suggested that in turbulent systems the dissipation of energy is confined to a fractal structure with features of all length scales. This behavior can be simulated by a simple forest-fire model [2.26]. Distribute randomly a number of trees (*green dots*) and a number of fires (*yellow dots*) on a two-dimensional rectangular lattice. Sites can also be empty. Update the system at each time t as follows: (1) grow new trees with probability p from sites that are empty at time $t-1$; (2) trees that were on fire at time $t-1$ die (become empty sites) and are removed at time t; (3) a tree that has a fire as a nearest neighbor at time $t-1$ catches fire at time t. Periodic boundaries are used. After a while the system evolves to a critical state with fire fronts of all sizes (Fig. 2.10). Drossel and Schwabl [2.27] have extended the model by adding a small probability f of igniting new fires at each time step. In the limit $f/p \to 0$ the ignitions create forest fires where the number of trees burned, s, follows a power law, $P(s) \sim s^{1-\tau}$, with $\tau \simeq 3$. (For percolation models of forest fires, see, e.g., Chap. 2 in [2.28].)

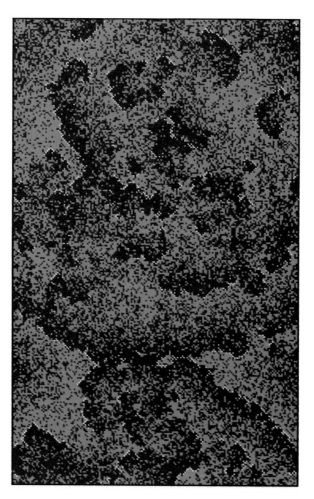

Fig. 2.10. A typical state in the dynamics of the forest-fire model described in the text. Trees are *green*, fires *yellow*, and empty sites *black*. The periodic lattice is 286 by 178 sites and new trees are born on empty sites with a probability of 1/32 per time step

The physics of earthquakes can be described in a similar language. The crust in a fault region is driven by imposing a force or a strain over a large length L. In the stationary state, the energy is dissipated in narrow fault structures forming a fractal set. The spatio-temporal correlation functions for the two phenomena are quite similar although the time scales are vastly different. In both cases, the energy enters the system uniformly (zero wave vector) and leaves the system locally. There are a lot of similarities between the earthquake model studied by Olami et al. [2.22] and the forest fire model. The analogy has been explored in some detail by Kagan [2.29]. Maybe it is useful to think of self-organized criticality and turbulence as one and the same phenomenon.

Acknowledgements. We are grateful for Kim Christensen for discussions and help with preparing figures. This work was supported by the U.S. Department of Energy under Contract No. DE-AC02-76CH00016. The color figures were produced on a Commodore Amiga.

References

2.1   B.B. Mandelbrot: *The Fractal Geometry of Nature* (Freeman, San Francisco 1982)
2.2   P. Bak, C. Tang, K. Wiesenfeld: Phys. Rev. Lett. **59**, 381 (1987); Phys. Rev. A **38**, 3645 (1988)
2.3   C. Tang, P. Bak: Phys. Rev. Lett. **60**, 2347 (1988)
2.4   K. Wiesenfeld, P. Bak, C. Tang: J. Stat. Phys. **54**, 1441 (1989)
2.5   P. Bak, K. Chen: Sci. Am. (January 1991) p. 46
2.6   K. Christensen, H.C. Fogedby, H.J. Jensen: J. Stat. Phys. **63**, 653 (1991)
2.7   K. Christensen: Thesis, University of Aarhus
2.8   D. Dhar: Phys. Rev. Lett. **64**, 1613 (1990)
2.9   D. Dhar, R. Ramaswamy: Phys. Rev. Lett. **63**, 1659 (1989)
2.10   D. Dhar, S. N. Majumdar: J. Phys. A **23**, 4333 (1990)
2.11   S.N. Majumdar, D. Dhar: Physica A **185**, 129 (1992)
2.12   M. Creutz: Comp. in Phys. **5**, 198 (1991)
2.13   M. Creutz: Nuc. Phys. B (Proc. Suppl.) **20**, 748 (1992)
2.14   J. Rosendahl, M. Vekii, J. Kelley: Phys. Rev. E. **47**, 1401 (1993)
2.15   M. Bretz, J.B. Cunningham, P.L. Kurczynski, F. Nori: Phys. Rev. Lett. **69**, 2431 (1992)
2.16   S.K. Grumbacher, K.M. McEwen, D.A. Halvorson, D.T. Jacobs, J. Lindler: Am. J. Phys. **61**, 329 (1993)
2.17   P. Diodati, F. Marchesoni, S. Piazza: Phys. Rev. Lett. **67**, 2239 (1991)
2.18   E.T. Lu, R.J. Hamilton: Astrophys. J. **L89**, 380 (1991)
2.19   P. Bak, C. Tang: J. Geophys. Res. B **94**, 15635 (1989)
2.20   K. Ito, M. Matsuzaki: J. Geophys. Res. B **95**, 6853 (1989)
2.21   K. Chen, P. Bak, S. P. Obukhov: Phys. Rev. A **43**, 625 (1991)
2.22   Z. Olami, H.J. Feder, K. Christensen: Phys. Rev. Lett. **68**, 1244 (1992)
2.23   P. Bak, K. Chen, M. Creutz: Nature **342**, 780 (1989)
2.24   K. Christensen, Z. Olami, P. Bak: Phys. Rev. Lett. **68**, 2417 (1992)
2.25   W. H. Press: Comments Astrophys. **7**, 103 (1978)
2.26   P. Bak, K. Chen, C. Tang: Phys. Lett. **147**, 297 (1990)
2.27   B. Drossel, F. Schwabl: Phys. Rev. Lett. **69**, 1629 (1992)
2.28   A. Bunde and S. Havlin, eds.: *Fractals and Disordered Systems* (Springer, Heidelberg 1991)
2.29   Y. Kagan: Nonlinear Science Today **2**, 1 (1992)

Self-Similarity of DNA Walks

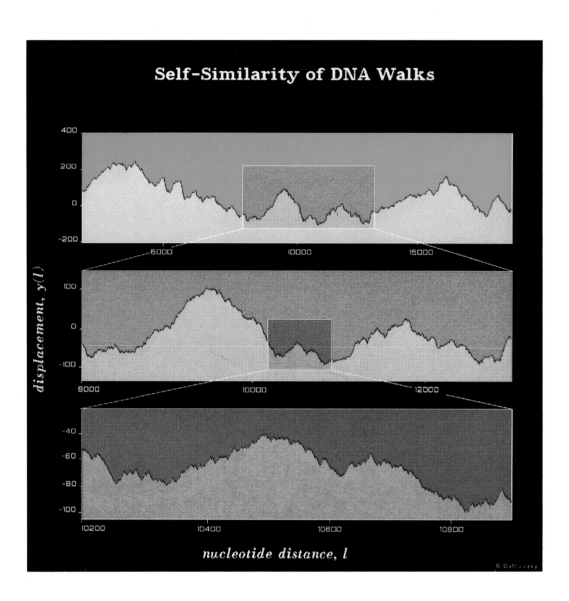

3 Fractals in Biology and Medicine: From DNA to the Heartbeat

Sergey V. Buldyrev, Ary L. Goldberger, Shlomo Havlin, C.-K. Peng, and H. Eugene Stanley

3.1 Introduction

The purpose of this chapter is to describe some recent progress in applying fractal concepts to systems of relevance to biology and medicine. We review several biological systems characterized by fractal geometry, with a particular focus on the long-range power-law correlations found recently in DNA sequences containing *noncoding* material. Furthermore, we discuss the finding that the exponent α quantifying these long-range correlations ("fractal complexity") is smaller for coding than for noncoding sequences. We also discuss the application of fractal scaling analysis to the dynamics of heartbeat regulation, and report the recent finding that the normal heart is characterized by long-range "anticorrelations" which are absent in the diseased heart.

In the last decade it was realized that some biological systems have no characteristic length or time scale, i.e., they have fractal—or, more generally, self-affine—properties [3.1,2] (see Chap. 1). However, the fractal properties in **different** biological systems, have quite different nature, origin, and appearance. In some cases, it is the geometrical shape of a biological object itself that exhibits obvious fractal features, while in other cases the fractal properties are more "hidden" and can only be perceived if data are studied as a function of time or mapped onto a graph in some special way. After an appropriate mapping, such a graph may resemble a mountain landscape, with jagged ridges of all length scales from very small bumps to enormous peaks (see Fig. 3.0). Mathematically, these landscapes can be quantified in terms of fractal concepts

---

◀ **Fig. 3.0.** The DNA walk representation for the rat embryonic skeletal myosin heavy chain gene ($\alpha = 0.63$). At the top the entire sequence is shown. In the middle the solid box shown in the top is magnified. At the bottom the solid box shown in the middle is magnified. The statistical self-similarity of these plots is consistent with the existence of a scale-free or fractal phenomenon which we call a fractal landscape. Note that one must magnify the segment by different factors along the ℓ (horizontal) direction and the y (vertical) direction; since F has the same units (dimension) as y, these magnification factors M_ℓ and M_y (along ℓ and y directions respectively) are related to the scaling exponent α by the simple relation $\alpha = \log(M_y)/\log(M_\ell)$ [e.g., from top to middle, $\log(M_y)/\log(M_\ell) = \log(2.07)/\log(3.2) = 0.63$]

such as self-affinity. The main part of the chapter is devoted to the study of such hidden fractal properties that have been recently discovered in DNA sequences and heartbeat activity.

3.2 Fractal Shapes

In contrast to compact objects, fractal objects have a very large *surface* area. In fact, they are composed almost entirely of "surface." This observation explains why fractals are ubiquitous in biology, where surface phenomena are of crucial importance (see also Chap. 6 (by B. Sapoval) of [3.3]).

Lungs exemplify this feature (see the picture on page II and Fig. 3.1). The surface area of a human lung is as large as a tennis court. The mammalian lung is made up of self-similar branches with many length scales, which is the defining attribute of a fractal surface. The efficiency of the lung is enhanced by this fractal property, since with each breath oxygen and carbon dioxide have to be exchanged at the lung surface. The structure of the bronchial tree has been quantitatively analyzed using fractal concepts [3.2,4]. In particular, fractal geometry could explain the power law decay of the average diameter of the bronchial tube with the generation number, in contrast to the classical model which predicts an exponential decay [3.6].

Not only the geometry of the respiratory tree is described by fractal geometry, but also the time-dependent features of inspiration. Specifically, Suki et al. [3.5] studied airway opening in isolated dog lungs. During constant flow inflations, they found that the lung volume changes in discrete jumps (Fig. 3.1), and that the probability distribution function of the relative size x of the jumps, $\Pi(x)$, and that of the time intervals t between these jumps, $\Pi(t)$, follow a power law over nearly two decades of x and t with exponents of 1.8 and 2.7, respectively. To interpret these findings, they developed a branching airway model in which airways, labeled ij, are closed with a uniform distribution of opening threshold pressures P. When the "airway opening" pressure P_{ao} exceeds P_{ij} of an airway, that airway opens along with one or both of its daughter branches if $P_{ij} < P_{ao}$ for the daughters. Thus, the model predicts "avalanches" of airway openings with a wide distribution of sizes, and the statistics of the jumps agree with those $\Pi(x)$ and $\Pi(t)$ measured experimentally. They concluded that power law distributions, arising from avalanches triggered by threshold phenomena, govern the recruitment of terminal airspaces.

A second example is the arterial system which delivers oxygen and nutrients to all the cells of the body. For this purpose blood vessels must have fractal properties [3.7,8]. The diameter distribution of blood vessels ranging from capillaries to arteries follows a power-law distribution which is one of the main characteristics of fractals. Sernetz et al. [3.9] have studied the branching patterns of arterial kidney vessels (see front cover). They analyzed the mass-radius relation and found that it can be characterized by fractal geometry, with fractal dimensions between 2.0 and 2.5. Similarly, the branching of trees and other

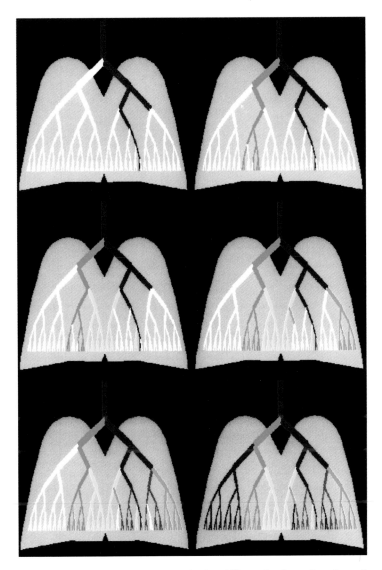

Fig. 3.1. The dynamic mechanism responsible for filling the lung involves "avalanches" or "bursts" of air that occur in all sizes—instead of an exponential distribution, one finds a power law distribution [3.5]. The underlying cause of this scale-free distribution of avalanches is the fact that every airway in the lung has its own threshold below which it is not inflated. Shown here is a diagram of the development of avalanches in the airways during airway opening. At first, almost all airways whose threshold value is smaller than the external pressure (red) are closed. Then the airway opening pressure increases until a second threshold is exceeded, and as a result all airways further up the tree whose thresholds are smaller become inflated (green). The airway opening pressure is successively increased until third, fourth, and fifth thresholds are exceeded (yellow, brown, and blue). The last threshold to be exceeded results in filling the airways colored violet; we notice that this last avalanche opens up over 25% of the total lung volume, thereby significantly increasing the total surface area available for gas exchange. After [3.5]

plants, as well as root systems have a fractal nature [3.10]. Moreover, the size distribution of plant-supported insects was found to be related to the fractal distribution of the leaves [3.11].

One of the most remarkable examples of a fractal object is the surface of a cauliflower, where every little head is an "almost" exact reduced copy of the whole head formed by intersecting Fibonacci spirals of smaller heads, which in turn consist of spirals of smaller and smaller heads, up to the fifth order of hierarchy (see Fig. 8.0 in [3.3]). West and Goldberger were first to describe such a "Fibonacci fractal" in the human lung [3.2]. (For a general review of fractals in physiology and medicine see also Goldberger, Rigney, and West [3.2])

Considerable interest in the biological community has also arisen from the possibility that neuron shape can be quantified using fractal concepts. For example, Smith et al. [3.12] studied the fractal features of vertebrate central nervous system neurons in culture and found that the fractal dimension is increased as the neuron becomes more developed. Caserta et al. [3.13] showed that the shapes of quasi-two-dimensional retinal neurons can be characterized by a fractal dimension d_f. They found for fully developed neurons *in vivo*, $d_f = 1.68 \pm 0.15$, and suggest that the growth mechanism for neurite outgrowth bears a direct analogy with the growth model called *diffusion limited aggregation* (DLA), see Chap. 1 (by H.E. Stanley) and Chap. 4 (by A. Aharony) in [3.3] and Fig. 3.2. The branching pattern of retinal vessels in a developed human eye is also similar to DLA [3.8]. The fractal dimension was estimated to be about 1.7, in good agreement with DLA for the case of two dimensions. For an alternative model for retinal growth see [3.14].

The DLA-type model governing viscous fingering may also serve to resolve the age-old paradox *"Why doesn't the stomach digest itself?"* [3.15]. Indeed, the concentration of hydrochloric acid in the mammalian stomach after each meal is sufficient to digest the stomach itself, yet the gastric epithelium normally remains undamaged in this harsh environment. One protective factor is gastric mucus, a viscous secretion of specialized cells, which forms a protective layer and acts as a diffusion barrier to acid. Bicarbonate ion secreted by the gastric

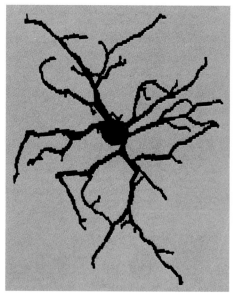

Fig. 3.2. Photograph of a retinal neuron (nerve cell), the morphology is similar to the DLA archetype. After [3.13]

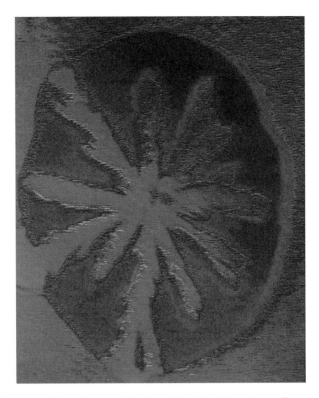

Fig. 3.3. Viscous fingers reflect the complex interface that develops when one fluid is pumped through another of higher viscosity. Shown is the formation of such viscous fingers or channels when hydrochloric acid is injected into solutions of gastric mucin. These channels may confine the acid and direct it to the lumen, thus protecting the gastric mucosa from acidification and ulceration; when the gastric glands contract, acid is ejected under high enough pressure to form viscous fingers. After [3.15]

epithelium is trapped in the mucus gel, establishing a gradient from pH 1-2 at the lumen to pH 6-7 at the cell surface. The puzzle, then, is how hydrochloric acid, secreted at the base of gastric glands by specialized parietal cells, traverses the mucus layer to reach the lumen without acidifying the mucus layer. Bhaskar et al. [3.15] resolved this puzzle by experiments that demonstrate the the possibility that flow of hydrochloric acid through mucus involves viscous fingering—the phenomenon that occurs when a fluid of lower viscosity is injected into a more viscous one (see Fig. 3.3). Specifically, Bhaskar et al. demonstrated that injection of hydrochloric acid through solutions of pig gastric mucin produces fingering patterns which are strongly dependent on pH, mucin concentration, and acid flow rate. Above pH 4, discrete fingers are observed, while below pH 4, hydrochloric acid neither penetrates the mucin solution nor forms fingers. These *in vitro* results suggest that hydrochloric acid secreted by the gastric gland can penetrate the mucus gel layer (pH 5-7) through narrow fingers, whereas hydrochloric acid in the lumen (pH 2) is prevented from diffusing back to the epithelium by the high viscosity of gastric mucus gel on the luminal side.

Yet another example of DLA-type growth is bacterial colony spread on the surface of agar (gel with nutrient) plates [3.16] (see Fig. 3.4). Vicsek et al. [3.17]

Fig. 3.4. A typical example of DLA-like colony patterns incubated at 35^OC for three weeks after inoculation on the surface of agar plates containing initially 1 g/ℓ of peptone as nutrient. This pattern has a fractal dimension of $d_f \cong 1.72$. After Matsushita and Fujikawa [3.16]

studied bacterial colony growth on a strip geometry which results in a self-affine surface (see Fig. 13.19 in [3.18]). They calculated the roughness exponent α for this surface and found $\alpha = 0.78 \pm 0.07$ (see also Chap. 4). The interfacial pattern formation of the growth of bacterial colonies was studied systematically by Ben-Jacob et al. [3.19]. They demonstrated that bacterial colonies can develop a pattern similar to morphologies in diffusion-limited growth observed in solidification and electro-chemical deposition. These include fractal growth, dense-branching growth, compact growth, dendritic growth and chiral growth. The results indicate that the interplay between the micro level (individual bacterium) and the macro level (the colony) play a major role in selecting the observed morphologies similar to those found in nonliving systems.

Another example of fractal interface appears in ecology, in the problem of the territory covered by N diffusing particles [3.20], see Fig. 3.5. As seen from the figure, the territory initially grows with the shape of a disk with a relatively smooth surface until it reaches a certain size, at which point the surface becomes increasingly rough. This phenomenon may have been observed by Skellam [3.21] who plotted contours delineating the advance of the muskrat population and noted that initially the contours were smooth but at later times they became rough (see Fig. 1 in [3.21]).

Other biological contexts in which fractal scaling seems to be relevant are the relation between brain size and body weight [3.22], between bone diameter and bone length [3.23], between muscle force and muscle mass [3.23], and between an organism's size and its rate of producing energy and consuming food [3.24].

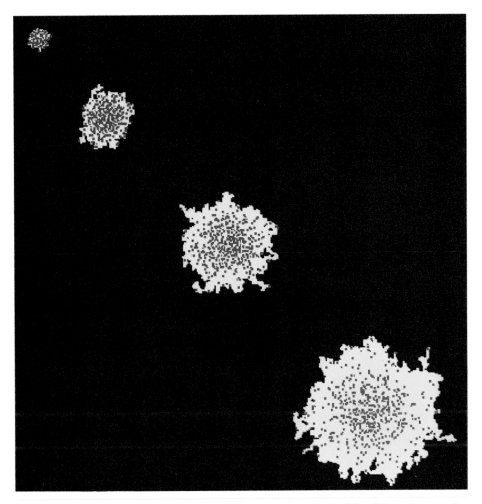

Fig. 3.5. Snapshots at successive times of the territory covered by N random walkers for the case $N = 500$ for a sequence of times. Note the roughening of the disc surface as time increases. The roughening is characteristic of the experimental findings for the diffusive spread of a population [3.21]. After [3.20], courtesy of P. Trunfio

3.3 Long-Range Power Law Correlations

In recent years long-range power-law correlations have been discovered in a remarkably wide variety of systems. Such long-range power-law correlations are a physical fact that in turn gives rise to the increasingly appreciated "fractal geometry of nature" [3.1,3,18,25–28]. So if fractals are indeed so widespread, it makes sense to anticipate that long-range power-law correlations may be similarly widespread. Indeed, recognizing the ubiquity of long-range power-law correlations can help us in our efforts to understand nature, since as soon as we find power-law correlations we can quantify them with a critical exponent. Quantification of this kind of scaling behavior for apparently unrelated systems allows us to recognize similarities between different systems, leading to underlying unifications that might otherwise have gone unnoticed.

Our intuition tells us that correlations should decay exponentially, not as power laws. Consider, e.g., a set of molecules in a linear chain ($d = 1$) with interaction between neighbors. If C_1 denotes the correlation function for two molecules that are nearest neighbors, then we expect that the correlation function for any two molecules separated by a distance r is [3.29]

$$C_r = (C_1)^r = e^{-r/\xi} \qquad (3.1a)$$

where the second equality in (3.1a) serves to define the correlation length $\xi \equiv -1/\log C_1$.

For $d = 1$ site percolation (see Chap. 2 in [3.3]), C_r denotes the pair connectedness, the probability that a site at position r is both occupied and also connected by a string of occupied sites to an occupied site at the origin. Again, (3.1a) holds, but now with $C_1 = p$, the probability a site is occupied.

Our initial prediction, that correlations decay exponentially because of the fashion in which order is "propagated," seems to work well with a major exception—at the critical point [3.29] the exponential decay of (3.1a) turns into to a power law decay

$$C_r \sim (1/r)^{d-2+\eta} \qquad (3.1b)$$

The difference between (3.1a) and (3.1b) is profound: (3.1a) states that there is a characteristic length ξ fixed by the strength of the nearest-neighbor correlation C_1, while (3.1b) states that there is *no* characteristic length at all.

Can we intuitively understand how it is possible to find a *nonexponential* decay of correlations? At first glance, it might appear that whenever we increment the distance between two molecules by one lattice constant, the correlation should decrease by roughly the same factor, but this intuition leads immediately to exponential decay. A possible resolution to this paradox stems from the fact that near a critical point, "information" propagates from a molecule at the origin to a molecule at position $\boldsymbol{r}$ *not via a single path* (as for $d = 1$), but *rather via an infinite number of paths*; some of these paths are explicitly enumerated in Fig. 9.4 of [3.29]. Ornstein and Zernike [3.30] recognized this fact, but approximated the fashion in which "order is propagated" and so obtained predictions that today we call "classical" (Fig. 7.5 of [3.29]). Exact enumeration methods, such as high-temperature series expansions, take into account exactly such paths up to a certain length k_{max}, where k_{max} is typically 20. To obtain power law correlations, the exact results for $k < 20$ are extrapolated to obtain an estimate of the behavior for all k. In some sense, although the correlation along each path *decreases* exponentially with the length of the path, the number of such path *increases* exponentially. Therefore, the net effect is that the dominant exponential decay is "magically canceled," leaving the sub-dominant but longer-range power-law correlations—which are in fact observed.

At one time, it was imagined that the "scale-free" case of (3.1b) was relevant to only a fairly narrow slice of physical phenomena—only to systems that had been "tuned" by exceedingly painstaking experimental work to be exactly at a critical point [3.29,31]. Now we appreciate the ubiquity of systems displaying such scale-invariant behavior. First of all, any system examined on length scales smaller than the correlation length is likely to display power-law behavior (because all paths between the origin and r are relevant up to the correlation length, and these cancel out the exponential decay for $r < \xi$). Moreover, the number and nature of systems displaying power law correlations has increased dramatically, including systems that no one might ever have suspected as falling under the umbrella of "critical phenomena." The latter part of the century has witnessed a veritable explosion in the study, both experimental and theoretical, of such systems. The *1991 Nobel Prize* was awarded to P.-G. de Gennes in part for his recognition that polymer systems behave analogously to systems near their critical points. The *1993 Wolf Prize* was awarded to Benoit Mandelbrot for the recognition of the "fractal geometry of nature."

Indeed, many systems drive themselves spontaneously toward critical points. One of the simplest models exhibiting such "self-organized criticality" (see Chap. 2 and [3.32]) is invasion percolation, a generic model that has recently found applicability to describing anomalous behavior of rough interfaces (see Chap. 4 and [3.18]). Instead of occupying all sites with random numbers below a pre-set parameter p, in invasion percolation one "grows" the incipient infinite cluster right at the percolation threshold by the trick of occupying always the perimeter site whose random number is smallest. Thus small clusters are certainly not scale-invariant and in fact contain sites with a wide distribution of random numbers. As the mass of the cluster increases, the cluster becomes closer and closer to being scale invariant or "fractal." Such a system is said to drive itself to a "self-organized critical state" [3.32].

In the following sections we will attempt to summarize the key findings of some recent work [3.33–57] suggesting that—under suitable conditions—the sequence of base pairs or "nucleotides" in DNA also displays power-law correlations. The underlying basis of such power law correlations is not understood at present, but this discovery has intriguing implications for molecular evolution and DNA structure, as well as potential practical applications for distinguishing coding and noncoding regions in long nucleotide chains.

3.4 Information Coding in DNA

The role of genomic DNA sequences in coding for protein structure is well known [3.58,59]. The human genome contains information for approximately 100,000 different proteins, which define all inheritable features of an individual. The genomic sequence is likely the most sophisticated information database created by nature through the dynamic process of evolution. Equally remarkable is the precise transformation of information (duplication, decoding, etc) that occurs in a relatively short time interval.

The building blocks for coding this information are called *nucleotides*. Each nucleotide contains a phosphate group, a deoxyribose sugar moiety and either a *purine* or a *pyrimidine base*. Two purines and two pyrimidines are found in

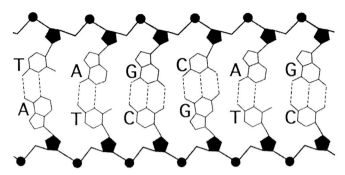

Fig. 3.6. The base pairing of two "double helix" DNA strands. The two chains of black pentagons and circles represent sugar-phosphate backbones of DNA strands linked by the hydrogen bonds (dashed lines) between complementary base pairs

DNA. The two purines are adenine (A) and guanine (G); the two pyrimidines are cytosine (C) and thymine (T). The nucleotides are linked end to end, by chemical bonds from the phosphate group of one nucleotide to the deoxyribose sugar group of the adjacent nucleotide, forming a long polymer (*polynucleotide*) chain. The information content is encoded in the sequential order of the bases on this chain. Therefore, as far as the information content is concerned, a DNA sequence can be most simply represented as a symbolic sequence of four letters: A, C, G and T, as shown in Fig. 3.6.

The double helix structure of DNA was discovered by James Watson and Francis Crick in 1953 (see Fig. 3.6). They deduced that two polynucleotide chains are helically interwined together such that the sugar-phosphate "backbones" are on the outside of the DNA molecule and that the purine and pyrimidine bases are on the inside. The bases on the opposing chains are bonded together by hydrogen bonds. The hydrogen bonds are very selective: Adenine (A) can pair only with thymine (T) by two hydrogen bonds (so called "weak bonded" base pair); guanine (G) can pair only with cytosine (C) by three hydrogen bonds (so called "strong bonded" base pair). This complementary property of DNA sequences is crucial for self-replication. The molecular mass of a purine is larger than molecular mass of a pyrimidine. A purine on one DNA strand is always coupled with a pyrimidine on the other strand, so that the total numbers of purines in the DNA molecule is always exactly equal to the number of pyrimidines. However, one strand may have large local excess of purines over pyrimidines which is called strand bias. Nevertheless, for long sequences, purine-pyrimidine content is roughly 50% even for one strand. The local concentration of each of four base pairs also have large fluctuations.

The genetic code directing amino acid assembly is now well understood. It was found that three nucleotides (*codon*) code for one amino acid, therefore, 64 different permutations are used to code for the 20 different amino acids that found in nature (with *degeneracy of the code*). To produce a protein, a segment of DNA is transcribed to an RNA molecule and the genetic code is read in a sequential order of non-overlapping triplets (codons) from the initiator methionine to the stop codon. The segment of DNA sequence that codes for one protein is called a *gene*.

However, in the genomes of high eukaryotic organisms only a small portion of the total genome length is used for protein coding (as low as 5% in the human genome). For example, genes are separated from each other by *intergenic sequences* which are not used for coding proteins and which (especially in mammalian genomes) can be several times longer than genes. Furthermore, in 1977 it was discovered that genes themselves have inclusions which are not used for coding proteins. A gene is transcribed to RNA (pre-mRNA) and then some segments of the pre-mRNA are "spliced out" during the formation of the smaller mRNA molecule. The mRNA then serves as the template for assembling protein. The segments of the chromosomal DNA that are spliced out during the formation of a mature mRNA are called *introns* (for intervening sequences). The coding sequences are called *exons* (for expressive sequences).

The role of introns and intergenomic sequences constituting large portions of the genome remains unknown. Furthermore, only a few quantitative methods are currently available for analyzing information which is possibly encrypted in the noncoding part of the genome.

3.5 Conventional Statistical Analysis of DNA Sequences

DNA sequences have been analyzed using a variety of models that can basically be considered in two categories. The first types are "local" analyses; they take into account the fact that DNA sequences are produced in sequential order; therefore, the neighboring nucleotides will affect the next attaching nucleotide. This type of analysis, represented by n-step Markov models, can indeed describe some observed short-range correlations in DNA sequences. The second category of analyses is more "global" in nature; they concentrate on the presence of repeated patterns (such as periodic repeats and interspersed base sequence repeats) that are chiefly found in eukaryotic genomic sequences. A typical example of analysis in this category is the Fourier transform, which can identify repeats of certain segments of the same length in nucleotide sequences [3.58].

However, DNA sequences are more complicated than these two standard types of analysis can describe. Therefore it is crucial to develop new tools for analysis with a view toward uncovering the mechanisms used to code other types of information. Promising techniques for genome studies may be derived from other fields of scientific research, including time-series analysis, statistical mechanics, fractal geometry, and even linguistics.

3.6 The "DNA Walk"

One interesting question that may be asked by statistical physicists would be whether the sequence of the nucleotides A,C,G, and T behaves like a one-dimensional "ideal gas", where the fluctuations of density of certain particles obey Gaussian law, or if there exist long range correlations in nucleotide content (as in the vicinity of a critical point). These result in domains of all size with different nucleotide concentrations. Such domains of various sizes were known for a long time but their origin and statistical properties remain unexplained. A natural language to describe heterogeneous DNA structure is long-range correlation analysis, borrowed from the theory of critical phenomena [3.29].

3.6.1 Graphical Representation

In order to study the scale-invariant long-range correlations of a DNA sequence, we first introduced a graphical representation of DNA sequences, which we term a *fractal landscape* or *DNA walk* [3.33]. For the conventional one-dimensional random walk model [3.60,61], a walker moves either "up" [$u(i) = +1$] or "down" [$u(i) = -1$] one unit length for each step i of the walk. For the case of an uncorrelated walk, the direction of each step is independent of the previous steps. For the case of a correlated random walk, the direction of each step depends on the history ("memory") of the walker [3.62–64].

One definition of the DNA walk is that the walker steps "up" [$u(i) = +1$] if a pyrimidine (C or T) occurs at position i along the DNA chain, while the walker steps "down" [$u(i) = -1$] if a purine (A or G) occurs at position i (see Fig. 3.7). The question we asked was whether such a walk displays only short-range correlations (as in an n-step Markov chain) or long-range correlations (as in critical phenomena and other scale-free "fractal" phenomena).

There are actually many possible rules of mapping of DNA sequence onto 1-dimensional random walk:

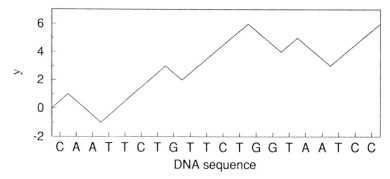

Fig. 3.7. Schematic illustration showing the definition of the "DNA walk"

Correlations of pairs of nucleotides:
(i) $u(i) = +1$ for C or T and $u(i) = -1$ otherwise ("purine-pyrimidine" rule);
(ii) $u(i) = +1$ for C or G and $u(i) = -1$ otherwise; ("hydrogen bond" rule)
(iii) $u(i) = +1$ for A or C and $u(i) = -1$ otherwise;

Correlations of one nucleotide with itself:
(iv) One can assign $u(i) = +1$ if nucleotide A occurs on the ith place and $u(i) = -1$ otherwise (in case of C,G, or T)
(v) $u(i) = +1$ for C and $u(i) = -1$ otherwise;
(vi) $u(i) = +1$ for G and $u(i) = -1$ otherwise;
(vii) $u(i) = +1$ for T and $u(i) = -1$ otherwise;

Correlations of some physical quantity:
(viii) $u(i) = 134$ for A, 110 for C, 150 for G, and 125 for T (molecular mass rule).

There have also been attempts to map DNA sequence onto multi-dimensional DNA walks [3.34,65]. However, recent work [3.57] indicates that the original purine-pyrimidine rule provides the most robust results, probably due to the purine-pyrimidine chemical complementarity.

The DNA walk allows one to visualize directly the fluctuations of the purine-pyrimidine content in DNA sequences: Positive slopes on Fig. 3.8 correspond to high concentration of pyrimidines, while negative slopes correspond to high concentration of purines. Visual observation of DNA walks suggests that the coding sequences and intron-containing noncoding sequences have quite different landscapes. Figure 3.8a shows a typical example of a gene that contains a significant fraction of base pairs that do *not* code for amino acids. Figure 3.8b shows the DNA walk for a sequence formed by splicing together the coding regions of the DNA sequence of this same gene (i.e., the cDNA). Figure 3.8c displays the DNA walk for a typical sequence with only coding regions. Landscapes for intron-containing sequences show very jagged contours which consist of patches of all length scales, reminiscent of the disordered state of matter near critical point. On the other hand, coding sequences typically consist of a few lengthy regions of different strand bias, resembling domains in the system in the ferromagnet state. These observations can be tested by rigorous statistical analysis. Figure 3.8 naturally motivates a quantification of these fluctuations by calculating the "net displacement" of the walker after ℓ steps, which is the sum of the unit steps $u(i)$ for each step i. Thus $y(\ell) \equiv \sum_{i=1}^{\ell} u(i)$.

3.6.2 Correlations and Fluctuations

An important statistical quantity characterizing any walk [3.60,61] is the root mean square fluctuation $F(\ell)$ about the average of the displacement; $F(\ell)$ is defined in terms of the difference between the average of the square and the square of the average,

$$F^2(\ell) \equiv \overline{[\Delta y(\ell) - \overline{\Delta y(\ell)}]^2} = \overline{[\Delta y(\ell)]^2} - \overline{\Delta y(\ell)}^2, \tag{3.2}$$

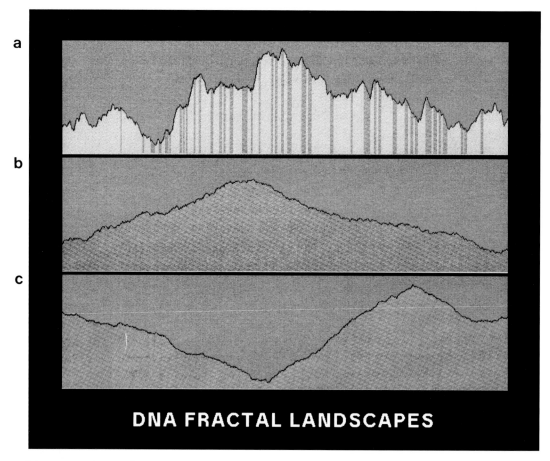

DNA FRACTAL LANDSCAPES

Fig. 3.8. The DNA walk representations of (a) human β-cardiac myosin heavy chain gene sequence, showing the coding regions as vertical golden bars, (b) the spliced together coding regions, and (c) the bacteriophage lambda DNA which contains only coding regions. Note the more complex fluctuations for (a) compared with the coding sequences (b) and (c). It is found that for almost all coding sequences studied that there appear regions with one strand bias, followed by regions of a different strand bias. In this presentation different step heights for purine and pyrimidine are used in order to align the end point with the starting point. This procedure is for graphical display purposes only (to allow one to visualize the fluctuations more easily) and is not used in any analytic calculations. After [3.33]

of a quantity $\Delta y(\ell)$ defined by $\Delta y(\ell) \equiv y(\ell_0 + \ell) - y(\ell_0)$ (see also Chaps. 1 and 5). Here the bars indicate an *average* over all positions ℓ_0 in the gene. Operationally, this is equivalent to (a) using calipers preset for a fixed distance ℓ, (b) moving the beginning point sequentially from $\ell_o = 1$ to $\ell_o = 2, \cdots$ and (c) calculating the quantity $\Delta y(\ell)$ (and its square) for each value of ℓ_o, and (d) averaging all of the calculated quantities to obtain $F^2(\ell)$.

The mean square fluctuation is related to the auto-correlation function

$$C(\ell) \equiv \overline{u(\ell_0)u(\ell_0 + \ell)} - \overline{u(\ell_0)}^2 \qquad (3.3a)$$

through the relation

$$F^2(\ell) = \sum_{i=1}^{\ell}\sum_{j=1}^{\ell} C(j - i). \qquad (3.3b)$$

The calculation of $F(\ell)$ can distinguish three possible types of behavior.

(i) If the base pair sequence were random, then $C(\ell)$ would be zero on average [except $C(0) = 1$], so $F(\ell) \sim \ell^{1/2}$ (as expected for a *normal* random walk).

(ii) If there were local correlations extending up to a characteristic range R (such as in Markov chains), then $C(\ell) \sim \exp(-\ell/R)$; nonetheless *the asymptotic ($\ell \gg R$) behavior $F(\ell) \sim \ell^{1/2}$ would be unchanged from the purely random case.*

(iii) If there is no characteristic length (i.e., if the correlation were "infinite-range"), then the scaling property of $C(\ell)$ would not be exponential, but would most likely to be a power law function, and the fluctuations will also be described by a power law

$$F(\ell) \sim \ell^{\alpha} \qquad (3.4a)$$

with $\alpha \neq 1/2$.

Figure 3.8a shows a typical example of a gene that contains a significant fraction of base pairs that do *not* code for amino acids. It is immediately apparent that the DNA walk has an extremely jagged contour which corresponds to long-range correlations. Figure 3.9 shows double logarithmic plots of the mean square fluctuation function $F(\ell)$ as a function of the linear distance ℓ along the DNA chain for a typical intron-containing gene.

The fact that the data for intron-containing and intergenic (i.e. noncoding) sequences are linear on this double logarithmic plot confirms that $F(\ell) \sim \ell^{\alpha}$. A least-squares fit produces a straight line with slope α substantially larger than the prediction for an uncorrelated walk, $\alpha = 1/2$, thus providing direct experimental evidence for the presence of long-range correlations.

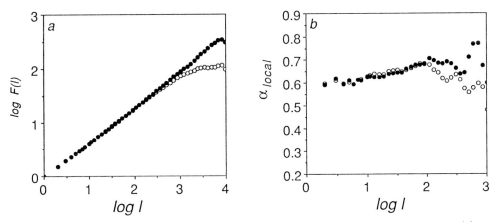

Fig. 3.9. (a) Double logarithmic plots of the mean square fluctuation function $F(\ell)$ as a function of the linear distance ℓ along the DNA chain for the rat embryonic skeletal myosin heavy chain gene (○) and its "intron-spliced sequence" (●). (b) The corresponding local slopes, α_{local}, based on pairs of successive data points of part (a). We see that the values of α are roughly constant. For this specific gene, the sequence with exons removed has an even broader scaling regime than the DNA sequence of the entire gene, indicated by the fact that part (a) is linear up to 10,000 nucleotides. After [3.42]

On the other hand, the dependence of $F(\ell)$ for coding sequences is not linear on the log-log plot: its slope undergoes a crossover from 0.5 for small ℓ to 1 for large ℓ. However, if a single patch is analyzed separately, the log-log plot of $F(\ell)$ is again a straight line with the slope close to 0.5. This suggests that within a large patch the coding sequence is almost uncorrelated.

It is known that functional proteins usually form a single compact three-dimensional conformation that corresponds to the global energy minimum in the conformational space. Recently, Shakhnovich and Gutin [3.66] found that in order to have such a minimum it is sufficient that an amino acid sequence forms an uncorrelated random sequence. The finding of Peng et al. [3.33] of the lack of long range correlations in the coding nucleotide sequences provides more evidence for this hypothesis, since there exist almost one-to-one correspondence between amino acid sequences and their nucleotide codes. Furthermore, this finding may also indicate that the *lack* of long range correlations in the amino acid sequences is, in fact, a necessary condition for a functional biologically active protein.

3.7 Other Methods
of Measuring Long-Range Correlations

An obvious question is whether the apparent long-range correlation behavior in noncoding sequences is simply an artifact of the DNA walk method itself. To compare the fluctuations of α in the DNA walk method with those found by other methods, two standard methods have been used [3.51] to study the correlation property of sequences, namely the power spectrum $S(f)$ and the correlation function $C(\ell)$. The power spectrum density, $S(f)$, is obtained by (a) Fourier transforming the sequence $\{u(i)\}$ and (b) taking the square of the Fourier component. For a stationary sequence, the power spectrum is the Fourier transform of the correlation function. If the correlation decays algebraically (not exponentially), i.e., there is no characteristic scale for the decay of the correlation (as we found in the noncoding DNA sequences) then we expect power-law behavior for both the power spectrum and the correlation function,

$$S(f) \sim (1/f)^{\beta}, \qquad (3.4b)$$

and

$$C(\ell) \sim (1/\ell)^{\gamma}. \qquad (3.4c)$$

The correlation exponents α, β and γ are not independent, since [3.62,63]

$$\alpha = \frac{1+\beta}{2} = \frac{2-\gamma}{2}. \qquad (3.5)$$

For a typical DNA sequence of *finite* length, both the correlation function and power spectrum are fairly noisy, but the estimates of β and γ obtained are consistent with those calculated from the DNA walk method. The reason for the smaller fluctuations of α in the DNA walk method is due to the fact that $F^2(\ell)$ is a double summation of $C(\ell)$. Thus it would seem that the original DNA walk method is more useful due to reduced noise. For a systematic analysis of the finite size effects on the exponent α, see [3.51] and [3.57].

Apart from the reduced noise mentioned above, one additional advantage of the DNA walk method [3.33] is that to find the exponent characterizing the long-range correlation one need not correct the data by subtracting the white noise, $S(\infty)$ [3.36]. Since there is no unambiguous method of estimating $S(\infty)$, this need to correct the data introduces an uncontrollable source of uncertainty.

3.8 Differences Between Correlation Properties of Coding and Noncoding Regions

The initial report [3.33] on long-range (scale-invariant) correlations only in non-coding DNA sequences has generated contradicting responses. Some [3.34,35, 38,39] support our initial finding, while some [3.35,40,44,50] disagree. However, the conclusions of Refs. [3.36] and [3.35,40,44,50] are inconsistent *with one another* in that [3.35] and [3.50] doubt the existence of long-range correlations (even in noncoding sequences) while [3.36] and [3.40,44] conclude that even coding regions display long-range correlations ($\alpha > 1/2$). Prabhu and Claverie [3.40] claim that their analysis of the putative *coding* regions of the yeast chromosome III [3.67] produces a *wide range of exponent values*, some larger than 0.5. The source of these contradicting claims may arise from the fact that, in addition to normal statistical fluctuations expected for analysis of rather short sequences, coding regions typically consist of only a few lengthy regions of alternating strand bias. Hence conventional scaling analyses cannot be applied reliably to the entire sequence but only to sub-sequences.

Peng et al. [3.56] have recently applied the "bridge method" to DNA, and have also developed a similar method specifically adapted to handle problems associated with non-stationary sequences which they term *detrended fluctuation analysis* (DFA).

The idea of the DFA method is to compute the dependence of the standard error of a linear interpolation of a DNA walk $F_d(\ell)$ on the size of the interpolation segment ℓ. The method takes into account differences in local nucleotide content and may be applied to the entire sequence which has lengthy patches. In contrast with the original $F(\ell)$ function, which has spurious crossovers even for ℓ much smaller than a typical patch size, the detrended function $F_d(\ell)$ shows linear behavior on the log-log plot for all length scales up to the characteristic patch size, which is of the order of a thousand nucleotides in the coding sequences. For ℓ close to the characteristic patch size the log-log plot of $F_d(\ell)$ has an abrupt change in its slope.

The DFA method clearly supports the difference between coding and noncoding sequences, showing that the coding sequences are less correlated than noncoding sequences for the length scales less than 1000, which is close to char-

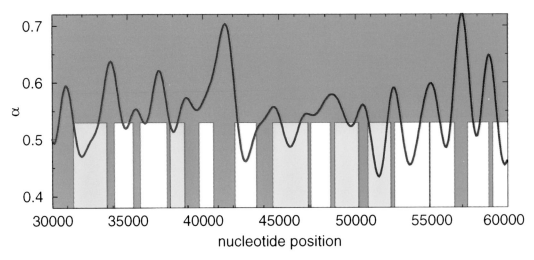

Fig. 3.10. Analysis of section of Yeast Chromosome III using the sliding box *Coding Sequence Finder* "CSF" algorithm. The value of the long-range correlation exponent α is shown as a function of position along the DNA chain. In this figure, the results for about 10% of the DNA are shown (from base pair #30,000 to base pair #60,000). Shown as vertical bars are the putative genes and open reading frames; denoted by the letter "G" are those genes that have been more firmly identified (March 1993 version of *GenBank*). Note that the local value of α displays **minima** where genes are suspected, while between the genes α displays **maxima**. This behavior corresponds to the fact that the DNA sequence of genes lacks long-range correlations ($\alpha = 0.5$ in the idealized limit), while the DNA sequence in between genes possesses long-range correlations ($\alpha \approx 0.6$). After [3.57]

acteristic patch size in the coding regions. One source of this difference is the tandem repeats (sequences such as AAAAAA...), which are quite frequent in noncoding sequences and absent in the coding sequences.

To provide an "unbiased" test of the thesis that noncoding regions possess but coding regions lack long-range correlations, Ossadnik et al. [3.57] analyzed several artificial uncorrelated and correlated "control sequences" of size 10^5 nucleotides using the GRAIL neural net algorithm [3.68]. The GRAIL algorithm identified about 60 putative exons in the uncorrelated sequences, but only about 5 putative exons in the correlated sequences.

Using the DFA method, we can measure the local value of the correlation exponent α along the sequence (see Fig 3.10) and find that the local minima of α as a function of a nucleotide position usually correspond to noncoding regions, while the local maxima correspond to noncoding regions. Statistical analysis using the DFA technique of the nucleotide sequence data for yeast chromosome III (315,338 nucleotides) shows that the probability that the observed correspondence between the positions of minima and coding regions is due to random coincidence is less than 0.0014. Thus, this method—which we called the "coding sequence finder" (CSF) algorithm—can be used for finding coding regions in the newly sequenced DNA, a potentially important application of DNA walk analysis.

Voss [3.36] has recently proposed that *coding* as well as noncoding DNA sequences display long-range power law correlations in their base pair (bp) sequences. This finding disagreed with the earlier claim [3.33], that coding

DNA sequences do not display power-law correlations. The discrepancy between [3.36] and [3.33] again may have arisen because the analysis in [3.33] was based on partitioning the entire coding sequence into a few large subsequences of constant overall compositional bias. It is important to resolve this discrepancy, since Voss based his scientific conclusion ("immunity to errors on all scales") on his claim of power-law correlations in *coding* sequences [3.69]. However, the finding in [3.33] and the results of the CSF analysis just described [3.57] suggest that this conjecture of Voss is unlikely to be valid.

Recently, additional evidence that the Voss proposal does not hold generally was presented [3.70]. Specifically, two counterexamples that clearly display *no* long-range correlations *when directly analyzed* (without partitioning into subsequences) were reported: (i) the complete genome of T7 bacteriophage (39,936 bp), which contains *only* coding regions, and (ii) the Ti plasmid fragment (24,595 bp), which is believed to consist almost entirely of coding regions.

Figure 3.11a shows the DNA walks for (i) and (ii). Figure 3.11b shows $F(\ell)$, the fluctuation in rms amplitude; the slopes of the log-log plots, *fit over 3 decades*, are 0.53 and 0.49, indicating the absence of long-range correlation for both cases. Figure 3.11c shows the power spectrum $S(f)$, which is almost perfectly flat (indicative of no correlation or "white noise"). The scaling behavior in Figs. 3.11b and 3.11c is markedly different than that found for genomic sequences containing substantial noncoding sub-regions, for which $F(\ell) \sim \ell^{\alpha}$ with $\alpha \approx 2/3$ and $S(f) \sim 1/f^{\beta}$ with $\beta \approx 1/3$ [3.33].

3.9 Long-Range Correlations and Evolution

What is the biological meaning of the finding of long-range correlations in DNA? If two nucleotides whose positions differ by 1000 base pairs were uncorrelated, then there might be no meaning. However, the finding that they are correlated suggests some underlying organizational property. The long-range correlations in DNA sequences are of interest because they may be an indirect clue to its three-dimensional structure [3.45,54] or a reflection of certain scale-invariant properties of long polymer chains [3.53,55]. In any case, the statistically meaningful long-range "scale-invariant" (see Fig. 3.0) correlations in noncoding regions and their absence in coding regions will need to be accounted for by future explanations of global properties in gene organization and evolution.

Molecular evolutionary relationships are usually inferred from comparison of coding sequences, conservation of intron/exon structure of related sequences, analysis of nucleotide substitutions, and construction of phylogenetic trees [3.71]. The changes observed are conventionally interpreted with respect to nucleotide sequence composition (mutations, deletions, substitutions, alternative splicing, transpositions, etc.) rather than overall genomic organization.

Very recently, Buldyrev et al. [3.55] sought to assess the utility of DNA correlation analysis as a complementary method of studying gene evolution.

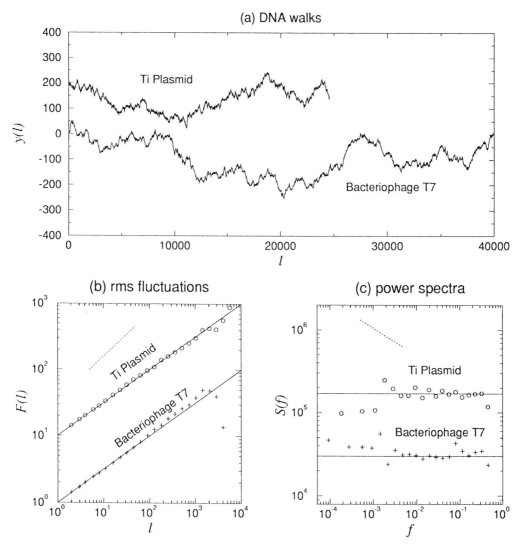

Fig. 3.11. (a) "DNA walks" [3.33] of nucleotide sequences; a random walker moves either up or down depending on whether the nucleotide at position ℓ is a pyrimidine or purine. (b) The rms fluctuation, $F(\ell)$, of the DNA walk displacement $y(\ell)$. (c) Power spectrum, $S(f)$, of the binary nucleotide sequences (pyrimidine$= 1$, purine$= -1$). Shown are two examples that display *no* long-range correlations: (i) the complete genome of T7 bacteriophage (GenBank name: PODOT7), which contains *only* coding regions, and (ii) the Ti plasmid fragment (ATACH5), which is believed to consist almost entirely of coding regions. The solid lines in (b) and (c) have slopes $\alpha = 1/2$ and $\beta = 0$ respectively; for comparison, dashed lines of slope 2/3 and -1/3 are also shown, corresponding to the typical behavior found for sequences containing noncoding regions [3.33]. The data for Ti (o) shifted on the plots for visual comparison. After [3.70]

In particular, they studied the changes in "fractal complexity" of nucleotide organization of a single gene family with evolution. A recent study by Voss [3.36] reported that the correlation exponent derived from Fourier analysis was lowest for sequences from organelles, but paradoxically higher for invertebrates than vertebrates. However, this analysis must be interpreted with caution since it was based on pooled data from different gene families rather than from the quantitative examination of any single gene family.

The hypothesis that the fractal complexity of genes from higher animals is greater than that of lower animals, using single gene family analysis was tested in [3.55]. This analysis focuses on the genome sequences from the conventional (Type II) myosin heavy chain (MHC) family. Such a choice limits potential bias that may arise secondary to non-uniform evolutionary pressures and differences in nucleotide content between unrelated genes. The MHC gene family was chosen because of the availability of completely sequenced genes from a phylogenetically diverse group of organisms, and the fact that their relatively long sequences are well-suited to statistical analysis.

The landscape produced by DNA walk analysis reveals that each MHC cDNA consists of two roughly equal parts with significant differences in nucleotide content (Fig. 3.12). The first part that codes for the heavy meromyosin or "head" of the protein molecule has a slight excess of purines (52% purines and 48% pyrimidines); the second part that codes for the light meromyosin or "tail" has about 63% purines and 37% pyrimidines. The *absolute nucleotide contents* are not shown in the graphical representation of Fig. 3.12a because we subtract the average slope from the landscape to make relative fluctuations around the average more visible. Indeed, one can easily see from Fig. 3.12a that the relative concentration of pyrimidines in the first part ("uphill" region) of the myosin cDNA is much higher than in the second ("downhill" region).

The landscapes of Fig. 3.12 show that the coding sequences of myosins remain practically unchanged with evolution, while the entire gene sequences become more heterogeneous and complex. The quantitative measurements of the exponent α by DFA method confirm this visual observation showing that for all coding sequences of MHC family $\alpha \approx 0.5$. In contrast, for entire genes of MHC family, the value of α monotonically increases from lower eukaryotes to invertebrates and from invertebrates to vertebrates [3.55].

3.10 Models of DNA Evolution

The finding described in the previous section suggests that the source of long-range correlations could be an evolutionary process specific to noncoding sequences. In contrast, the coding sequences should preserve their uncorrelated structure in order to maintain the functional properties of the encoded proteins.

Li [3.72] suggested a simple model of a biologically plausible process of duplication and mutation that can produce a sequence with any given value of α. His model can be used to explain certain features of highly repetitive DNA

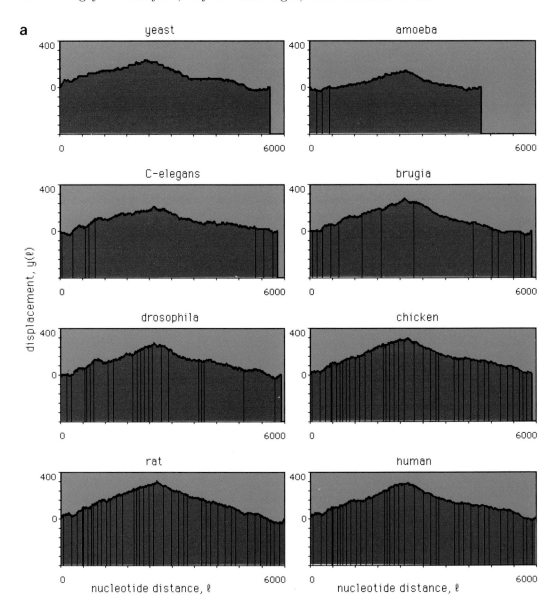

Fig. 3.12. The DNA walk representations of (a) 8 cDNA sequences from the MHC family and (b) the corresponding genes. DNA landscapes are plotted so that the end points have the same vertical displacement as the starting points [3.33]. The graphs are for yeast, amoeba, worms: *C. elegans*, *Brugia malayi*, drosophila, chicken, rat and human (from top to bottom, left to right). The shaded areas in (b) denote coding regions of the genes. The DNA walks

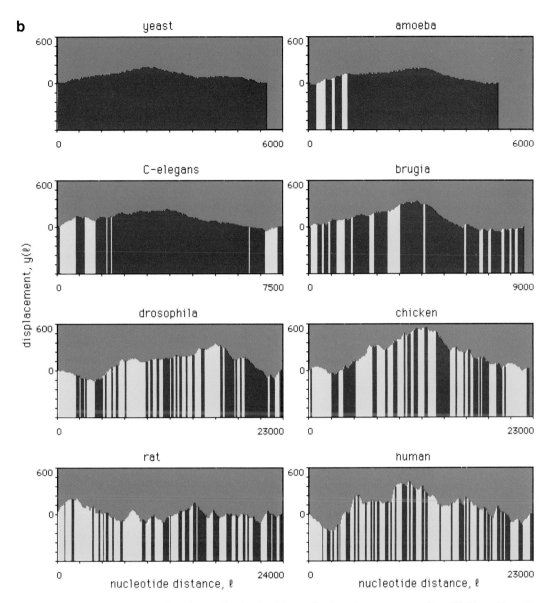

for the genes show increasing "complexity" with evolution. In contrast, the cDNA walks all show remarkably similar crossover patterns due to sequential "up-hill" and "down-hill" slopes representing different purine/pyrimidine strand biases in the regions coding for the head and tail of the MHC molecule, respectively. After [3.55]

(such as tandem repeats), but does not take into account many other important processes of DNA evolution like retroviral insertions and deletions, which are probably the main source of the rapid evolution of DNA sequences.

An example is the LINE-1 sequence which consists of 6,139 base pairs and is believed to contain a code for a functional protein [3.73]. In agreement with this it is found that the LINE-1 sequence has value of α close to 0.5, indicating the lack of long-range correlations [3.53]. Moreover, the LINE-1 sequence has a strong strand bias of about 59% of purines, which is also typical for coding sequences. The total number of LINE-1 sequences and fragments in the human genome is estimated to be 107,000, while in the genome of the chimpanzee there are only 51,000 copies of the LINE-1 sequence [3.74]. This dramatic difference indicates that thousands of insertions or deletions of LINE-1 sequences took place over a relatively short evolutionary time scale. LINE-1 sequences are found on both strands of DNA and therefore produce large local fluctuations of nucleotide content. Another frequent repetitive element is the ALU sequence [3.75], which is also statistically similar to protein coding DNA, but, in contrast with the LINE-1 sequence, is only 290 base pairs long.

The central idea of the insertion model [3.53,55] is based on the assumption that the insertion of retroelements, formed by the inverse-transcribed RNA, plays a major role in DNA evolution. The statistical properties of retroelements are similar to those of protein coding sequences. In order to be inserted into DNA, a retroelement must form a loop. The probability to find a loop of certain size ℓ in a long polymer chain in a solvent is given [3.76] by the formula

$$P(\ell) \propto (1/\ell)^{\mu}, \tag{3.6}$$

where μ is a critical exponent with a value close to 2.2. Thus we assume

(i) that DNA sequences are comprised of subsequences distributed according to Eq. (3.6), and

(ii) that these subsequences are statistically similar to protein coding sequences which (a) usually have a significant excess of purines over pyrimidines (or vice versa because of DNA two-strand complementarity) and (b) can be modeled by a Markovian process with short range correlations.

This biological evolution model is mathematically equivalent to the generalized Lévy walk which gives rise to a landscape with a well defined power-law long-range correlation exponent α that depends upon the Lévy walk parameter μ [3.53,64]

$$\alpha = \begin{cases} 1 & \mu \leq 2 \\ 2 - \mu/2 & 2 < \mu < 3 \\ 1/2 & \mu \geq 3. \end{cases} \tag{3.7}$$

Thus nontrivial behavior of α corresponds to the case $2 < \mu < 3$, where the first moment of $P(l)$ converges while the second moment diverges. The long-range correlation property for the Lévy walk, in this case, is related to the broad distribution of Eq. (3.6) that lacks a characteristic length scale. Equation (3.7) is valid only asymptotically for large values of ℓ. For small ℓ the slope of the log-log plot of the function $F(\ell)$ for the generalized Lévy walk model increases monotonically from a value defined by short range Markovian correlations of the inserted subsequences to a value $\alpha = 0.9$ predicted by (3.7). However, this limiting value can be achieved only for very long sequences of about 10^6 base pairs, and has a large standard error for finite sequences [3.51].

To test the insertion model [3.53], we have adjusted its parameters, to best approximate features of actual DNA sequences and found a good agreement between the model and the actual data on successive slopes of the $F(\ell)$ function for all sequences, that contain a substantial percentage of noncoding material.

In order to create a more realistic model of DNA evolution which includes also deletion of certain DNA subsequences, intron insertion [3.77] and the exchange of genomic material between DNA strands and chromosomes, Buldyrev et al. [3.55] modified the generalized Lévy walk model by allowing random deletion and re-insertions of subsequences with length distribution defined by Eq. (3.6). Starting with an uncorrelated sequence , statistically similar to mRNA of MHC, the model generates, with each new iteration, more and more heterogeneous sequences, and reproduces the increase of α with evolution observed for MHC gene family.

Note that the values of α for the model and actual DNA sequences are measured in [3.55] using the DFA procedure for relatively small values of $\ell < 100$, i.e., in the fitting range where the differences between coding and noncoding sequences are the most pronounced. Voss [3.36] measured α by fitting the low frequency part of power spectrum, which corresponds to much larger values of ℓ. That is why Voss does not find any significant difference between coding and noncoding sequences. Moreover, in this fitting range the standard error of α for finite sequences is very large and may be a reason for his paradoxical observations of evolutionary changes of α. However, the high values of α that Voss finds for this fitting range are generally consistent with the expected behavior of the generalized Lévy walk model.

Two major theories have been advanced to explain the origin and evolution of introns. One suggests that precursor genes consisted entirely of coding sequences and introns were inserted later in the course of evolution to help facilitate development of new structures in response to selective pressure, perhaps, by means of "exon shuffling" [3.78]. The alternative theory suggests that precursor genes were highly segmented and subsequently organisms not requiring extensive adaptation or new development or, perhaps, facing the high energetic costs of replicating unnecessary sequences, lost their introns [3.79,80]. Support for these hypotheses has remained largely conjectural; no models have been brought forward to support either process. The landscape analysis of the MHC gene family and the stochastic model reviewed here are more consistent with the former view.

3.11 Long-Range Correlations
and DNA Spatial Structure

From the point of view of statistical mechanics long-range correlations cannot exist in a one-dimensional system at equilibrium. One possible explanation of their existence is the type of non-equilibrium evolutionary process described above. However, another source of these correlations could be the actual physical interactions between parts of DNA molecule in three dimensional space. It is known that the DNA polymer chain bound by histones folds itself into the chromosome according to a hierarchical structure of loops of many length scales. Moreover, it has an ability to unwind without forming knots. This complex behavior suggests that the sequence of DNA base pairs may contain a "code" that defines DNA packaging and unwinding. At the very least, the local nucleotide composition may be affected by the way DNA is packaged.

Recently, Grosberg et al. [3.45] proposed a theoretical model which relates the long-range correlations in the nucleotide sequence to the spatial structure of the chromosome; they argue that this structure has to be self-similar. Simple estimates based on packing considerations (packing an object of length of up to $1m$ into a region of roughly $1\mu m$ size), show that in any real biological system, from virus to chromosome, the native spatial structure of DNA has to be of a globular type, rather than that of an expanded coil (for a review on DNA behavior in the globular phase, see [3.46]).

There are a large number of different globular structures which have been investigated in polymer physics [3.47]. For a sufficiently long "simple" polymer, most of the conformations of any equilibrium globule contain a vast number of complicated knots, so that the number of entanglements is comparable to the number of chain segments. Such complex knotted conformations cannot dominate the native state of a functioning biopolymer since entanglements will dramatically reduce its ability to respond to biochemical influences. Indeed, globular proteins, including complex ones which have quaternary structure, are free of any knots. A DNA globule is more complicated than a protein globule. For example, DNA involves dramatically larger length scales and is governed by volume interactions of tremendous complexity which are mediated by proteins, and include the property of recognition of particular sequences by protein. Grosberg et al. [3.45] justify the assumption that, *in a statistical sense*, the DNA globule is practically unknotted.

For a sufficiently long polymer, the prohibition of knot formation leads to a nontrivial self-similar fractal spatial structure, the so-called "crumpled globule" [3.48] (see Fig. 3.13). The key property of this structure is that each part of the chain with an arbitrary length l has to be folded into a compact self-similar globular state, i.e., its spatial size should scale statistically as $l^{1/3}$. It differs dramatically from the equilibrium (with respect to formation of knots) globule where any chain part which is small compared to the size of the whole globule looks like a chain in a polymer melt, i.e., its size is of order $l^{1/2}$. Such a globule of linear size R can be imagined as built of blobs formed by chain segments of

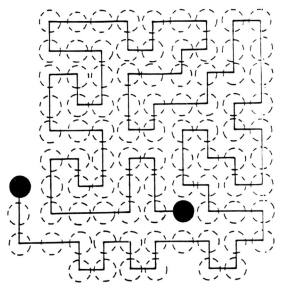

Fig. 3.13. Schematic representation of "crumpled globule" structure in two dimensions. The monomers are represented by dashed circles (the ends of the chain are given by solid circles), and the chain contour is indicated by the solid line. Notice that while *all* random collapsed configurations of a polymer in two dimensions are of the crumpled globule type, this does not hold, in general, in three dimensions where a typical configuration of a random globule will contain a large number of knots. After [3.45]

length $(R/a)^2$, a being some microscopic scale; these blobs strongly penetrate each other and in fact are all placed in the same region of size R. In the crumpled globule, on the other hand, each crumple of an arbitrary length l, having a size of order $l^{1/3}$, does *not* penetrate other crumples of the same scale, which therefore remain segregated in space as can be visualized in two dimensions, by considering a rope that is densely folded between two flat surfaces.

For eukaryotic DNA, the existence of several levels of self-similarity of spatial structure is well-established [3.49]. Since we know the spatial structure of eukaryotic DNA, we conclude that the shapes and relative positions of crumples must be well-defined too. One can now answer the question: how many units in the primary structure of DNA are needed to specify a particular realization of crumpled spatial structure on some length-scale l? Since each chain part of contour length l forms a globular crumple of spatial size $l^{1/3}$, it has the surface of order $l^{2/3}$. Larger-scale structures are formed as the result of surface interactions between the crumples (since interpenetration does not occur). Therefore, the number of units which participate in the fixation of spatial structure on any arbitrary scale l is of order $l^{2/3}$. This results in a predicted universal value of $\alpha = 2/3$ for all eukaryotic DNA sequences that form chromosomes, which agrees with the measurements discussed above. In fact, DNA spatial structure as well as evolutionary processes in DNA may both play a certain role in forming long-range correlations in the nucleotide sequences. Further studies are needed to resolve this question.

3.12 Other Biological Systems
with Long-Range Correlations

The catalog of systems in which power law correlations appear has grown rapidly in recent years [3.32,81,82]; see also Chap. 2. What do we anticipate for other biological systems? Generally speaking, when "entropy wins over energy"—i.e., randomness dominates the behavior—we find power laws and scale invariance. The absence of characteristic length (or time) scales may confer important biological advantages, related to adaptability of response [3.2]. Biological systems sometimes are described in language that makes one think of a Swiss watch. Such mechanistic or "Rube Goldberg" descriptions must in some sense be incomplete, since it is only some appropriately-chosen averages that appear to behave in a regular fashion. The trajectory of each individual biological molecule is of necessity random—albeit correlated. Thus one might hope that recent advances in understanding "correlated randomness" [3.83,62–64] could be relevant to biological phenomena.

3.12.1 The Human Heartbeat

Traditionally, clinicians describe the normal electrical activity of the heart as "regular sinus rhythm." However, cardiac interbeat intervals fluctuate in a complex, apparently erratic manner in healthy subjects even at rest. Analysis of heart rate variability has focused primarily on short time oscillations associated with breathing (0.15–0.40 Hz) and blood pressure control (~ 0.1 Hz) [3.80]. Fourier analysis of longer heart rate data sets from healthy individuals typically reveals a $1/f$-like spectrum for frequencies < 0.1 Hz [3.84–87].

Peng et al. [3.88] recently studied scale-invariant properties of the human heartbeat time series, the output of a complicated integrative control system. The analysis is based on the digitized electrocardiograms of beat-to-beat heart rate fluctuations over very long time intervals (up to 24 h $\approx 10^5$ beats) recorded with an ambulatory monitor. The time series obtained by plotting the sequential intervals between beat n and beat $n+1$, denoted by $B(n)$, typically reveals a complex type of variability. The mechanism underlying such fluctuations is related to competing neuroautonomic inputs. Parasympathetic (vagal) stimulation decreases the firing rate of pacemaker cells in the heart's sinus node; sympathetic stimulation has the opposite effect. The nonlinear interaction (competition) between these two branches of the involuntary nervous system is the postulated mechanism for much of the erratic heart rate variability recorded in healthy subjects, although non-autonomic factors may also be important.

To study these dynamics over large time scales, the time series is passed through a digital filter that removes fluctuations of frequencies > 0.005 beat^{-1}, and plot the result, denoted by $B_L(n)$, in Fig. 3.14. One observes a more complex pattern of fluctuations for a representative healthy adult (Fig. 3.14a) compared to the "smoother" pattern of interbeat intervals for a subject with severe heart disease (Fig. 3.14b). These heartbeat time series produce a contour rem-

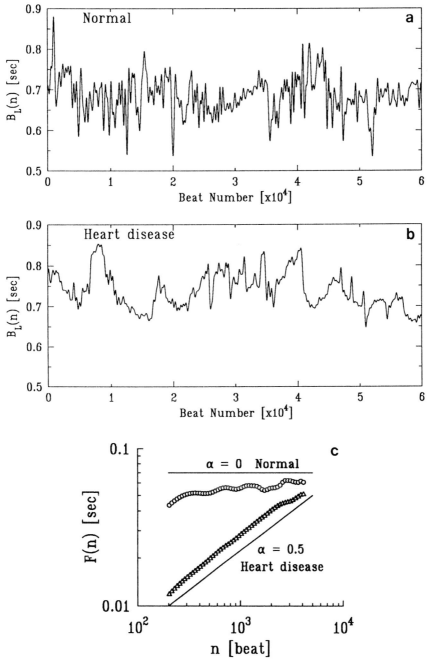

Fig. 3.14. The interbeat interval $B_L(n)$ after low-pass filtering for (a) a healthy subject and (b) a patient with severe cardiac disease (dilated cardiomyopathy). The healthy heartbeat time series shows more complex fluctuations compared to the diseased heart rate fluctuation pattern that is close to random walk ("brown") noise. (c) Log-log plot of $F(n)$ vs n. The circles represent $F(n)$ calculated from data in (a) and the triangles from data in (b). The two best-fit lines have slope $\alpha = 0.07$ and $\alpha = 0.49$ (fit from 200 to 4000 beats). The two lines with slopes $\alpha = 0$ and $\alpha = 0.5$ correspond to "$1/f$ noise" and "brown noise," respectively. We observe that $F(n)$ saturates for large n (of the order of 5000 beats), because the heartbeat interval are subjected to physiological constraints that cannot be arbitrarily large or small. The low-pass filter removes all Fourier components for $f \geq f_c$. The results shown here correspond to $f_c = 0.005$ beat^{-1}, but similar findings are obtained for other choices of $f_c \leq 0.005$. This cut-off frequency f_c is selected to remove components of heart rate variability associated with physiologic respiration or pathologic Cheyne-Stokes breathing as well as oscillations associated with baroreflex activation (Mayer waves). After [3.88]

iniscent of the irregular landscapes that have been widely studied in physical systems.

To quantitatively characterize such a "landscape," Peng et al. introduce a mean fluctuation function $F(n)$, defined as

$$F(n) \equiv \overline{|B_L(n' + n) - B_L(n')|}, \qquad (3.8)$$

where the bar denotes an average over all values of n'. Since $F(n)$ measures the average difference between two interbeat intervals separated by a time lag n, $F(n)$ quantifies the magnitude of the fluctuation over different time scales n.

Figure 3.14c is a log-log plot of $F(n)$ vs n for the data in Figs. 3.14a and 3.14b. This plot is approximately linear over a broad physiologically-relevant time scale $(200 - 4000$ beats$)$ implying that

$$F(n) \sim n^{\alpha}. \qquad (3.9)$$

It is found that the scaling exponent α is markedly different for the healthy and diseased states: for the healthy heartbeat data, α is close to 0, while α is close to 0.5 for the diseased case. Note that $\alpha = 0.5$ corresponds to a random walk (a Brownian motion), thus the low-frequency heartbeat fluctuations for a diseased state can be interpreted as a stochastic process, in which the heartbeat intervals $I(n) \equiv B(n + 1) - B(n)$ are uncorrelated for $n \geq 200$.

To investigate these dynamical differences, it is helpful to study further the correlation properties of the time series. It is useful to study $I(n)$ because it is the appropriate variable for the aforementioned reason. Since $I(n)$ is stationary, one can apply standard spectral analysis techniques [3.3] (see also Chap. 1). Figures 3.15a and 3.15b show the power spectra $S_I(f)$, the square of the

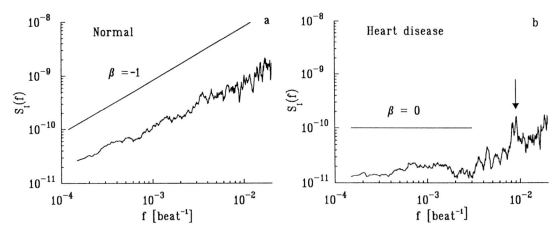

Fig. 3.15. The power spectrum $S_I(f)$ for the interbeat interval increment sequences over ~ 24 hours for the same subjects in Fig. 3.14. (a) Data from a healthy adult. The best-fit line for the low frequency region has a slope $\beta = -0.93$. The heart rate spectrum is plotted as a function of "inverse beat number" (beat^{-1}) rather than frequency (time^{-1}) to obviate the need to interpolate data points. The spectral data are smoothed by averaging over 50 values. (b) Data from a patient with severe heart failure. The best-fit line has slope 0.14 for the low frequency region, $f < f_c = 0.005$ beat^{-1}. The appearance of a pathologic, characteristic time scale is associated with a spectral peak (arrow) at about 10^{-2} beat^{-1} (corresponding to Cheyne-Stokes respiration). After [3.88]

Fourier transform amplitudes for $I(n)$, derived from the same data sets (without filtering) used in Fig. 3.14. The fact that the log-log plot of $S_I(f)$ vs f is linear implies

$$S_I(f) \sim \frac{1}{f^\beta}. \tag{3.10}$$

The exponent β is related to α by $\beta = 2\alpha - 1$ [3.62]. Furthermore, β can serve as an indicator of the presence and type of correlations:

(i) If $\beta = 0$, there is no correlation in the time series $I(n)$ ("white noise").

(ii) If $0 < \beta < 1$, then $I(n)$ is correlated such that positive values of I are likely to be close (in time) to each other, and the same is true for negative I values.

(iii) If $-1 < \beta < 0$, then $I(n)$ is also correlated; however, the values of I are organized such that positive and negative values are more likely to alternate in time ("anti-correlation") [3.62].

For the diseased data set, we observe a flat spectrum ($\beta \approx 0$) in the low frequency region (Fig. 3.15b) confirming that $I(n)$ are not correlated over long time scales (low frequencies). Therefore, $I(n)$, the first derivative of $B(n)$, can be interpreted as being analogous to the *velocity* of a random walker, which is uncorrelated on long time scales, while $B(n)$—corresponding to the *position* of the random walker—are correlated. However, this correlation is of a trivial nature since it is simply due to the summation of uncorrelated random variables.

In contrast, for the data set from the healthy subject (Fig. 3.15a), we obtain $\beta \approx -1$, indicating *nontrivial* long-range correlations in $B(n)$—these correlations are not the consequence of summation over random variables or artifacts of non-stationarity. Furthermore, the "anti-correlation" properties of $I(n)$ indicated by the negative β value are consistent with a nonlinear feedback system that "kicks" the heart rate away from extremes. This tendency, however, does not only operate on a beat-to-beat basis (local effect) but on a wide range of time scales. To our knowledge, this is the first explicit description of long-range anticorrelations in a fundamental biological variable, namely the interbeat interval increments.

Furthermore, the histogram for the heartbeat intervals increments is found to be well-described by a Lévy stable distribution. For a group of subjects with severe heart disease, it is found that the distribution is unchanged, but the long-range correlations vanish. Therefore, the different scaling behavior in health and disease must relate to the underlying dynamics of the heartbeat. Applications of this analysis may lead to new diagnostics for patients at high risk of cardiac disease and sudden death.

3.12.2 Physiological Implications

The finding of nontrivial long-range correlations in healthy heart rate dynamics is consistent with the observation of long-range correlations in other biological systems that do not have a characteristic scale of time or length. Such behavior may be adaptive for at least two reasons. (i) The long-range correlations serve as an organizing principle for highly complex, nonlinear processes that generate fluctuations on a wide range of time scales. (ii) The lack of a characteristic scale helps prevent excessive *mode-locking* that would restrict the functional responsiveness of the organism. Support for these related conjectures is provided by observations from severe diseased states such as heart failure where the breakdown of long-range correlations is often accompanied by the emergence of a dominant frequency mode (e.g., the Cheyne-Stokes frequency). Analogous transitions to highly periodic regimes have been observed in a wide range of other disease states including certain malignancies, sudden cardiac death, epilepsy and fetal distress syndromes.

The complete breakdown of normal long-range correlations in any physiological system could theoretically lead to three possible diseased states: (i) a random walk (brown noise), (ii) highly periodic behavior, or (iii) completely uncorrelated behavior (white noise). Cases (i) and (ii) both indicate only "trivial" long-range correlations of the types observed in severe heart failure. Case (iii) may correspond to certain cardiac arrhythmias such as fibrillation. More subtle or intermittent degradation of long-range correlation properties may provide an early warning of incipient pathology. Finally, we note that the long-range correlations present in the healthy heartbeat indicate that the neuroautonomic control mechanism actually drives the system away from a single steady state. Therefore, the classical theory of homeostasis, according to which stable physiological processes seek to maintain "constancy" [3.89], should be extended to account for this dynamical, far from equilibrium, behavior.

3.12.3 Human Writings

Long-range correlations have been found recently in human writings [3.81]. A novel, a piece of music or a computer program can be regarded as a one-dimensional string of symbols. These strings can be mapped to a one-dimensional random walk model similar to the DNA walk (Sect. 3.6) allowing calculation of the correlation exponent α using (3.4a). Values of α between 0.6 and 0.9 were found for various texts.

An interesting fractal feature of languages was found in 1949 by Zipf [3.90]. He observed that the frequency of words as a function of the word order decays as a power law (with a power close to -1) for more than four orders of magnitude. A theory for this empirical finding, based on assumptions of coding words in the brain, was given by Mandelbrot [3.1,91]. A related interesting statistical measure of short-range correlations in languages and in general series sequences is the entropy and redundancy defined by Shannon [3.92].

3.12.4 Dynamics of Membrane Channel Openings

Ions, such as potassium and sodium, cannot cross the lipid cell membrane. They can, however, enter or exit the cell through ion channel proteins that are distributed on the cell membrane. These proteins spontaneously fluctuate between open or closed states. Liebovitch [3.93] found that the histograms of the open and closed duration times of some channels are self-similar and behave as power laws. This approach may provide new models for the ion channel gating mechanisms.

3.12.5 Fractal Music and the Heartbeat

Fourier analysis of instantaneous fluctuations in amplitude as well as internote intervals for certain classical music pieces (e.g., Bach's First Brandenburg Concerto) reveals a $1/f$ distribution over a broad frequency range [3.94-96]. Voss and Clarke [3.95] used an algorithm for generating $1/f$-noise to "compose" music.

Based on the observation of different scaling patterns for healthy and pathologic heartbeat time series [3.88], it was very recently postulated [3.97] that (i) actual biological rhythms such as the heartbeat might serve as a more natural template for musical compositions than artificially-generated noises, and (ii) audibly appreciable differences between the note series of healthy and diseased hearts could potentially serve as the basis for a clinically useful diagnostic test. Accordingly, a computer program was devised to map heart rate fluctuations onto intervals of the diatonic musical scale [3.97]. As anticipated, the *normal* ($1/f$-like) heartbeat obtained from the low pass filtered time series reported in [3.88] generated a more variable (complex) type of music than that generated by the *abnormal* times series (Fig. 3.16).

The "musicality" of these transcriptions is intriguing and supports speculations about the brain's possible role as a translator/manipulator of biological $1/f$-like noise into aesthetically pleasing art works. Current investigations are aimed at extending these preliminary observations by (i) comparing the "musicality" of note sequences generated by natural (biological) vs. artificial (computer simulated) correlated and uncorrelated noises, and (ii) using heartbeat time series as the template for simultaneously generating fluctuations in musical rhythm and intensity, not only pitch.

3.12.6 Fractal Approach to Biological Evolution

Fossil data suggest that evolution of biological species takes place as intermittent bursts of activity, separated by relatively long periods of quiescence [3.98]. Recently Bak and Sneppen [3.99] suggested that these spontaneous catastrophic extinctions may be related to the power law distribution of avalanches of growth observed in a model of self-organized criticality (SOC). Such SOC models are reminiscent of recent surface growth models based on the concept of directed invasion percolation [3.100].

82   Sergey V. Buldyrev, Ary L. Goldberger, Shlomo Havlin et al.

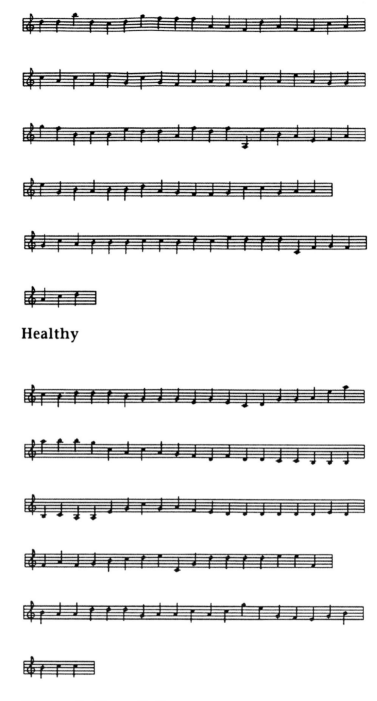

Healthy

Diseased: Heart Failure

Fig. 3.16. Musical mapping of two heartbeat times series, derived from normal (top) and pathologic (bottom) data sets. The original heart beat time series were obtained from 24 hour recordings consisting of about 10^6 heartbeats. The heartbeat time series were then low-pass filtered to remove fluctuations > 0.05 (beat^{-1}), roughly equivalent to averaging every 200 beats. The pattern of fluctuations in the normal is more complex than that of the "music" generated from the abnormal data sets. Musical compositions based on these times series are available on cassette by request along with the "scores"; contact Zachary D. Goldberger (e-mail: ary "at" astro.bih.harvard.edu). There is a nominal charge for copying and mailing. After [3.97], courtesy of Z.D. Goldberger

Acknowledgements. We are grateful to R. Bansil, A.-L. Barabasi, K.R. Bhaskar, F. Caserta, G. Daccord, W. Eldred, P. Garik, Z.D. Goldberger, Z. Hantos, J.M. Hausdorff, R.E. Hausman, T.J. LaMont, H. Larralde, R.N. Mantegna, M.E. Matsa, J. Mietus, A. Neer, J. Nittmann, S.M. Ossadnik, F. Péták, I. Rabin, F. Sciortino, A. Shehter, M.H.R. Stanley, B. Suki, P. Trunfio, M. Ukleja, G.H. Weiss, and especially M. Simons for major contributions to those results reviewed here that represent collaborative research efforts. We also wish to thank C. Cantor, C. DeLisi, M. Frank-Kamenetskii, A.Yu. Grosberg, G. Huber, I. Labat, L. Liebovitch, G.S. Michaels, P. Munson, R. Nossal, R. Nussinov, R.D. Rosenberg, J.J. Schwartz, M. Schwartz, E.I. Shakhnovich, M.F. Shlesinger, N. Shworak, and E.N. Trifonov for valuable discussions. Partial support was provided to to SVB and HES by the National Science Foundation, to ALG by the G. Harold and Leila Y. Mathers Charitable Foundation, the National Heart, Lung and Blood Institute and the National Aeronautics and Space Administration, to SH by Israel-USA Binational Science foundation, and to C-KP by an NIH/NIMH Postdoctoral NRSA Fellowship.

References

3.1   B.B. Mandelbrot: *The Fractal Geometry of Nature* (W.H. Freeman, San Francisco 1982)
3.2   B.J. West, A.L. Goldberger: J. Appl. Physiol., **60**, 189 (1986);
B.J. West, A.L. Goldberger: Am. Sci., **75**, 354 (1987);
A.L. Goldberger, B.J. West: Yale J. Biol. Med. **60**, 421 (1987);
A.L. Goldberger, D.R. Rigney, B.J. West: Sci. Am. **262**, 42 (1990);
B.J. West, M.F. Shlesinger: Am. Sci. **78**, 40 (1990);
B.J. West: *Fractal Physiology and Chaos in Medicine* (World Scientific, Singapore 1990);
B.J. West, W. Deering: Phys. Reports **xx**, xxx (1994)
3.3   A. Bunde, S. Havlin, eds.: *Fractals and Disordered Systems* (Springer-Verlag, Berlin 1991)
3.4   M.F. Shlesinger, B.J. West: Phys. Rev. Lett. **67**, 2106 (1991)
3.5   B. Suki, A.-L. Barabasi, Z. Hantos, F. Peták, H.E. Stanley: Nature **368**, xxx (1994)
3.6   E.R. Weibel, D.M. Gomez: Science **137**, 577 (1962)
3.7   A.A. Tsonis, P.A. Tsonis: Perspectives in Biology and Medicine **30**, 355 (1987)
3.8   F. Family, B.R. Masters, D.E. Platt: Physica D **38**, 98 (1989);
B.R. Masters, F. Family, D.E. Platt: Biophys. J. (Suppl.) **55**, 575 (1989);
B.R. Masters, D.E. Platt: Invest. Ophthalmol. Vis. Sci. (Suppl.) **30**, 391 (1989)
3.9   M. Sernetz, J. Wübbeke, P. Wlczek: Physica A **191**, 13 (1992)
3.10 H. Takayasu: *Fractals in the Physical Sciences* (Manchester University Press, Manchester 1990)
3.11 D.R. Morse: Nature **314**, 731 (1985)
3.12 T.G. Smith, W.B. Marks, G.D. Lange, W.H. Sheriff Jr., E.A. Neale: J. Neuroscience Methods **27**, 173 (1989)
3.13 H.E. Stanley: Bull. Am. Phys. Soc. **34**, 716 (1989);
F. Caserta, H.E. Stanley, W.D. Eldred, G. Daccord, R.E. Hausman, J. Nittmann: Phys. Rev. Lett. **64**, 95 (1990);
F. Caserta, R.E. Hausman, W.D. Eldred, H.E. Stanley, C. Kimmel: Neurosci. Letters **136**, 198 (1992);
F. Caserta, W.D. Eldred, E. Fernandez, R.E. Hausman, L.R. Stanford, S.V. Buldyrev, S. Schwarzer, H.E. Stanley: J. Neurosci. Methods (in press)

84   Sergey V. Buldyrev, Ary L. Goldberger, Shlomo Havlin et al.

3.14 D.R. Kayser, L.K. Aberle, R.D. Pochy, L. Lam: Physica A **191**, 17 (1992)
3.15 K.R. Bhaskar, B.S. Turner, P. Garik, J.D. Bradley, R. Bansil, H.E. Stanley, J.T. La-
      Mont: Nature **360**, 458 (1992)
3.16 T. Matsuyama, M. Sugawa, Y. Nakagawa: FEMS Microb. Lett. **61**, 243 (1989);
      H. Fujikawa, M. Matsushita: J. Phys. Soc. Japan **58**, 387 (1989);
      M. Matsushita, H. Fujikawa: Physica A **168**, 498 (1990)
3.17 T. Vicsek, M. Cserzö, V.K. Horváth: Physica A **167**, 315 (1990);
      S. Matsuura, S. Miyazima: Physica A **191**, 30 (1992)
3.18 T. Vicsek: *Fractal Growth Phenomena, Second Edition* (World Scientific, Singapore
      1992)
3.19 E. Ben-Jacob, H. Shmueli, O. Shochet, A Tenenbaum: Physica A **187**, 378 (1992);
      E. Ben-Jacob, A. Tenenbaum, O. Shochet, O. Avidan: Physica A **202**, 1 (1994)
3.20 H. Larralde, P. Trunfio, S. Havlin, H.E. Stanley, G.H. Weiss: Nature **355**, 423 (1992);
      M.F. Shlesinger: Nature **355**, 396 (1992);
      H. Larralde, P. Trunfio, S. Havlin, H.E. Stanley, G.H. Weiss: Phys. Rev. A **45**, 7128
      (1992);
      S. Havlin, H. Larralde, P. Trunfio, J.E. Kiefer, H.E. Stanley, G.H. Weiss: Phys. Rev.
      A **46**, R-1717 (1992)
3.21 J.G. Skellam: Biometrika **38**, 196 (1951)
3.22 P.H. Harvey, J.R. Krebs: Science **249**, 140 (1990)
3.23 R.H. Peters: *The Ecological Implications of Body Size* (Cambridge University Press,
      Cambridge 1983);
      J.E.I. Hokkanen: J. Theor. Biol. **499**, 499 (1956);
      A.A. Biewener: Science **250**, 1097 (1990)
3.24 M. Sernetz, B. Gelleri, J. Hofmann: J. Theor. Biol. **187**, 209 (1985);
      R.R. Strathmann: Science **250**, 1091 (1990)
3.25 J. Feder: *Fractals* (Plenum, NY, 1988)
3.26 D. Stauffer, H.E. Stanley: *From Newton to Mandelbrot: A Primer in Theoretical
      Physics* (Springer-Verlag, Heidelberg & N.Y. 1990)
3.27 E. Guyon, H.E. Stanley: *Les Formes Fractales* (Palais de la Découverte, Paris 1991);
      English translation: *Fractal Forms* (Elsevier North Holland, Amsterdam 1991)
3.28 H.E. Stanley, N. Ostrowsky, eds.: *On Growth and Form: Fractal and Non-Fractal Pat-
      terns in Physics*, Proceedings 1985 Cargèse NATO ASI, Series E: Applied Sciences
      (Martinus Nijhoff, Dordrecht 1985)
3.29 H.E. Stanley: *Introduction to Phase Transitions and Critical Phenomena* (Oxford Uni-
      versity Press, London 1971)
3.30 F. Zernike: Physica **7**, 565 (1940)
3.31 H.E. Stanley, N. Ostrowsky, eds.: *Correlations and Connectivity: Geometric Aspects
      of Physics, Chemistry and Biology*, Proceedings 1990 Cargèse Nato ASI, Series E:
      Applied Sciences (Kluwer, Dordrecht 1990)
3.32 P. Bak, C. Tang, K. Wiesenfeld: Phys. Rev. Lett. **59**, 381 (1987);
      P. Bak, C. Tang, K. Wiesenfeld: Phys. Rev. A **38**, 3645 (1988)
3.33 C.-K. Peng, S.V. Buldyrev, A.L. Goldberger, S. Havlin, F. Sciortino, M. Simons, H.E.
      Stanley: Nature **356**, 168 (1992)
3.34 W. Li, K. Kaneko: Europhys. Lett. **17**, 655 (1992)
3.35 S. Nee: Nature **357**, 450 (1992)
3.36 R. Voss: Phys. Rev. Lett. **68**, 3805 (1992)
3.37 J. Maddox: Nature **358**, 103 (1992)
3.38 P.J. Munson, R.C. Taylor, G.S. Michaels: Nature **360**, 636 (1992)
3.39 I. Amato: Science **257**, 747 (1992)
3.40 V.V. Prabhu, J.-M. Claverie: Nature **357**, 782 (1992)
3.41 P. Yam: Sci. Am. **267**[3], 23 (1992)
3.42 C.-K. Peng, S.V. Buldyrev, A.L. Goldberger, S. Havlin, F. Sciortino, M. Simons, H.E.
      Stanley: Physica A **191**, 25 (1992)
3.43 H.E. Stanley, S.V. Buldyrev, A.L. Goldberger, J.M. Hausdorff, S. Havlin, J. Mietus,
      C.-K. Peng, F. Sciortino, M. Simons: Physica A **191**, 1 (1992)
3.44 C.A. Chatzidimitriou-Dreismann, D. Larhammar: Nature **361**, 212 (1993);
      D. Larhammar, C.A. Chatzidimitriou-Dreismann: Nucleic Acids Res. **21**, 5167 (1993)
3.45 A.Yu. Grosberg, Y. Rabin, S. Havlin, A. Neer: Biofisika (Russia) **38**, 75 (1993);
      A.Yu. Grosberg, Y. Rabin, S. Havlin, A. Neer: Europhys. Lett. **23**, 373 (1993)

3.46   J.-L. Sikorav, G.M. Church: J. Mol. Biol. **222**, 1085 (1991)

3.47   A.Yu. Grosberg, A.R. Khokhlov: *Statistical Physics of Macromolecules* (Nauka Publishers, Moscow 1989)

3.48   A.Yu. Grosberg, S.K. Nechaev, E.I. Shakhnovich: J. Physique **49**, 2095 (1988)

3.49   K.W. Adolph, in: *Chromosome Structure and Function*, ed. by M.S. Risley (Van Nostrand, New York 1986);
       M. Takahashi: J. Theor. Biol. **141**, 117 (1989)

3.50   S. Karlin, V. Brendel: Science **259**, 677 (1993)

3.51   C.-K. Peng, S.V. Buldyrev, A.L. Goldberger, S. Havlin, M. Simons, H.E. Stanley: Phys. Rev. E **47**, 3730 (1993)

3.52   N. Shnerb, E. Eisenberg: Phys. Rev. E **49**, xxx (1994)

3.53   S.V. Buldyrev, A.L. Goldberger, S. Havlin, C.-K. Peng, M. Simons, H.E. Stanley: Phys. Rev. E **47**, 4514 (1993)

3.54   A.S. Borovik, A.Yu. Grosberg, M.D. Frank-Kamenetskii: preprint

3.55   S.V. Buldyrev, A.L. Goldberger, S. Havlin, C.-K. Peng, H.E. Stanley, M.H.R. Stanley, M. Simons: Biophys. J. **65**, 2675 (1993)

3.56   C.-K. Peng, S.V. Buldyrev, S. Havlin, M. Simons, H.E. Stanley, A.L. Goldberger: Phys. Rev. E **49**, 1968 (1994)

3.57   S.M. Ossadnik, S.V. Buldyrev, A.L. Goldberger, S. Havlin, C.-K. Peng, M. Simons, H.E. Stanley: preprint;
       H.E. Stanley, S.V. Buldyrev, A.L. Goldberger, S. Havlin, C.-K.Peng, M. Simons: [Proceedings of Internat'l Conf. on Condensed Matter Physics, Bar-Ilan], Physica A **200**, 4 (1993);
       H.E. Stanley, S.V. Buldyrev, A.L. Goldberger, S. Havlin, S.M. Ossadnik, C.-K. Peng, M. Simons: Fractals **1**, 238 (1993)

3.58   S. Tavaré, B.W. Giddings, in: *Mathematical Methods for DNA Sequences*, Eds. M.S. Waterman (CRC Press, Boca Raton 1989), pp. 117-132

3.59   J.D. Watson, M. Gilman, J. Witkowski, M. Zoller: *Recombinant DNA* (Scientific American Books, New York 1992).

3.60   E.W. Montroll, M.F. Shlesinger: "The Wonderful World of Random Walks" in: *Nonequilibrium Phenomena II. From Stochastics to Hydrodynamics*, ed. by J.L. Lebowitz, E.W. Montroll (North-Holland, Amsterdam 1984), pp. 1–121

3.61   G.H. Weiss: *Random Walks* (North-Holland, Amsterdam 1994)

3.62   S. Havlin, R. Selinger, M. Schwartz, H.E. Stanley, A. Bunde: Phys. Rev. Lett. **61**, 1438 (1988);
       S. Havlin, M. Schwartz, R. Blumberg Selinger, A. Bunde, H.E. Stanley: Phys. Rev. A **40**, 1717 (1989);
       R.B. Selinger, S. Havlin, F. Leyvraz, M. Schwartz, H.E. Stanley: Phys. Rev. A **40**, 6755 (1989)

3.63   C.-K. Peng, S. Havlin, M. Schwartz, H.E. Stanley, G.H. Weiss: Physica A **178**, 401 (1991);
       C.-K. Peng, S. Havlin, M. Schwartz, H.E. Stanley: Phys. Rev. A **44**, 2239 (1991)

3.64   M. Araujo, S. Havlin, G.H. Weiss, H.E. Stanley: Phys. Rev. A **43**, 5207 (1991);
       S. Havlin, S.V. Buldyrev, H.E. Stanley, G.H. Weiss: J. Phys. A **24**, L925 (1991);
       S. Prakash, S. Havlin, M. Schwartz, H.E. Stanley: Phys. Rev. A **46**, R1724 (1992)

3.65   C.L. Berthelsen, J.A. Glazier, M.H. Skolnick: Phys. Rev. A **45**, 8902 (1992)

3.66   E.I. Shakhnovich, A.M. Gutin: Nature **346**, 773 (1990)

3.67   S.G. Oliver et al.: Nature **357**, 38 (1992)

3.68   E.C. Uberbacher, R.J. Mural: Proc. Natl. Acad. Sci. USA **88**, 11261 (1991)

3.69   In the power spectrum analysis for those sequences containing noncoding regions, subtracting of the white noise, $S(\infty)$, as performed in [3.36], gives more weight to the noncoding segments (correlated) than the coding segments (uncorrelated). See also H.E. Stanley, S.V. Buldyrev, A.L. Goldberger, S. Havlin, C-K Peng, M. Simons: Physica A **200**, 4 (1993)

3.70   S.V. Buldyrev, A. Goldberger, S. Havlin, C.-K. Peng, F. Sciortino, M. Simons, H.E. Stanley: Phys. Rev. Lett. **71**, 1776 (1993);
       R.F. Voss, Phys. Rev. Lett. **71**, 1777 (1993)

3.71   W.-H. Li, D. Graur: *Fundamentals of Molecular Evolution* (Sinauer Associates, Sunderland MA 1991)

3.72 W. Li: International Journal of Bifurcation and Chaos **2**, 137 (1992)

3.73 J. Jurka: J. Mol. Evol. **29**, 496 (1989)

3.74 R.H. Hwu, J.W. Roberts, E.H. Davidson, R.J. Britten: Proc. Natl. Acad. Sci. USA. **83**, 3875 (1986)

3.75 J. Jurka, T. Walichiewicz, A. Milosavljevic: J. Mol. Evol. **35**, 286 (1992)

3.76 J. Des Cloizeaux: J. Physique (Paris) **41**, 223 (1980);
P.G. de Gennes: *Scaling Concepts in Polymer Physics* (Cornell Univ. Press, Ithaca 1979)

3.77 G.F. Joyce: Nature **338**, 217 (1989)

3.78 W. Gilbert: Nature **271**, 501 (1978)

3.79 P. Hagerman: Ann. Rev. Biochem. **59**, 755 (1990)

3.80 W.F. Doolitle, in: *Intervening Sequences in Evolution and Development*, ed. by E. Stone, R. Schwartz (Oxford University Press, NY 1990), p. 42

3.81 A. Schenkel, J. Zhang, Y-C. Zhang: Fractals **1**, 47 (1993);
M. Amit, Y. Shmerler, E. Eisenberg, M. Abraham, N. Shnerb: Fractals **2**, xxx (1994)

3.82 R.N. Mantegna: Physica A **179**, 23 (1991)

3.83 H.E. Stanley, N. Ostrowsky, eds.: *Random Fluctuations and Pattern Growth: Experiments and Models*, Proceedings 1988 NATO ASI, Cargèse (Kluwer Academic Publishers, Dordrecht 1988)

3.84 R.I. Kitney, O. Rompelman: *The Study of Heart-Rate Variability* (Oxford University Press, London 1980);
S. Akselrod, D. Gordon, F.A. Ubel, D.C. Shannon, A.C. Barger, R.J. Cohen: Science **213**, 220 (1981)

3.85 M. Kobayashi, T. Musha: IEEE Trans. Biomed. Eng. **29**, 456 (1982)

3.86 A.L. Goldberger, D.R. Rigney, J. Mietus, E.M. Antman, S. Greenwald: Experientia **44**, 983 (1988)

3.87 J.P. Saul, P. Albrecht, D. Berger, R.J. Cohen: *Computers in Cardiology* (IEEE Computer Society Press, Washington, D.C. 1987), pp. 419–422;
D.T. Kaplan, M. Talajic: Chaos **1**, 251 (1991)

3.88 C.-K. Peng, J.E. Mietus, J.M. Hausdorff, S. Havlin, H.E. Stanley, A.L. Goldberger: Phys. Rev. Lett. **70**, 1343 (1993);
C.K. Peng, Ph.D. Thesis, Boston University (1993);
C.K. Peng, S.V. Buldyrev, J.M. Hausdorff, S. Havlin, J.E. Mietus, M. Simons, H.E. Stanley, A.L. Goldberger, in: *Fractals in Biology and Medicine*, ed. by G.A. Losa, T.F. Nonnenmacher, E.R. Weibel (Birkhauser Verlag, Boston 1994)

3.89 W.B. Cannon: Physiol. Rev. **9**, 399 (1929)

3.90 G.K. Zipf: *Human Behavior and the Principle of "Least Effort"* (Addison-Wesley, New York 1949)

3.91 L. Brillouin: *Science and Information Theory* (Academic Press, New York 1956)

3.92 C.E. Shannon: Bell Systems Tech. J. **80**, 50 (1951)

3.93 L.S. Liebovitch, J. Freichbarg, J.P. Koniarek: Math. Biosci. **89**, 36 (1987);
L.S. Liebovitch: Biophys. J. **55**, 373 (1989)

3.94 R.V. Voss, J. Clarke: Nature **238**, 317 (1975)

3.95 R.V. Voss, J. Clarke: J. Acous. Soc. Am. **63**, 258 (1978)

3.96 M. Schroeder: *Fractals, Chaos, Power Laws: Minutes from an Infinite Paradise* (W. H. Freeman, New York 1991)

3.97 H.E. Stanley, S.V. Buldyrev, A.L. Goldberger, Z.D. Goldberger, S. Havlin, R.N. Mantegna, S.M. Ossadnik, C.-K. Peng, M. Simons: Physica A **205**, xxx (1994)

3.98 N. Eldredge, S. J. Gould, in: *Models in Paleobiology*, edited by T. J. M. Schopf (Freeman and Cooper Inc., San Francisco 1972);
S. J. Gould, N. Eldredge: *Nature* **366**, 223 (1993)

3.99 P. Bak, K. Sneppen: Phys. Rev. Lett. **71**, 4083 (1993)

3.100 S. Havlin, A.-L. Barabási, S.V. Buldyrev, C.-K. Peng, M. Schwartz, H.E. Stanley, T. Vicsek, in: *Growth Patterns in Physical Sciences and Biology* [Proc. NATO Advanced Research Workshop, Granada, Spain, October 1991], J.M. Garcia-Ruiz, E. Louis, P. Meakin, L. Sander, eds. (Plenum, New York 1993), pp. 85–98;
L.-H. Tang, H. Leschhorn: Phys. Rev. A **46**, R-8309 (1992);
S.V. Buldyrev, A.-L. Barabási, F. Caserta, S. Havlin, H.E. Stanley, T. Vicsek: Phys. Rev. A **46**, R-8313 (1992);

A.-L. Barabási, S. V. Buldyrev, S. Havlin, G. Huber, H. E. Stanley, T. Vicsek, in: R. Jullien, J. Kertész, P. Meakin, D. E. Wolf (eds.), *Surface Disordering: Growth, Roughening, and Phase Transitions* [Proc. of the Les Houches Workshop 1992] (Nova Science, New York 1992), pp. 193–204;

S.V. Buldyrev, S. Havlin, J. Kertesz, A. Shehter, H.E. Stanley: Fractals **1**, 827 (1993);

L.A.N. Amaral, A.L. Barabási, S.V. Buldyrev, S. Havlin, H.E. Stanley: Phys. Rev. Lett. **72**, 641 (1994)

4 Self-Affine Interfaces

János Kertész and Tamás Vicsek

4.1 Introduction

Self-similar objects are the simplest examples for scale invariance. Scaling is maintained, although in an anisotropic form, when the invariance is required for the general affine transformation and not for pure dilation only. Examples include critical directed percolation clusters, directed lattice animals, or the record of a random walk. These objects are called *self-affine.*

For an interesting and quite common class of scaling objects the invariance transformation can be represented by a diagonal matrix with all but one of the eigenvalues equal to λ and one to λ^α with $0 < \alpha < 1$. Surfaces of homogeneous (non-fractal) systems are often *rough* and show such self-affine structure. Further examples of *random self-affine functions* can be found in fluctuating time dependent phenomena ($1/f^\phi$ noise).

The concept of self-affinity was introduced by Mandelbrot, who studied the fractal properties of self-affine functions [4.1]. There are several reviews describing self-affine geometry [4.2-4] including Chap. 7 (by J.-F. Gouyet, M. Rosso, and B. Sapoval) in [4.5]. Therefore, we restrict ourselves in this chapter to summarizing only the fundamental features and the important new developments in this very active field of research.

The simplest example for the self-affine random function is the record $h(x)$ of a one-dimensional Brownian motion. The time is put on the x axis and the position is denoted by h. In analogy with the case of self-similar fractal objects the statistical scale invariance is reflected in the correlation functions. It is convenient to consider the *height-difference correlation function $C(x)$*. For a

◀ **Fig. 4.0.** An example of a model interface describing imbibition. The front propagates in the presence of quenched randomness and the pinning of the interface is governed by a mechanism related to directed percolation. Different colors mean different heights. Courtesy of S.V. Buldyrev

long record of the walk $(0 \leq x \leq X)$

$$C(x) = \langle [h(x' + x) - h(x')]^2 \rangle_{x'}^{1/2}, \tag{4.1}$$

where $\langle \cdots \rangle_{x'}$ means averaging over the positions x'. Since the mean square displacement of a Brownian particle is proportional to time, we have (see Chap. 5)

$$C(x) \sim x^{1/2}. \tag{4.2}$$

For an arbitrary self-affine function of one variable

$$C(x) \sim x^\alpha, \tag{4.3}$$

with α being the Hurst exponent, known in the case of surfaces as the *roughness exponent*. Thus the Hurst exponent is $1/2$ for the record of the Brownian motion (see also Chap. 1).

The fractal dimension of a self-affine function can be determined by box counting. Choosing the box length as ε, the number $N(\varepsilon)$ of boxes needed to cover the graph of the function on the interval X is

$$N(\varepsilon) \sim \frac{X}{\varepsilon} \times \frac{\varepsilon^\alpha}{\varepsilon} \sim \varepsilon^{\alpha - 2}, \tag{4.4}$$

since the first term on the right-hand side is the number of subintervals of length ε and the second term is the number of boxes covering the graph in a subinterval. The fractal dimension is therefore

$$d_f = 2 - \alpha, \tag{4.5a}$$

since $d_f = -\lim_{\varepsilon \to 0}(\log N(\varepsilon)/\log \varepsilon)$. This d_f is the *local fractal dimension*.

In the case of self-similar fractals the dimension can also be calculated by introducing a lower cut-off and covering the object with growing boxes. For self-affine functions this procedure obviously leads to a different result, namely to the trivial global dimension 1. The change in the behavior occurs at a crossover point x_c defined by $C(x_c) \simeq x_c$. When $\varepsilon \ll x_c$, the second term on the right-hand side of (4.4) is > 1. In the opposite limit, when $\varepsilon \gg x_c$, this term is < 1, thus one box is sufficient for covering the graph of the function in a subinterval ε. The generalization of these results to self-affine functions of d variables is straightforward. The local fractal dimension is

$$d_f = d + 1 - \alpha, \tag{4.5b}$$

while the global dimension is equal to d.

The position (h) and the time (x) are physically different quantities and their units are independent of each other. Nevertheless, if the record as a geometric object is considered, a length scale has to be set onto the "time axis" as well. Since the ratio of the time and length units can be chosen arbitrarily, the position of the crossover point x_c can be changed.

If the self-affine function is discrete, e.g., it represents the surface of a lattice system, the units are fixed in all directions and the range below the crossover point is usually not present. Self-affine surfaces are in this sense not fractals. Nevertheless, they can be transformed by scale transformations into objects with nontrivial local fractal dimensions.

The fractal aspect of self-affine surfaces also becomes apparent when the surface is cut by a d-dimensional (hyper-)plane with an orientation parallel to the average orientation of the surface. The section is a self-similar object with the fractal dimension $d_{\mathrm{cut}} = d_f - 1 = d - \alpha$.

A concept analog to multifractality (see [4.2] and [4.5]), called *multiaffinity* [4.6] can be introduced for the objects obeying affine scaling. Introducing the generalized qth-order correlation functions

$$C_q(x) = \langle |h(x + x') - h(x')|^q \rangle_{x'}^{1/q}, \tag{4.6}$$

the expected behavior is

$$C_q(x) \sim x^{\alpha_q}. \tag{4.7}$$

In simple cases α_q does not depend on q. For multiaffine objects the function α_q can be related to D_q or $f(\alpha)$ in the multifractal formalism [4.6].

In the following we shall concentrate on self-affine *surfaces or interfaces* [4.3,7-13]. From the microscopic point of view it is a difficult problem to decide which phase a particular point in the surface region belongs to. In a realistic lattice model this feature is reflected in the occurrence of holes, high steps, and overhangs, which makes it cumbersome to define the surface. A self-affine single-valued function results from a realistic surface by *coarsening*. Of course, there are models particularly tailored to investigate such surface problems, like the so-called solid-on-solid (SOS) models, where just these overhangs and holes are forbidden by definition. For these models the equivalence of the simplified surfaces with the real ones has to be shown.

The term "surface region" of the former paragraph needs further specification [4.14]. The scaling fluctuations of physical self-affine surfaces (which are either already defined on a lattice or have a lower cutoff anyway) is observable in the long distance limit ($x \to \infty$ in (4.3)). These fluctuations result in wandering of the surface and can be characterized by the surface width:

$$w^2 = \langle (h - \langle h \rangle)^2 \rangle = \langle h^2 \rangle - \langle h \rangle^2, \tag{4.8}$$

where, in analogy with (4.1), $\langle \cdots \rangle$ means spatial average over the range of $\boldsymbol{x}$ above which the (coarsened) surface positions $h(\boldsymbol{x})$ are defined. The scaling of the surface can be described by the dependence of the width w on the linear extent L of the system:

$$w \sim L^\alpha. \tag{4.9}$$

Relation (4.9) is valid for asymptotically large L values. Over short distances the behavior is characterized by the so called *intrinsic surface width*, the range over which one phase crosses over into the other (Fig. 4.1). For equilibrium interfaces this range can be identified with the bulk correlation length.

What is the origin of the roughness of surfaces? One source can be destruction, such as *fracture* [4.15-17], *erosion* [4.18,19] or *burning* [4.20]. In many cases

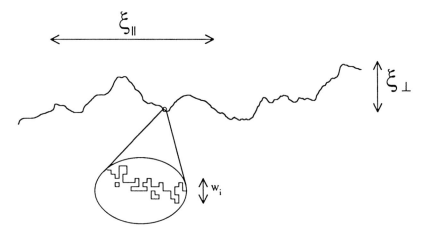

Fig. 4.1. The schematic structure of an interface. The intrinsic surface width w_i is indicated in the inset. For the meaning of $\xi_\parallel$ and $\xi_\perp$ see (4.38)

rough surfaces are generated by a *growth* [4.2,3,10]. Rough interfaces often occur under *equilibrium* circumstances as well [4.7,8]. It seems to be a general feature that (time dependent or quenched) randomness is essential in the development of self-affine surfaces.

4.2 Roughness and Pinning in Equilibrium

Let us consider a $d+1$ dimensional Ising model in zero external field below the transition temperature with anti-periodic boundary conditions in one specific direction and periodic boundary conditions in the other d directions. Owing to the boundary conditions the system cannot order according to one of the ground states and a d-dimensional *interface* is created. Using the lattice-gas representation of the Ising model, where spins pointing in one direction are identified with occupied sites and the other ones with empty sites, the above configuration can be regarded as a model of a solid–vapor interface.

The interface is not a unique d-dimensional object because the configurations contain overhangs, high steps, and holes; thus one phase goes into the other one over a range characterized by the bulk correlation length ξ_b. Therefore, the intrinsic width of the surface can be identified with ξ_b. It is generally accepted that the behavior of the interface over distances much larger than ξ_b can be adequately described by SOS models, where overhangs and holes are disregarded and the interface can be represented by a single valued height variable h of d spatial variables. An appropriate Hamiltonian is [4.7]

$$\mathcal{H} = \frac{J}{2}\sum_{j,\text{nn}}(h_j - h_{j+\text{nn}})^2 \tag{4.10a}$$

where nn means nearest neighbour. Equation (4.10a) has the continuum limit

$$\mathcal{H} = \frac{J}{2} \int_{L^d} d^d x |\nabla h|^2. \tag{4.10b}$$

This *Gaussian model* has the advantages of being solvable and it is general in the sense that its long-wavelength limit describes the scaling properties of an interface characterized by a surface tension.

Taking the Fourier transform of this equation one has

$$\mathcal{H} = \frac{J}{2} \sum_k k^2 |\bar{h}_k|^2, \tag{4.11}$$

where

$$\bar{h}_k = (1/L)^{d/2} \int_{L^d} h(\boldsymbol{x}) e^{-i\boldsymbol{k}\boldsymbol{x}} \tag{4.12}$$

and L is the linear extent of the d dimensional surface. Owing to the equipartition theorem $Jk^2|\bar{h}_k|^2 = T$. The fluctuation of the interface over a distance $\boldsymbol{x} - \boldsymbol{x}'$ is therefore given as

$$C^2(\boldsymbol{x}-\boldsymbol{x}') = 2\langle h^2\rangle - 2\langle h(\boldsymbol{x})h(\boldsymbol{x}')\rangle \sim (T/J) \int \frac{d^d k}{k^2}[1 - \cos(\boldsymbol{k}(\boldsymbol{x}-\boldsymbol{x}'))], \tag{4.13}$$

where the k integral is limited from above by the inverse of the lower cutoff in the system (lattice constant or interatomic distance). The evaluation of the integral leads to $C^2(\boldsymbol{x} - \boldsymbol{x}') \sim (x-x')^{2-d}$ for all dimensions except $d = 2$, where $C^2(\boldsymbol{x} - \boldsymbol{x}') \sim \log(x - x')$ [4.7-9]. This result can be summarized as

$$\alpha = \frac{2-d}{2}. \tag{4.14}$$

The one dimensional interface of a two dimensional system is rough with a Hurst exponent $1/2$. The interface is marginally rough for the $d = 2$ interface because of the logarithmic divergence – this is what is meant by the zero exponent in (4.14). In higher dimensions the interface is flat, i.e., its width does not diverge with increasing system size.

The possible *lattice effects* do not occur in the above consideration. It turns out that the periodic potential due to the lattice is irrelevant for the one-dimensional interface while it leads to the so called *roughening transition* for the two dimensional interface. Below the roughening temperature T_R the interface is smooth: the free energy of forming a step is larger than zero. At T_R this free energy contribution vanishes and for $T \geq T_R$ the surface becomes logarithmically rough. Important properties such as equilibrium crystal shapes or the growth rate depend heavily on whether the temperature is above or below T_R.

The nature of the roughening transition has been intensely studied [4.7,8]. The two dimensional Hamiltonian to be considered is a modification of (4.10a):

$$\mathcal{H} = \frac{J}{2} \sum_{j,\text{nn}}(h_j - h_{j+\text{nn}})^2 - 2y_0 J \sum_j \cos(2\pi h_j), \tag{4.15}$$

where the lattice effects are taken into account by the periodic potential. From the point of view of critical behavior this Hamiltonian is dual to the two-dimensional Coulomb gas or, equivalently, to the planar XY model; thus the roughening transition is of the Kosterlitz-Thouless type. The logarithmic roughness for $T \geq T_R$ corresponds to the low-temperature correlations in the Coulomb gas. This picture of the roughening transition based on the mapping and on renormalization group calculations has been reinforced by exact solutions and numerical results [4.8,9].

The roughness of the equilibrium interface is generated by thermal fluctuations. *Quenched, time-independent noise* also leads to the roughening of the interface. The disorder can be in the couplings as represented by a zero-field Ising Hamiltonian with random interactions so that the ferromagnetic ground state is not destroyed. The continuum equation describing the interface is [4.21]:

$$\mathcal{H} = \int_V d^d x [\frac{J}{2} |\nabla h|^2 + \eta(\boldsymbol{x}, h(\boldsymbol{x}))], \qquad (4.16)$$

where the random potential can be approximated by Gaussian white noise. The important difference in the thermal noise is that η in (4.16) is time independent, i.e., quenched. There are two exponents characterizing the interface: H, the roughness exponent, and ω, defined by

$$\Delta E \sim L^\omega, \qquad (4.17)$$

where ΔE is the fluctuation of the energy along the interface in the ensemble of different noise configurations. There is a scaling relation between the two exponents:

$$2\alpha - \omega = 2 - d. \qquad (4.18)$$

In 1+1 dimensions the weight W of an interface (or a directed path) in a random medium starting at $(0,0)$ and ending at (x,y) satisfies the equation [4.22]:

$$\partial_x W(x,y) = (T/J)\partial_y^2 W(x,y) + (1/T)V(x,y)W(x,y). \qquad (4.19)$$

As we shall discuss later, at zero temperature this is equivalent to a *global optimization problem*. Equation (4.19) can be transformed into the Burgers equation with random force and can be solved exactly in 1+1 dimensions. The result is $\alpha = 2/3$ and $\omega = 1/3$. Therefore the interface roughness with disorder is larger than in the pure case because of the pinning effect of the impurities.

Enhanced roughening is caused by another type of quenched disorder when the spins of an Ising model feel *random, time-independent magnetic field* η_i:

$$\mathcal{H} = -J \sum_{<i,j>} s_i s_j + \sum_i \eta_i s_i, \qquad (4.20)$$

where the sum runs over nearest-neighbor pairs of spins. An interesting *dynamic* effect in this model is that there is a *depinning transition* as a function of an additional homogeneous external field H [4.23]. A finite $H \geq H_c$ field is needed to get the interface to move away from its pinned position. We shall come back later to this problem of a moving interface in the presence of quenched disorder.

4.3 Dynamic Scaling and Growth Models

In order to study the time dependent fluctuations of the interface it is useful to introduce the *Langevin equation* corresponding to the Hamiltonian (4.10) [4.7,8]:

$$\partial_t h_j = -\frac{\Gamma}{T}\frac{\delta\mathcal{H}}{\delta h_j} + \eta_j(t), \tag{4.21}$$

which leads to

$$\partial_t h_j(t) = -(\Gamma 2J/T)\sum_{\mathrm{nn}}(h_j(t) - h_{j+\mathrm{nn}}(t)) + \eta_j(t). \tag{4.22}$$

Here η_j is a Gaussian fluctuating white noise satisfying the fluctuation dissipation theorem:

$$\langle\eta_j(t)\rangle = 0, \tag{4.23}$$

$$\langle\eta_i(t)\eta_j(t')\rangle = 2\Gamma\delta_{i,j}\delta(t - t'). \tag{4.24}$$

The continuum limit of this equation is

$$\partial_t h(\boldsymbol{x},t) = \nu\nabla^2 h(\boldsymbol{x},t) + \eta(\boldsymbol{x},t). \tag{4.25}$$

It is simple to obtain the Fourier transform of the correlation function from (4.25):

$$\tilde{C}^2(k,\omega) = \langle\bar{h}_k(\omega)\bar{h}_{-k}(\omega)\rangle \tag{4.26}$$

and

$$\tilde{C}^2(k,\omega) = \frac{1}{\omega^2 + \nu^2 k^4}. \tag{4.27}$$

The result (4.14) could have been calculated from this expression as well. Moreover, another exponent, *the dynamic exponent z*, can be obtained using (4.27). The right-hand side of (4.27) can be rewritten so that

$$\tilde{C}^2(k,\omega) = \omega^\varphi f(k^z\omega), \tag{4.28}$$

where f is a scaling function and $z = 2$ in all dimensions for the above problem [4.7].

What happens if the interface is *driven* by an external force? In this case the position of the mean $\langle h\rangle$ changes with time and the fluctuating surface is a result of a growth process:

$$\partial_t h(\boldsymbol{x},t) = F + \nu\nabla^2 h(\boldsymbol{x},t) + \eta(\boldsymbol{x},t). \tag{4.29}$$

This is the Edwards-Wilkinson (EW) model of sedimentation [4.24]. Since both the Laplace and the noise terms disappear if an average over the whole system is taken, $\langle h\rangle = F$ and (4.29) can be transformed to (4.25) using a co-moving frame. Therefore, the exponents α and z of this growth model are the same as those for the thermal roughening problem.

Much of the information about growth phenomena has been obtained by studying simple computer models which cannot be related to Hamiltonians. By analyzing such a model, Family and Vicsek postulated [4.25] the following *dynamic-scaling* form of the width (4.8) of an initially flat surface developing in a $d + 1$ dimensional space:

$$w(L, t) = L^\alpha f(t/L^z), \tag{4.30}$$

where L is the linear size of the substrate. In the limit where the argument of the scaling function $f(a)$ is small, $a \ll 1$, the width depends only on t and this implies that for small a the scaling function $f(a)$ has to be of the form $f(a) \sim a^\beta$ with $\beta = \alpha/z$. At large times the width saturates, i.e., $f(a)$ goes to a constant in that limit. Thus,

$$f(a) \sim \begin{cases} a^\beta, & \text{if } a \ll 1 \\ \text{const}, & \text{if } a \gg 1. \end{cases} \tag{4.31}$$

For short times ($t \ll L^z$) the width depends only on the time

$$w \sim t^\beta, \tag{4.32}$$

while for ($t \gg L^z$)

$$w \sim L^\alpha. \tag{4.33}$$

Stationarity is reached for times $t \gg \tau$ where the relaxation time $\tau \sim L^z$ is measured from the beginning of the growth.

Similar behavior is expected in the stationary state for the *time-dependent correlation function* $\tilde{C}(r, t)$, the generalization of (4.1)

$$\tilde{C}(r, t) = \langle (h(\boldsymbol{x}', t') - h(\boldsymbol{x}' + \boldsymbol{x}, t' + t))^2 \rangle_{\boldsymbol{x}', t'}^{1/2} \sim r^\alpha \tilde{f}\left(\frac{t}{r^z}\right), \tag{4.34}$$

with an $\tilde{f}(a)$ analogous to (4.31); however, in this case L is replaced by $r = |\boldsymbol{x}|$. For $r \ll L$, $\tilde{C}(r, 0)$ behaves as

$$\tilde{C}(r, 0) \sim r^\alpha, \tag{4.35}$$

while for times $t \ll \tau$

$$\tilde{C}(0, t) \sim t^\beta. \tag{4.36}$$

In the transient region, before stationarity sets in, the scaling can be studied by the *equal time correlation function* $\hat{C}(r, t)$:

$$\hat{C}(r, t) = \langle (h(\boldsymbol{x}', t) - h(\boldsymbol{x}' + \boldsymbol{x}, t))^2 \rangle_{\boldsymbol{x}'}^{1/2} \sim r^\alpha \hat{f}(t/r^z). \tag{4.37}$$

The spread of fluctuations in the direction parallel to the substrate is characterized by the correlation length $\xi_\parallel(t)$. This shows how far the effect of a perturbation can get after some time t along the surface (Fig. 4.1).

$$\xi_\parallel \sim t^{1/z}, \quad \text{for} \quad t \ll \tau \quad \text{and} \quad \xi_\parallel = L, \quad \text{for} \quad t \gg \tau. \tag{4.38a}$$

The average height difference $\xi_\perp$ over the distance $\xi_\parallel$ is

$$\xi_\perp = \xi_\parallel^\alpha \tag{4.38b}$$

reflecting the affine relationship between directions parallel and perpendicular to the substrate. According to (4.34–38), the surface has self-affine properties up to the scale $\xi_\parallel(t)$ which grows with time (4.38). Dynamic scaling in self-affine surface growth has been verified in many cases by different methods. The scaling properties exist for large system sizes and long times as compared to the microscopic length and time units. For intermediate times and lengths corrections to the asymptotic scaling often have to be taken into account.

One of the simplest related system is the *Eden growth process* [4.26]. In this model all perimeter sites of the cluster are growth sites with equal growth probability and at one time one of the growth sites gets part of the cluster. The initial stage is an occupied d-dimensional flat (hyper-)plane. Of course, other initial geometries, such as a single seed, can also be considered. This model was originally invented to describe biological growth, such as that of cancer cells, but it can also be related to crystal growth.

Deep in the bulk, Eden clusters have no holes. The surface region is rather complex with holes, high steps and overhangs. As in the situation of Ising-type interfaces, these contribute to an *intrinsic width* of the surface which does not scale with the system size. The effect on the scaling of the intrinsic width w_i is well described by the *convolution approximation*:

$$w_{total}^2 = w_{scaling}^2 + w_i^2 \tag{4.39}$$

where $w_{scaling}$ is described by (4.30). Equation (4.39) expresses the idea that the fluctuations resulting in the intrinsic width and the scaling long wavelength fluctuations can be considered independent. Reasonable scaling for the Eden model in $d + 1 = 2, 3$ and 4 dimensions could be obtained [4.27] when the corrections to scaling due to the intrinsic width were taken into account. The results are: $\beta \simeq 0.33$ ($d = 1$), $\beta \simeq 0.22$ ($d = 2$), and $\beta \simeq 0.15$ ($d = 3$), and a scaling relation $\alpha(1 + 1/\beta) = \alpha + z = 2$, to be discussed later, seems to hold.

Another model which shows similar behavior is *ballistic deposition* [4.28], which describes vapor deposition [4.29]. Particles – one after the other – fall down at random positions on parallel straight paths and stick to the cluster or deposit when they get into contact with it either by falling on the top of a particle of the deposit or by touching one at the side. Again, the kinetic roughening process starts with a flat surface. In contrast to the Eden model, here holes survive forever since a particle stuck to the cluster screens all the sites below it.

The numerical results on ballistic aggregation lead to exponents close to those of the Eden model [4.30]; these models can therefore be assumed to belong to the same universality class. As discussed above, the determination of the exponents is hindered by the corrections due to the intrinsic width. There are two ways to overcome this difficulty. One is to suppress the intrinsic width.

This can be done for the Eden model by noise reduction [4.31] – in fact, the results mentioned above were obtained using this method [4.27]. The other way is to introduce models which belong to the same universality class but have minimal intrinsic width by definition. Such are the *single-step* models. The surface configurations are SOS-like without any overhangs, holes or high steps.

In one of these models [4.32] the particles are assumed to rain down along randomly positioned vertical trajectories as in ordinary ballistic deposition, but they are incorporated into the deposit only if (a) the height difference between any two columns is smaller than or equal to unity and (b) no overhangs occur, i.e., a particle always has to sit on the top of another one. This model leads to perfectly compact clusters, which have a surface with slopes 0 or ±1.

The extensive simulations of the above restricted solid-on-solid model has resulted in accurate estimates of the exponent $\beta(d)$ in dimensions $d+1 = 2-5$ [4.32]:

$$\beta(1) = 0.332 \pm 0.005, \qquad \beta(2) = 0.25 \pm 0.005,$$

$$\beta(3) = 0.20 \pm 0.01, \qquad \beta(4) \simeq 0.17. \tag{4.40}$$

Now we turn to the description of the results obtained by the *hypercube stacking* model [4.30,33,34]. In this model the surface of a stack of (hyper-)cubes on a $(11\cdots)$ substrate is considered. This means that the hypercubes are piled up in such a way that one of their corners points downwards. The model is defined as a sequence of stacking events; each time a hypercube is added to the surface at a deposition rate p^+ or removed with a rate p^- [4.33,34]. The single step solid-on-solid condition is assumed at all stages. Forrest and Tang used a multispin coding, parallel-processing algorithm to obtain accurate estimates for this model.

The careful evaluation of the simulations carried out on systems of sizes $L = 11520^2$ in $d+1 = 3$ and $L = 2 \times 192^3$ in $d+1 = 4$ dimensions has led to the values

$$\beta(2) = 0.24 \pm 0.001 \qquad \text{and} \qquad \beta(3) = 0.180 \pm 0.005 \tag{4.41}$$

which together with the scaling law $\alpha + z = 2$ uniquely determine the other exponent as well. Two comments are in order concerning (4.41). (a) The studied system sizes are considerably larger than in any of the previous investigations, and (b) the estimates (4.41) are in *disagreement* with earlier conjectures based on numerical results [4.27,32].

4.4 Continuum Equations, Directed Polymers, and Morphological Transitions

Why are the diverse models like Eden growth, ballistic deposition and the different single-step models expected to belong to the same universality class? Long-wavelength fluctuations, which are responsible for scaling can be described by stochastic hydrodynamic (continuum) equations. It turns out that all these models are described by the same continuum equation because they share a relevant common property: the growth velocity depends on the tilt of the substrate.

We have already seen the simplest continuum equation describing kinetic roughening, namely the EW equation (4.29). The exponents of this equation are different from the Eden universality class (4.39,40) (except for $\alpha = 1/2$ for the one dimensional surface). Edwards and Wilkinson have already pointed out [4.24] that taking into account nonlinearities may change their result. Kardar, Parisi and Zhang (KPZ) [4.35] suggested the following *nonlinear* equation:

$$\partial_t h = \nu \nabla^2 h + \lambda/2 (\nabla h)^2 + \eta(\boldsymbol{x}, t), \qquad (4.42)$$

where $\lambda/2$ is the coupling of the nonlinear term. The constant force F has already been transformed out. The noise $\eta(\boldsymbol{x}, t)$ in (4.42) was assumed to be uncorrelated and white and to have a Gaussian distribution, so that $\langle \eta(\boldsymbol{x}, t) \rangle = 0$, and

$$\langle \eta(\boldsymbol{x}, t) \eta(\boldsymbol{x}', t') \rangle = 2 D \delta^d (\boldsymbol{x} - \boldsymbol{x}') \delta(t - t'), \qquad (4.43)$$

where D is a constant. The origin of the nonlinearity can be understood if the Eden-type growth is considered; in the continuum approximation, this is just growth normal to the surface in the presence of noise. The surface is described by a height function and therefore normal growth Δ at positions where $\nabla h \neq 0$ leads to an increment $\Delta \sqrt{1 + (\nabla h)^2} \simeq \Delta(1 + (\nabla h)^2/2)$. It is clear from this argument that a nonlinear coupling $\lambda \neq 0$ is generated whenever the *growth velocity depends on the tilt of the surface*. The coupling is positive if the velocity increases (as in the Eden model) and negative if it decreases (as in the single-step model) with the tilt. However, the scaling behavior is independent of the sign of λ since the relevant combination of the couplings is $\lambda^2 d_f / \nu^3$. If surface relaxation is allowed in the ballistic deposition model, i.e., the particles roll down and fill the holes, there will be no tilt dependence of the growth velocity and the EW (or equilibrium) exponents are recovered [4.36]. The actual value of the nonlinear coupling constant can be measured by investigating the tilt dependence of the growth velocity [4.37] or by using inhomogeneous deposition [4.38].

We have already referred to the scaling relation [4.30]

$$\alpha + z = 2. \qquad (4.44)$$

(In $1 + 1$ dimensions, where the equivalence to the equilibrium roughening problem is valid, the corresponding equation is (4.18).) This equation (4.44) is a consequence of the invariance of the KPZ equation with respect to an infinitesimal tilt [4.39]. It can be understood in simple geometric terms if growth normal to the surface is assumed [4.40]. Let us approximate the fluctuation of size $\xi_{\|}$ in the surface by a (hyper-)spherical calotte with a height and diameter equal to $\xi_\perp$ and $\xi_{\|}$, respectively. The radius of the sphere grows proportionally to the time t. Therefore,

$$t^2 - (t - \xi_\perp)^2 = \xi_{\|}^2. \qquad (4.45)$$

The term $\xi_\perp^2$ can be neglected for long times. The scaling relation (4.44) follows if relations (4.38) are used in (4.45).

Using the so called Cole-Hopf transformation $W(\boldsymbol{x},t) = \exp[(\lambda/2\nu)h(\boldsymbol{x},t)]$ the KPZ equation can be transformed into the diffusion equation with multiplicative noise (the $d+1$ dimensional generalization of (4.19)) describing the problem of $d+1$ dimensional directed polymers. At zero temperature this is equivalent to finding the optimal directed path in a random potential, i.e., the path along which the sum of the potential is minimal. As mentioned before, the solution to (4.19) in $1+1$ dimensions is known, thus the exponents of the KPZ or Eden-type growth can be determined – but the meaning of the exponents here is different. The roughening exponent for the equilibrium-interface problem with quenched disorder is equal to the inverse of the dynamic exponent in the surface growth process $1/z = 2/3$ while the exponent describing energy fluctuations is equal to $\beta = 1/3$.

The relation between directed polymers and surface growth becomes more explicit on the microscopic level using waiting time growth models [4.41], moreover, the mapping can be made exact [4.42]. Imagine a hypercubic stacking model where random waiting times are assigned to each growth site when they are born. The time when a particular site at $\boldsymbol{R}$ becomes occupied satisfies the equation

$$t(\boldsymbol{R}) = \tau(\boldsymbol{R}) + \max_t(\boldsymbol{R}'), \tag{4.46}$$

where $\boldsymbol{R}'$ denotes sites preceding $\boldsymbol{R}$. Applying (4.46) recursively it turns out that the sites on the surface are connected to the substrate by ancestor lines for which

$$t(\boldsymbol{R}) = \max_{\text{Path } \boldsymbol{0} \to \boldsymbol{R}} \sum_{\boldsymbol{R}' \in \text{Path}} \tau(\boldsymbol{R}'). \tag{4.47}$$

Thus the path is equivalent to a zero temperature directed polymer if $-\tau(\boldsymbol{R})$ is identified with the random potential at site $\boldsymbol{R}$. Besides its theoretical interest, the model of directed polymers can be helpful in understanding the self-affine structure of rupture lines, for example, in paper sheets [4.17].

The KPZ equation can be mapped to the Burgers equation [4.35] and this allows us to apply the renormalization group analysis of Forster et al. [4.43] to this problem. The following picture emerged [4.35,39] (see Fig. 4.2). Up to a critical dimension $d_c + 1 = 3$ (two-dimensional substrate) the nonlinear term dominates except if it is not present at all ($\lambda = 0$). This means that for $\lambda = 0$ the growth is described by the EW linear theory (thus, in $d = 2$ $\beta = 1/4$, $z = 2$), while for any nonzero λ the strong-coupling fixed point dominates the dynamics, leading to $\beta = 1/3$ and $z = 3/2$ in $d = 2$.

There are no exact results related to the KPZ equation for dimensions higher than $1+1$. However, the RG results and the computer simulations seem to be in quite good agreement. The direct numerical solution of the nonlinear stochastic differential equation [4.44,45] has been shown to be an effective tool and the results support the picture obtained by the other methods. Recently alternative approximate methods have been put forward by Schwartz and Edwards [4.46]

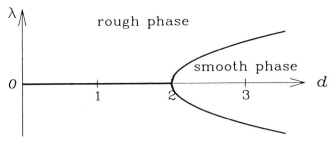

Fig. 4.2. A phase diagram of KPZ-type surface growth [4.42]

as well as by Bouchoud and Cates [4.47]. In these similar but not equivalent studies either a variational treatment of the Fokker-Planck equation is proposed [4.46] or a self-consistent approach is applied to the KPZ equation [4.47]. The numerical results for $2+1$ dimensions are rather encouraging ($\alpha = 0.29$ and 0.37, respectively). For higher dimensions the calculated values do not agree well with the numerical results. It is not clear whether the conclusion of the theories about the existence of an upper critical dimension d_u is correct but if d_u exists at all it must be higher than predicted [4.47]. (For $d > d_u$ the strong coupling phase should vanish.)

For $d > d_c$ there is a finite, dimension dependent critical value of $\lambda^2 d_f / \nu^3$ at which the smooth linear phase changes over to the rough phase. Less is known about what happens right at d_c. In fact, most of the computer simulations devoted to detecting a transition at d_c have been subject to criticisms of various kinds. The renormalization group in this case suggests an *exponentially slow* logarithmic to power-law crossover [4.48], and some of the simulations seem to support this result.

Before reviewing the numerical results on morphological transitions in surface growth let us discuss what are some of the major expected features of a non-equilibrium transition of this kind. In the general case the transition takes place between two morphologies characterized by different scaling exponents. A well understood specific situation is when the transition is triggered by a diverging length scale [4.49] $\xi_r \sim |p - p_c|^{-\nu_p}$, where ν_p is some exponent and the transition takes place at p_c as the parameter p is approaching its critical value. Under such circumstances the dynamic scaling (4.30) for the width has to be modified according to the scaling ansatz

$$w \sim \xi_r^{\alpha'} F(L/\xi_r, t/\xi_r^{z'}), \qquad (4.48)$$

where the anomalous values of the roughening and the dynamic exponents are denoted by α' and z', respectively. The usual dynamic scaling is recovered only in the limit $L \gg \xi_r$ and $t \gg \xi_r^{z'}$, when (4.48) becomes equivalent to (4.30). At $p = p_c$ the length ξ_r diverges and the width scales as

$$w \sim L^{\alpha'} \tilde{f}(t/L^{z'}). \qquad (4.49)$$

The above phenomenological scaling theory has been tested using a polynucleation growth model. In short, this model corresponds to a simultaneous deposition of particles followed by a relaxation step when a particle is added at each kink position along the surface. This model can be treated in terms of *directed percolation* [4.49] and the exact values of the anomalous exponents can be obtained. The result $\alpha' = 0$ has been verified by computer simulations which indicate that

$$w(L,t) \sim (\log t)^{1/2} \quad \text{and} \quad w(L, t = \infty) \sim (\log L)^{1/2}. \qquad (4.50)$$

The morphological transition has an important effect on the shape of the clusters grown from a central seed. Again, the relation to directed percolation turned out to be helpful in describing this non-equilibrium roughening transition [4.50].

Several communications reported on morphological phase transitions in computer models of surface growth. The existence of a transition from the smooth to the rough phase in dimensions $d + 1 > 3$ seems to be well established (see, e.g., [4.34,45,51,52]).

The situation is less clear in $d + 1 = 3$, although this case is physically relevant. In some papers the existence of a phase transition is indicated. The problem of interest here is the question whether the observed transition is a trivial one or whether it takes place for a parameter value of the models which corresponds to a nonzero value of λ in the language of the KPZ equation. By a trivial transition here one means that by tuning the parameters of a model it is in principle possible to achieve a situation in which the behaviour of the model becomes linear ($\lambda = 0$). (For example, $\lambda > 0$ for ballistic deposition, but $\lambda < 0$ for the single-step models.) Then the smooth phase is reduced to a single point as is predicted by the renormalization group for $d \le 2$.

Morphology transitions were reported to occur in the following models. Amar and Family [4.53] considered a modified solid-on-solid model in which the probability of growing at a given site depended on the change of energy (depending on a temperature-like variable and the number of neighbours) upon adding a site to the deposit. By varying the temperature a sharp transition was observed between two phases described by different exponents accompanied by a flat phase at the point of the transition. Arguments based on the KPZ equation suggested [4.53-55] that this transition is trivial in the above sense of making λ equal to zero by tuning the temperature. The difference in the values of the exponents is not explained by these arguments; it may be due to lattice effects or to an *extremely slow crossover*.

Yan et al. [4.51] studied a ballistic deposition model in which the particles were allowed to diffuse. The effect of diffusion was taken into account in a manner which destroyed the discrete nature of the model in the growth direction. As the parameter p of the model (controlling the rate of aggregation versus diffusion) was changed, a marked decrease in the effective exponents was observed at a critical value of p signaling a non-trivial phase transition since the exponents remained small beyond p_c. Similar results were reported by Pellegrini and Jullien [4.52], who investigated ballistic deposition of sticky and nonsticky particles. However, it was concluded in a later publication by these authors that no nontrivial phase transition is present [4.56].

The complexity of the situation is demonstrated in Fig. 4.3, where results obtained for the hypercube stacking model are shown [4.34]. The quantity

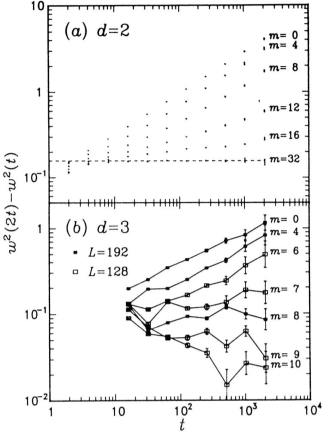

Fig. 4.3a,b. Crossover in the hypercube stacking model in $2+1$ dimensions and morphological transition in $3+1$ dimensions [4.34]. The strength of the nonlinear coupling is controlled by p^- at fixed $p^+ = 1/2$. The EW (linear) case is $p^- = p^+$

$w^2(2t) - w^2(t)$ is plotted versus the time t in order to exclude the effect of the intrinsic width (4.39). The data are for various $p^- = m/64$ values for fixed $p^+ = 1/2$. It can be clearly seen that as we approach the equilibrium case ($m = 32$, $\lambda = 0$) the crossover time, at which the asymptotic behavior sets in, diverges rapidly. As a result, effective exponents different from the asymptotic value can be associated with large fractions of the data. The situation in $2+1$ dimensions is different (bottom); $m = 8$ seems to be a critical value: for larger m the fluctuations of the interface decay in time, the surface remains flat. It seems most probable that in $2+1$ dimensions there is no nontrivial phase transition and the data need vary careful analysis owing to the extremely slow crossover at $\lambda = 0$.

According to Villain [4.57] and Wolf [4.58], in some limit surfaces grown by molecular beam epitaxy (MBE) can be described by the anisotropic KPZ equation where the couplings ν and λ in (4.42) are substituted by direction dependent, *anisotropic couplings*. For $d = 2$, a transition from the KPZ behavior to the EW one is predicted [4.58] if the nonlinear coupling in one direction gets a sign different from the sign of the coupling in the other direction.

An interesting question is whether there is an analog of the roughening transition in a driven system. We have already mentioned the transition triggered by directed percolation [4.49], but in this case a general constraint is responsible for the smooth phase. The KPZ transition for $d+1 \geq 3$ describes a change from a smooth to a rough phase, but it is not justified to call it a roughening transition since the smooth phase is not a consequence of the lattice.

Noziéres and Gallet [4.59] investigated the effect of the lattice periodicity by introducing a sine term into the EW equation (4.27). Hwa et al. [4.60] considered the analog extension of the KPZ equation:

$$\mu^{-1}\partial_t h = F + \nu\nabla^2 h - y\sin(2\pi h) + \lambda/2(\nabla h)^2 + \eta(\boldsymbol{x}, t). \qquad (4.51)$$

Both studies concluded that any finite F is relevant and destroys the smooth faceted state. Although this is true in principle, for a finite system at finite times it may have no practical meaning. Therefore, the effects should be studied close to T_R. It turns out that the terms investigated by Noziéres and Gallet explain very well the experiments on He IV while the roughening due to the nonlinear term is not accessible experimentally because of the extremely large crossover scales [4.61].

The experimental systems where KPZ-type kinetic surface roughening has been expected include biological growth, vapor deposition (see Fig. 4.4), roughening due to sputter erosion, and the motion of a fluid interface in a random medium.

In *biological growth* the evolution of colonies of bacteria [4.62] and of fungi [4.63] has been investigated. In these quasi two-dimensional systems exponents $\alpha \simeq 0.78$ and $\alpha \simeq 0.62$, respectively, were measured, which are significantly larger than the KPZ value (1/2). In these experiments the interface roughness occurs on the mm–cm scale and the patterns are evaluated by digital-image processing.

The roughness exponents measured in *vapor deposited* silver experiments by Krim's group (see [4.64]) seem to depend on the material of the substrate and they are also larger than the values predicted by KPZ (~ 0.4). The roughness

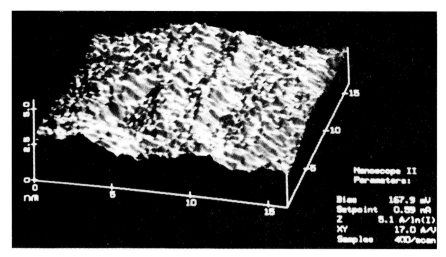

Fig. 4.4. An image of a vapor deposited surface. The picture was generated by using STM. Courtesy of J. Krim

of an iron surface bombarded by Ar ions can be characterized by $\alpha \simeq 0.53$. To our knowledge, there are two experimental results compatible with the KPZ theory: one is a measurement on heteroepitaxial film of Fe [4.65] and the other is by the group of Williams on graphite *erosion by sputtering* [4.18,66]. The recently developed powerful tools such as STM (Scanning Tunneling Microscope) or AFM (Atomic Force Microscope) are essential in the investigation of the microscopic roughness of these systems.

In fluid displacement experiments the roughening of an initially flat interface is investigated. Here two fluids fill the gap between two parallel plates and the more viscous one propagates. The static randomness in the prepared Hele-Shaw cell causes the roughening. The exponents are again larger than predicted by the KPZ theory [4.67,68]. Large exponents were also measured in the imbibition experiments using paper as 1+1 dimensional and oasis brick as 2+1 dimensional porous random media [4.69-71]. In Sect. 4.5 we shall discuss these experiments in some detail.

At the end of this brief overview of experimental results we mention some "exotic" systems: Zhang et al. investigated the propagation of the burning front of paper [4.20] and obtained $\alpha \simeq 0.7$ for the roughness. By analyzing the self-affine scaling of rupture lines in paper [4.17] (see Fig. 4.5) α close to 2/3 was obtained and this result was interpreted in terms of the directed polymer model. In an experiment on mountain formation by erosion Vicsek et al. produced [4.19] self-affine structures with a roughness exponent close to 0.8.

The KPZ equation reflects some symmetry properties of the system. In fact, this is the simplest nonlinear equation of surface growth which is invariant under constant shift and infinitesimal tilting. Other stochastic equations have been proposed to describe kinetic surface roughening under different cir-

Fig. 4.5. A self-affine rupture line in a paper sheet. The roughness exponent is close to 2/3 [4.17]

cumstances. Two equations of this type received special attention. The first is a *conservative version* of the KPZ equation [4.72]:

$$\partial_t h = -\nabla^2[\nu\nabla^2 h + \lambda/2(\nabla h)^2] + \eta(\boldsymbol{x}, t), \qquad (4.52)$$

where the noise is also conservative:

$$\langle \eta(\boldsymbol{x}, t)\eta(\boldsymbol{x}', t')\rangle = -2D\nabla^2\delta^d(\boldsymbol{x} - \boldsymbol{x}')\delta(t - t'). \qquad (4.53)$$

Owing to conservation the total volume below the interface is constant. The equation has been studied by RG and nontrivial results were obtained. The scaling relation (4.44) is replaced by $\alpha + z = 4$ and the roughness exponent can be calculated $\alpha = (2 - d)/3$. The upper critical dimension is $d_u = 2$. The nonlinearity arises from breaking the detailed balance of the dynamics, which, however, generally induces a term $\nabla^2 h$ [4.73]. Then the scaling is different; it is characterized by the Edwards Wilkinson exponent, since the Laplace term is the most relevant one. Therefore, the nontrivial behavior is just at a special value of the parameters and cannot be considered generic. An interesting feature of this model is the existence of a *dynamic roughening transition* [4.74].

The situation seems to be somewhat similar in the case of the equation proposed to describe growth in Molecular Beam Epitaxy (MBE). Wolf and Villain [4.75] as well as Das Sarma and Tamborenea [4.76] suggested an equation similar to (4.52) but – according to the growth due to deposition – *with a non conservative noise term*. The fourth-order derivative term is due to *surface diffusion* [4.77] while the nonlinearity is the most relevant term if the KPZ type squared gradient term reflecting the tilt-dependent growth velocity is suppressed. In MBE this seems to be the case since no holes are present in the deposit, and it is difficult to imagine a realistic mechanism leading to tilt-dependent growth velocity without holes.

It turns out, however, that a Laplacian term is usually *generated by the dynamics* [4.78-80], either driving the behavior to the EW one or leading to surface instability. It depends on the sign of the coefficient of the Laplacian term which of these possibilities is realized. The so called Schwoebel barrier [4.81] in some systems hinders the atoms in rolling down at the edge of a terrace, leading to a negative sign in the coefficient, and therefore an instability sets in.

Complicated crossovers often hinder the study of the asymptotic behavior, but, from the theoretical point of view, it seems that there are three different generic cases of kinetic roughening of growing surfaces. One is the KPZ universality class, where the dominant term is the nonlinearity $(\nabla h)^2$. If this is suppressed, i.e., the growth velocity is tilt-independent, the behavior is governed by the (often dynamically generated) Laplacian term. If the coefficient of the Laplacian is positive, EW exponents are observed, while for negative coefficients instabilities occur. It should be emphasized that the very weak, logarithmic roughness in $2 + 1$ dimensional EW-type growth means a practically flat surface. The instability corresponds to $\alpha = 1$ and in this case the surface is not self-affine.

4.5 Effects of Correlated, Power-Law, and Quenched Noise: Nonuniversal Roughening and Pinning

4.5.1 Correlated Noise

The computer simulations and the continuum-equation approach are more or less in agreement and provide a coherent picture of kinetic roughening. As indicated in the previous section, the situation is much worse from the experimental point of view. In most experiments the observed exponents are larger than those one would expect by applying the theoretical considerations naively. Therefore the assumptions of the theories have to be checked.

A very important input into the theories is the form of the noise in the stochastic differential equations describing the evolution of the surface. Usually Gaussian white noise (with or without conservation) is assumed. This, however, may be an oversimplification. Instead of (4.43), Medina et al. [4.39] considered the cases when the noise term is allowed to have *long-range correlations* in space and/or time. The spectrum $D(k, \omega)$ of this kind of noise can be defined through the noise correlator

$$\langle \eta(\boldsymbol{k}, \omega) \eta(\boldsymbol{k}', \omega') \rangle = 2D(k, \omega) \delta^d(\boldsymbol{k} + \boldsymbol{k}') \delta(\omega + \omega'), \tag{4.54}$$

which has power-law singularities of the form

$$D(k, \omega) \sim |\boldsymbol{k}|^{-2\rho} \omega^{-2\theta}. \tag{4.55}$$

The results of a detailed study can be summarized as follows. In dimensions higher than a critical d_c weak and strong noise lead to different scaling exponents, while for $d < d_c$ any amount of noise is relevant, resulting in the strong-coupling behavior. For weak noise ($\lambda \simeq 0$) the exponent α has the modified value $\alpha(\lambda \simeq 0) = (2 - d)/2 + \rho + 2\theta$, while z is unchanged $z(\lambda \simeq 0) = 2$. For $d < d_c = 2 + 2\rho + 4\theta$ these results are not valid as λ becomes a relevant parameter. If $\theta = 0$ but $\rho > 0$, two regimes are found. For small ρ the exponents α_w and z_w are the same as those obtained for white noise. When ρ becomes larger than $\rho_c(d) = \alpha_w + (d - z_w)/2$ the long-range part of $D(k)$ takes over and the new exponents can be obtained by a simple Flory type scaling

$$\alpha(d) = (4 - d + 2\rho)/3 \quad \text{and} \quad z(d) = (2 + d - 2\rho)/3. \tag{4.56}$$

Furthermore, the scaling law $\alpha + z = 2$ remains valid. Simulations with correlated noise [4.82] are in agreement with the theoretical predictions. When long-time temporal correlations are present the scaling behavior becomes rather complicated and several regimes appear. Moreover, in this case the invariance against infinitesimal tilting of the surface is lost and $\alpha + z = 2$ breaks down. A possible source of long time correlations is nucleation: the first nucleated terrace is likely to be the place where nucleation of the next generation starts and so on [4.83].

4.5.2 Noise with Power-Law Distributed Amplitudes

An effect due to another nontrivial form of the noise was investigated first by Zhang [4.84]. He suggested white noise but a non-Gaussian distribution of the noise amplitudes. This was shown to lead to nonuniversal exponents in a surface-growth model.

Following Zhang let us consider uncorrelated noise with the amplitudes according to the distribution density

$$P(\eta) = \frac{\mu}{\eta^{1+\mu}} \quad \text{for} \quad \eta > 1 \quad \text{and} \quad P(\eta) = 0 \quad \text{otherwise.} \quad (4.57)$$

This distribution has finite first and second moments (the second moment diverges as $\mu \to 2$), but it is not bounded like a Gaussian or exponential noise where all moments exist. Correspondingly, *rare events*, i.e., very large values of η, may appear with a relatively high probability (see Fig. 4.6). For independent variables the conditions of the central-limit theorem would be fulfilled and no effect should be expected. This is the case if the noise characterized by (4.57) is introduced in the EW equation. The correlations due to the nonlinearity of the KPZ equation change the situation dramatically.

If the noise (4.57) was introduced into models which belonged with Gaussian or bounded noise to the KPZ universality class, nonuniversal, μ-dependent exponents were obtained (with $\alpha + z = 2$) [4.84]. Extensive simulations have been carried out in $1+1$ [4.85-87] and $2+1$ [4.88] dimensions. There has been some controversy about the existence of a critical μ_c above which the universal KPZ exponents should set in. A simple theory [4.89], based on the assumption that for $\mu \leq \mu_c$ the roughening is dominated by the rare events, leads to:

Fig. 4.6. Ballistic deposition of rods with a size distrinbution according to (4.57). The effect of rare events is apparent [4.86]

$$\alpha = \frac{d+2}{\mu+1}. \tag{4.58}$$

Since (4.44) is valid (4.58) determines both exponents. The critical value μ_c is given by the condition:

$$\alpha_{KPZ} = \frac{d+2}{\mu_c+1}, \tag{4.59}$$

where α_{KPZ} is the exponent from the KPZ equation with bounded noise. Recently Lam and Sander argued that (4.58) with (4.59) should be exact [4.90].

It is remarkable that an analysis of the fluid-flow experiments by Horváth et al. [4.91] supports the Zhang theory. It was found that the distribution of the local jumps of the experimental interfaces (associated with noise) followed a power law with an exponent $\mu \simeq 2.67$. For the measured roughness exponent $\alpha \simeq 0.81$, (4.58) gives $\mu = 2.7$ in surprisingly good agreement with the experimental value. Similar results were obtained in the simulations of wetting in a system of disks [4.92].

The Zhang model has interesting multiaffine aspects as well (c.f. (4.6,7)) [4.93]. Figure 4.7 presents data for a Zhang-type model, $L = 2^{16}$ at time $t = 602890$, i.e., deep in the saturation regime; an average over five runs was taken. The important conclusions one can draw from these log-log plots showing the qth-order correlation function as a function of q are the following: (i) The initial part of the data sets for each q exhibits scaling behavior with a unique slope depending on q, i.e., multifractal scaling is present, and (ii) this kind of scaling crosses over into the uniform scaling behavior for x exceeding some characteristic crossover length $x_\times$.

Feature (ii) means that in addition to the characteristic length $\sim t^{\beta/\alpha}$ always present in kinetic roughening before the system size L is reached at saturation, a new characteristic length $x_\times$ occurs in surface growth dominated by rare events. For $x < x_\times$ the qth-order correlations show multifractal scaling be-

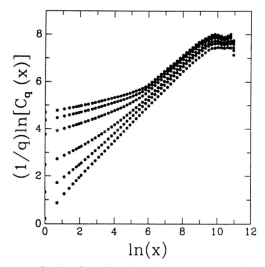

Fig. 4.7. The qth order correlation function in a rare-events-dominated growth [4.93]

haviour. For $x_\times < x < t^{\beta/\alpha}$ conventional scaling sets in, while no correlations are present for $x > t^{\beta/\alpha}$. The crossover length $x_\times$ depends on L. Estimation of the size dependence of the crossover length demonstrates that as L is increased, the relative size of the region over which the multifractal scaling behavior is observed becomes larger.

4.5.3 Quenched Noise

In the above examples time-dependent fluctuations induce kinetic roughening. If an interface propagates in a random medium, the noise is quenched. The dynamics of interfaces whose motion is dominated by the pinning forces present in a random medium is not yet solved and it represents one of the most actively investigated areas in the field of rough surfaces. Since this case is relevant to many experimental situations such as two phase fluid flows in porous materials [4.67,68,94] the motion of domain walls in magnetically ordered systems [4.23,95] or the pinning of charge density waves [4.96], it is of great importance to understand the dynamics of such moving interfaces.

The main effect of the above mentioned pinning forces is that they are able to dramatically slow down the motion of the interface. In particular, at the so called pinning transition the pinning forces completely halt the interface.

The interest in the pinning dominated surface growth has been greatly motivated by the fact that this process is expected to play a relevant role in the development of wetting fronts which can be regarded as a typical experimental realization of surface roughening. In these imbibition experiments a viscous fluid driven by wetting forces invades a porous material. A possible realization is glycerine entering a layer of glass beads or water penetrating a sheet of paper. The values obtained in these experiments for the exponents α and β have been found to be considerably different from the prediction of the KPZ theory. In the most studied 1+1 dimensional case the observed wetting fronts gave estimates $0.63 < \alpha < 0.81$ and $\beta \simeq 0.65$.

The main questions to be answered are: (i) Does a well-defined, distinct universality class exist for kinetic roughening in disordered media close to pinning? (ii) Do the surfaces exhibit any unusual behavior related to the appearance of new exponents?

In the rest of this section we present the results of numerical and theoretical investigations of various equations relevant from the perspective of the above questions.

A natural approach to the study of the role of disorder is to examine the behavior of the EW and KPZ equations (including the driving force F) with a noise term $\langle \eta(x,h) \rangle = 0$ representing quenched noise with a Gaussian distribution of amplitudes and a correlator, which in the continuum limit in $1+1$ dimensions is formally given by

$$\langle \eta(x_0, h_0)\eta(x_0 + x, h_0 + h') \rangle = D\delta(h')\delta(x), \qquad (4.60)$$

where D is a constant. For a large diving force the universal behaviour is expected but pinning the quenched disorder may have serious effects for small values of F.

Let us next investigate the properties of the solutions of the KPZ equation by comparing the scaling behavior of its various relevant terms.

The key point in such an analysis is the appropriate choice for the form of the noise term. In other words, the results are determined by the assumed dependence of the noise on h and x. If we intend to compare our theoretical predictions with the results of the experiments and numerical integration (simulation) of the above equations, we need to consider a noise term with correlations corresponding to the conditions of the experiments and the numerical methods. In the numerical studies the surface is discretized in the horizontal direction (along the substrate) but is assumed to move continuously in the vertical direction. Similarly, in the experiments there always exists a typical size of the inhomogeneities in the disordered medium (for example, the diameter of the glass beads) which plays the role of the lattice constant. Correspondingly, we are led to study the effects of noise having the correlator [4.97,98]:

$$\langle \eta(\boldsymbol{x}_0, h_0)\eta(\boldsymbol{x}_0 + \boldsymbol{x}, h_0 + h') \rangle = \Delta(|h'|)\delta^d(\boldsymbol{x}). \qquad (4.61)$$

In $\Delta(|h'|)$ is a monotonically decreasing function with a cut-off at a small characteristic size a (corresponding to the lattice constant) playing the role of the typical correlation length of the fluctuations of noise in the h direction.

From dynamic scaling it follows that the change Δh over an interval Δx is proportional to Δx^α. The scaling of the noise term with Δx can be written in the form $\Delta x^{-d/2}$ since the integral of the correlations $\langle \eta\eta \rangle$ over the interval Δx is normalized to 1. In the $\Delta x \to 0$ limit the relevant terms in the equation should scale the same way. Equating the $(\nabla h)^2$ and the η terms we get

$$\frac{(\Delta h)^2}{(\Delta x)^2} = \Delta x^{-d/2}. \qquad (4.62).$$

From the condition $\Delta h \sim \Delta x^\alpha$ we immediately obtain $\alpha = \frac{4-d}{4}$. For $d = 1$ (4.62) predicts $\alpha = 0.75$. Similarly, we can obtain for β the following expression: $\beta = (4 - d)/(4 + d)$, giving for the 1+1 dimensional case $\beta = 0.6$. These results are in good agreement with the experiments.

The EW equation with quenched noise has been theoretically studied by dimensional analysis [4.99] and dynamic-renormalization-group calculations [4.100,101]. Since the critical dimension of the problem is $d = 4$, the renormalization-group results are given in terms of ϵ expansion around $d = 4$. This theory predicts $\alpha = (4 - d)/3$, and there are arguments that the result $\alpha = 1$ for the $d = 1$ dimensional case is exact to all orders in ϵ. The prediction $\alpha = 2/3$ for $d = 2$ has been seen in Monte Carlo simulations of the motion of rough domain walls in Ising ferromagnets. The 1+1 dimensional case is more controversial; the experiments are not yet entirely conclusive. One of the reasons for the complexity of the situation is that the scaling behavior of the resting interface can be qualitatively different from the moving one, and thus very severe crossover effects are expected close to the transition.

The numerical integration of the quenched version of the KPZ and EW equations led to the following results:

Additive noise, KPZ case. Just above the pinning threshold a scaling regime of $w(t) \sim t^\beta$ was observed with an exponent $\beta = 0.61 \pm 0.06$. A further important aspect of the growth is the geometry of the surface in the stationary regime. Furthermore *as the system size becomes larger* the behavior of the surface approaches self-affine scaling with a static exponent close to $\alpha = 0.71 \pm 0.06$. This value is in agreement with our result $\alpha = 3/4$ predicted by dimensional analysis.

Additive noise, EW case. The latest, unpublished results indicate that in the moving regime β is close to $3/4$, in good agreement with the theoretical predictions [4.99-101]. The resting interface can be characterized by a roughness exponent $\alpha = 1$. This is also in agreement with the theory; however, the moving interface already cannot be described in terms of $\alpha = 1$: it becomes rougher.

Detailed data for the scaling of the asymptotic surface velocity $v_a = \lim_{t\to\infty} \partial_t h$ as a function of the distance from the critical driving force F_c have also been obtained. It was found that v_a vanishes at $F = F_c$ and for $F - F_c > 0 \ll 1$ it scales as $F_a \sim (F - F_c)^\theta$ [4.98]. To analyze the data it is useful to assume that the combined scaling behavior of v_a in the two regimes (close and far from pinning) can be well represented by the expression

$$v_a = F - F_c \frac{1}{1 + A(F/F_c - 1)^\theta},\tag{4.63}$$

where A is some constant providing the best fit (together with the appropriately chosen F_c and θ) to the data provided by the numerical integration. Since such an analysis takes into account crossover effects, it is expected to provide a considerably better estimate for θ than a simple fit of a straight line to the data for relatively (but not sufficiently) small $F - F_c$. Figure 4.8 shows the rescaled data for various ν and $L = 1300$. On the basis of Fig. 4.8 we conclude that θ is close to 0.5.

The detailed dynamics of the growing surface can be characterized by the distribution of time intervals or *waiting times* the interface spends between two grid points. Let us consider the development of the surfaces obtained for large times (in the stationary regime) and determine the number of places $N(\tau)$ where the surface spends a given amount of time τ, where τ is the so-called waiting time. This time strongly depends on the value of η as well as on the local geometry of the surface. The simulations showed that the uniform (Gaussian-type) input noise is transformed by the growth mechanism into a power-law distribution of the waiting times. In this respect the results for the exponents α and β are consistent with the predictions of Tang et al. [4.42], who showed that in models with a power law distribution of waiting times the statics and the dynamics of the interface are described by nonuniversal exponents. In an even wider perspective, we can conclude that the three seemingly different facts (i) nonuniversal exponents, (ii) power law noise, and (iii) Gaussian-quenched noise are intimately related and the uncovering of their connection is likely to add greatly to the understanding of kinetic roughening.

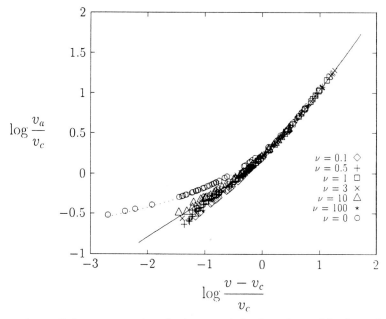

Fig. 4.8. A scaling of the asymptotic velocity v_a of the interface with the reduced driving force (denoted here by $(v - v_c)/v_c$)

Multiplicative noise. It has been proposed [4.102] that the development of the interface $h(x, t)$ in the experiments on quasi $1+1$ dimensional viscous flows is described by the equation

$$\partial_t h = \left(\nabla^2 h + v(1 + (\nabla h)^2)^{1/2} \right) (p + \eta), \qquad (4.64)$$

where $p > 0$ is some constant, v is the normal velocity and the term $\eta > 0$ corresponds to quenched noise with no correlations as in (4.60). In this case η is assumed to follow some other simple distribution, for example, the Poisson or the uniform distribution.

The main idea behind incorporating noise in a multiplicative form (4.64) is that the motion of the flowing phase in an inhomogeneous medium is determined by the simultaneous effects of surface tension, capillary forces, and local flow properties (permeabilities of the channels), so that the local velocity depends on the local properties of the medium, $v_s = F\epsilon$, where v_s is the velocity of the surface in the vertical direction, ϵ (noise) is the randomly changing local permeability, and F denotes a general driving force.

Figure 4.9 shows the calculated surfaces for the following sets of parameters: $L = 1500$, $p = 0.0001$, $v = 0.5$, and $\lambda = 0$. The simulated surfaces look very similar to the observed ones. The corresponding exponent β was found to be very close to the experimentally observed value of 0.65.

Growth models with percolative pinning. An alternative approach to the interpretation of imbibition experiments is to study lattice growth models with built-in quenched randomness. In the model of Buldyrev et al. the surface advances in a system of available and blocked sites in a manner similar to percolation and invasion percolation (see Chaps. 1 and 9 and Chaps. 1 and 2 in [4.5]). The main new feature of the process is that the overhangs which would necessarily appear in a standard growth regime are eliminated by an "erosion"

Fig. 4.9. Successive interfaces resulting from the numerical solution of (4.64) [4.102]

type rule [4.69]. The simulations lead to the exponents $\alpha = 0.633 \pm 0.001$ for a pinned (static) interface and $\alpha_{\mathrm{dyn}} = 0.70 \pm 0.05$ for a moving (dynamic) interface in $1 + 1$ dimensions. In addition, it has been numerically observed that for this model $\beta \simeq \alpha$. In this context Tang and Leschhorn investigated independently a similar model [4.103] and later Sneppen [4.104] considered a version corresponding to invasion-directed percolation [4.17]. In the model of Sneppen a power-law distribution of the activity of the surface sites has also been demonstrated [4.105].

The $2 + 1$ dimensional analog problem raises the question of percolating directed surfaces [4.70,71]. (For a computer image of such an interface see Fig. 4.0.) Recently the relation of the random resistor–diode networks to the percolating directed (hyper-)surfaces has been elucidated [4.106]. The exponents measured in imbibition experiments on a three-dimensional porous medium [4.71] are in reasonable agreement with the numerical simulations ($\alpha \simeq 0.5$).

4.6 Summary

In this chapter we have tried to show the less-often emphasized relations between equilibrium and kinetic roughening, pointed out the main mechanisms and the corresponding models leading to self-affine surfaces, and discussed some recent developments. It is clear that we could not cover all the important results and we apologize to colleagues whose contributions have not been included. The main conclusions are the following:

(i) Roughness of surfaces is the consequence of noise, either time dependent or quenched.

(ii) In equilibrium the roughness exponent α is $1/2$ in $1 + 1$ dimensions and 0 (logarithmic roughness) in $2 + 1$ dimensions for the pure systems. There is a Kosterlitz-Thouless type roughening transition in $2 + 1$ dimensions owing to the lattice periodicity.

(iii) For kinetic roughening in surface growth with bounded white noise there seem to be three generic cases: the Edwards-Wilkinson universality class, the Kardar-Parisi-Zhang universality class, and the case where instability occurs, i.e., the assumption $\alpha < 1$ breaks down. The dynamic scaling hypothesis (4.30) for kinetic roughening of self-affine surfaces has been established and verified in many different cases.

(iv) The different special forms of noise (correlated, power law amplitude, multiplicative, and quenched) may lead to interesting effects including new universality classes. The relation to other models of statistical physics, such as directed percolation, can be helpful in understanding roughening.

There are several unsolved problems in the field which deserve more effort directed to their resolution. The question of pinning is certainly one of them. Another problem is the variety of measured exponents in deposition experiments. Possibly both are closely related to the topic of *crossovers*. Theoreticians often concentrate on the asymptotics in order to sort out the universal behavior. On the other hand, in complicated, real systems with finite sizes and in finite times the measured data can be heavily influenced by the (sometimes extremely slow) crossovers.

More experiments are definitely needed for further progress. A challenge for theoreticians is to construct realistic models based on microscopic interactions in order to explain experimental findings. Surface instabilities during growth are expected to attract increasing attention in the near future.

Acknowledgements. Collaboration with many individuals has been essential for our understanding of the field. It is a pleasure to thank them, including A.-L. Barabási, R. Bourbonnais, S.V. Buldyrev, Z. Csahók, F. Czirók, F. Family, S. Havlin, K. Honda, V.K. Horváth, G. Huber, M.H. Jensen, J. Krug, K. Moser, E. Somfai, H.E. Stanley, L.-H. Tang, M. Vicsek, D.E. Wolf, and Y.-C. Zhang. The support of the Hungarian Science Foundation OTKA 1218 is acknowledged.

References

4.1   B.B. Mandelbrot: Physica Scripta **32**, 257 (1985); B.B. Mandelbrot, in: *Fractals in Physics*, ed. by L. Pietronero, E. Tosatti (Elsevier, Amsterdam 1986)

4.2   T. Vicsek: *Fractal Growth Phenomena*, 2nd ed. (World Scientific, Singapore 1991)

4.3   F. Family, T. Vicsek, eds.: *Dynamics of Fractal Surfaces* (World Scientific, Singapore 1991)

4.4   J. Feder: *Fractals* (Plenum, New York 1988)

4.5   A. Bunde, S. Havlin, eds.: *Fractals and Disordered Systems* (Springer, Berlin 1991)

4.6   A.-L. Barabási, T. Vicsek, P. Szépfalusy: Physica A **178**, 17 (1990)

4.7   J.D. Weeks, in: *Ordering in Strongly Fluctuating Condensed Matter Systems*, ed. by T. Riste (Plenum, New York 1980),

4.8   H. van Beijeren, I. Nolde, in: *Structures and Dynamics of Surfaces II*, ed. by W. Schommers and P. von Blanckenhagen (Springer, Berlin, 1987)

4.9   P. Nozières, in: *Solids far from Equilibrium: Growth, Morphology and Defects*, ed. by C. Godrèche (Plenum, New York, 1990)

4.10   J. Krug, H. Spohn, in: *Solids far from Equilibrium: Growth, Morphology and Defects*, cd. by C. Godrèche (Plenum, New York, 1990)

4.11   D.E. Wolf, in: *Kinetics of Ordering and Growth on Surfaces*, ed. by. M. Lagally (Plenum, New York 1990)

4.12   R. Jullien, J. Kertész, P. Meakin, D.E. Wolf, eds.: *Surface Disordering: Growth, Roughening and Phase Transitions* (Nova, Commack N.Y. 1992)

4.13   P. Meakin: Physics Reports **235**, 189 (1993)

4.14   J. Kertész, D.E. Wolf: Physica D **38**, 221 (1989)

4.15   H.J. Herrmann, S. Roux, eds.: *Statistical Models for the Fracture of Disordered Media* (North Holland, Amsterdam 1990)

4.16   R.H. Daushardt, F. Haubensack, R.O. Ritchie: Acta Metall **38**, 143 (1990)

4.17   J. Kertész, V.K. Horváth, F. Weber: Fractals **1**, 67 (1993)

4.18   R. Bruinsma, in: *Surface Disordering: Growth, Roughening and Phase Transitions*, ed. by R. Jullien, J. Kertész, P. Meakin and D.E. Wolf (Nova, Commack N.Y. 1992)

4.19   A. Czirók, E. Somfai, T. Vicsek: *preprint*

4.20   J. Zhang, Y.-C. Zhang, P. Alstrom, M.T. Levinsen: in *Surface Disordering: Growth, Roughening and Phase Transitions* ed. by R. Jullien, J. Kertész, P. Meakin and D.E. Wolf (Nova, Commack N.Y. 1992)

4.21   D.A. Huse and C.L. Henley: Phys. Rev. Lett. **54**, 2708 (1985)

4.22   D.A. Huse, C.L. Henley, D. Fisher: Phys. Rev. Lett. **55**, 2924 (1985)

4.23   R. Bruinsma, G. Aeppli: Phys. Rev. Lett. **52**, 1547 (1985)

4.24   S.F. Edwards, D.R. Wilkinson: Proc. Roy. Soc. London, Ser A **381**, 17 (1982)

4.25   F. Family, T. Vicsek: J. Phys. A **18**, L75 (1985)

4.26   M. Eden, in: *Proc. of the IV-th Berkeley Symposium on Mathematics and Probability, Vol. IV.*, ed. by J. Neyman (University of California Press, Berkeley 1961)

4.27   D.E. Wolf, J. Kertész: Europhys. Lett. **4**, 656 (1987)

4.28   M.J. Vold: J. Coplloid Interface Sci. **14**, 168 (1959)

4.29   H.J. Leamy, G.H. Gilmer, S. Liang, in: *Current Topics in Mat. Sci., Vol. 6.*, ed. by E. Kaldis (North Holland, Amsterdam 1980)

4.30   P. Meakin, P. Ramanlal, L. Sander, R.C. Ball: Phys. Rev. A **34**, 5091 (1986)

4.31   J. Kertész, D.E. Wolf: J. Phys. A **21**, 747 (1988)

4.32   J.M. Kim, J.M. Kosterlitz: Phys. Rev. Lett. **62**, 2289 (1989)

4.33   M. Plischke, Z. Rácz, D. Liu: Phys. Rev. B **35**, 3485 (1987)

4.34   B.M. Forrest, L.-H. Tang: Phys. Rev. Lett. **64**, 1405 (1990)

4.35   M. Kardar, G. Parisi, Y.-C. Zhang: Phys. Rev. Lett. **56**, 889 (1986)

4.36   F. Family: J.Phys. A **19**, L441 (1986)

4.37   J. Krug: J. Phys. A **22**, L769 (1989)

4.38   D.E. Wolf, L.-H. Tang: Phys. Rev. Lett. **65**, 1591 (1990)

4.39   E. Medina, T. Hwa, M. Kardar, Y.-C. Zhang: Phys. Rev. A **29**, 3053 (1989)

4.40   D.E. Wolf, J. Kertész: Phys. Rev. Lett. **63**, 1191 (1989)

4.41   S. Roux, A. Hansen, E.L. Hinrichsen: J. Phys. A **24**, L295 (1991)

4.42   L.-H. Tang, J. Kertész, D.E. Wolf: J. Phys. A **24**, L1193 (1991)

4.43   D. Forster, D. Nelson, M.J. Stephen: Phys. Rev. A **16**, 732 (1977)

4.44   J. Amar, F. Family: Phys. Rev. A **41**, 3399 (1990)

4.45   K. Moser, J. Kertész, D.E. Wolf: Physica A **178**, 215 (1991)

4.46   M. Schwartz, S.F. Edwards: Europhys. Lett. **20**, 301 (1992)

4.47   J.P. Bouchaud, M.E. Cates: Phys. Rev. E **47**, R1455 (1993) and *erratum*

4.48   L.-H. Tang, T. Nattermann, T. and B.M. Forrest: Phys. Rev. Lett. **66**, 2422 (1990)

4.49   J. Kertész, D.E. Wolf: Phys. Rev. Lett. **62**, 2571 (1989)

4.50   J. Krug, J. Kertész, D.E. Wolf: Europhys. Lett. **12**, 113 (1990)

4.51   H. Yan, D. Kessler and L.M. Sander: Phys. Rev. Lett. **64**, 926 (1990)

4.52   Y.P. Pellegrini, R. Jullien Phys. Rev. Lett. **64**, 1745 (1990)

4.53   J. G. Amar, F. Family: Phys. Rev. Lett. **64**, 543 (1990);
       D.A. Huse, J.G. Amar, F. Family: Phys. Rev. A **41**, 7075 (1990)

4.54   J. Krug, H. Spohn: Phys. Rev. Lett. **64**, 2332 (1990)

4.55   J.M. Kim, T. Ala-Nissila, J.M. Kosterlitz: Phys. Rev. Lett. **64**, 2333 (1990)

4.56   R. Jullien, Y.P. Pellegrini, P. Meakin, in: *Surface Disordering: Growth, Roughening and Phase Transitions*, ed. by R. Jullien, J. Kertész, P. Meakin and D.E. Wolf (Nova, Commack N.Y. 1992)

4.57   J. Villain: J. Physique I **1**, 19 (1991)

4.58   D.E. Wolf: Phys. Rev. Lett. **67**, 1783 (1991)

4.59   P. Nozières, F. Gallet: J. Physique **48**, 353 (1987)

4.60   T. Hwa, M. Kardar, M. Paczuski: Phys. Rev. Lett. **66**, 441 (1991)

4.61 S. Balibar, J.P.Bouchaud, F. Gallet, C. Guthman, E. Rolley, in: *Surface Disordering: Growth, Roughening and Phase Transitions*, ed. by R. Jullien, J. Kertész, P. Meakin and D.E. Wolf (Nova, Commack N.Y. 1992)

4.62 T. Vicsek, M. Cserző and V.K. Horváth: Physica **A167**, 315 (1990)

4.63 S. Matsuura, S. Miyazima: Fractals **1**, 11 (1993)

4.64 J. Krim, in: *Surface Disordering: Growth, Roughening and Phase Transitions*, ed. by R. Jullien, J. Kertész, P. Meakin and D.E. Wolf (Nova, Commack N.Y. 1992)

4.65 J. Chevrier, V. Le Thanh, R. Buys, J. Derrien: Europhys. Lett. **16**, 737 (1991)

4.66 E. Eklund, R. Bruinsma, J. Rudnick, R. S. Williams: Phys. Rev. Lett. **67**, 1759 (1991)

4.67 M.A. Rubio, C.A. Edwards, A. Dougherty, J.P. Gollub: Phys. Rev. Lett. **63**, 1685 (1989)

4.68 V.K. Horváth, F. Family, T. Vicsek: J. Phys. A **24**, L25 (1991)

4.69 S.V. Buldyrev, A.-L. Barabási, F. Caserta, S. Havlin, H.E. Stanley, T. Vicsek: Phys. Rev. A **45**, R8313 (1991)

4.70 A.-L. Barabási, S.V. Buldyrev, S. Havlin, G. Huber, H.E. Stanley, T. Vicsek, in: *Surface Disordering: Growth, Roughening and Phase Transitions*, ed. by R. Jullien, J. Kertész, P. Meakin and D.E. Wolf (Nova, Commack N.Y. 1992)

4.71 S.V. Buldyrev, A.-l. Barabási, S. Havlin, J. Kertész H.E. Stanley, H.S. Xenias: Physica A **191**, 220 (1992)

4.72 T. Sun, M. Guo, M. Grant: Phys. Rev. A **40**, 6763 (1989)

4.73 Z. Rácz, M. Siegert, D. Liu, M. Plischke: Phys. Rev. A **43**, 5275 (1991)

4.74 T. Sun, B. Morin, H. Guo, M. Grant, in: *Surface Disordering: Growth, Roughening and Phase Transitions*, ed. by R. Jullien, J. Kertész, P. Meakin and D.E. Wolf (Nova, Commack N.Y. 1992)

4.75 D.E. Wolf, J. Villain: Europhys. Lett. **13**, 389 (1990)

4.76 S. Das Sarma, P. Tamborenea: Phys. Rev. Lett. **66**, 325 (1991)

4.77 W.W. Mullins: J. Appl. Phys. **28**, 333 (1957)

4.78 M. Siegert, M. Plischke: Phys. Rev. Lett. **68**, 2035 (1992)

4.79 J. Krug, M. Plischke, M. Siegert: Phys. Rev. Lett. **70**, 3271 (1993)

4.80 D. Kessler: Physica A *preprint*; talk presented at the Bar Ilan Conference on Frontiers in Condensed Matter Physics (March, 1993)

4.81 R.L. Schwoebel: J. Appl. Phys. **40**, 614 (1969)

4.82 P. Meakin, R. Jullien: Europhys. Lett. **9**, 71 (1989)

4.83 D.E. Wolf, J. Kertész *to be published*

4.84 Y.-C. Zhang: J. Physique **51**, 2129 (1990)

4.85 J. Amar, F. Family: J. Phys. A **24**, L79 (1991)

4.86 S.V. Buldyrev, S. Havlin, J. Kertész, H.E. Stanley, T. Vicsek: Phys. Rev. A **43**, 7113 (1991)

4.87 R. Bourbonnais, H.J. Herrmann, T. Vicsek: Intl. J. Mod. Phys. C **2**, 719 (1991)

4.88 R. Bourbonnais, J. Kertész, D.E. Wolf: J. Physique II. **1**, 493 (1991)

4.89 J. Krug: *J. Physique I.* **1**, 9 (1991);
      Y.-C. Zhang: Physica A **170**, 1 (1990)

4.90 C.-H. Lam, L.M. Sander: Phys. Rev. Lett. **69**, 3338 (1992)

4.91 V.K. Horváth, F. Family, T. Vicsek: Phys. Rev. Lett. **67**, 3207 (1991)

4.92 N. Martys, M. Cieplak, M.O. Robbins: Phys. Rev. Lett. **66**, 1058 (1991);
      M.O. Robbins: *private communication*

4.93 A.-L. Barabási, R. Bourbonnais, M. Jensen, J. Kertész, T. Vicsek, Y.-C. Zhang: Phys. Rev. A **45**, R6949 (1991)

4.94 S. He, G.L.M.K.S. Kahanda, P.-Z. Wong: Phys. Rev. Lett. **69**, 3731 (1992)

4.95 H. Ji, M. O. Robbins, *Phys. Rev.* A44, 2538 (1991)

4.96 L.P. Gorkov, G. Grüner, eds.: *Charge Density Waves* (Elsevier, New York 1989)

4.97 D. A. Kessler, H. Levine, Y. Tu: Phys. Rev. **A43**, R4551 (1991)

4.98 Z. Csahók, K. Honda, T. Vicsek: J. Phys. A **26**, L171 (1993)

4.99 G. Parisi: Europhys. Lett. **17**, 673 (1992)

4.100 T. Nattermann, S. Stepanow, L-H. Tang, H. Leschhorn: J. Physique II, **2**, 1483 (1992)

4.101 G. Narayan, D. Fisher: *preprint*

4.102 T. Vicsek, E. Somfai, M. Vicsek: J. Phys. A **25**, L763 (1992)

4.103 L.-H. Tang, H. Leschhorn: Phys. Rev. A **45**, R8309 (1992)

4.104 K. Sneppen: Phys. Rev. Lett. **69**, 3537 (1993)

4.105 K. Sneppen, M.H. Jensen: Phys. Rev. Lett. **71**, 101 (1993)

4.106 S.V. Buldyrev, S. Havlin, J. Kertész, A. Shehter, H.E. Stanley: Fractals **1**, 827 (1993)

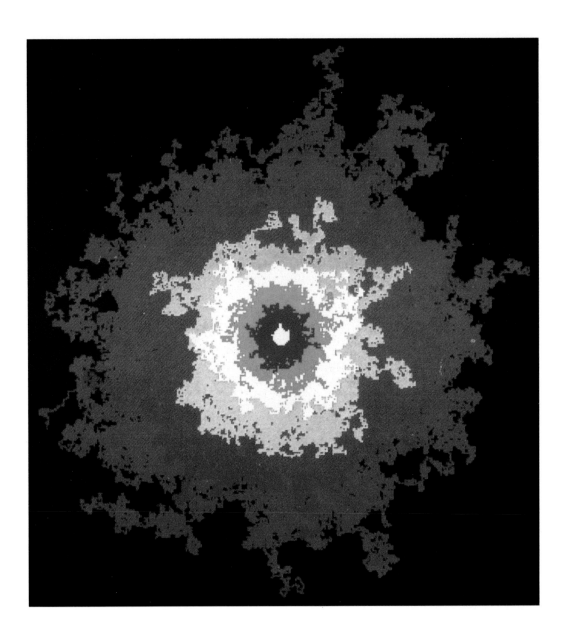

5 A Primer of Random Walkology

George H. Weiss

5.1 Introduction

The phenomenon of Brownian motion has been known since the time that van
Leeuwenhoek first peered through a microscope. Although it must have been
regarded as a nuisance by early microscopists, Brownian motion has played a
significant role in the development of our understanding of the physical world.
Indeed, Einstein's successful theoretical explanation of Brownian motion led
to a universal acceptance of the atomic hypothesis, which was by no means
evident to scientists as eminent as Ostwald and Mach in the early part of the
twentieth century. The basis of Einstein's explanation of Brownian motion is a
form of diffusion theory, although not termed as such in his original papers on
the subject.

Diffusion and diffusive effects are both observable and apparent to almost
anyone, and on the simplest level, the principal ideas underlying the theory
are a part of the training of every contemporary physical scientist. The fact
that diffusive phenomena very often exhibits a high degree of regularity masks
underlying fluctuations that occur at a microscopic level. These fluctuations
can be modeled in terms of properties of the random walk. More importantly,
an understanding of the kinetics of transport in fractal and disordered media is
almost necessarily phrased in terms of the random walk. The usual concepts of
transport by ideal diffusive motion are too narrow to encompass the diversity
of kinetic behavior possible for transport in disordered media.

One of the earliest applications of what is effectively diffusion theory is
to describe flow of liquids through porous media [5.1]. However, to engineers
working in this area, it is also well known that D'Arcy's law and the associated
theoretical framework based on this postulate are inadequate in a significant

◄ **Fig. 5.0.** Contours of the two-dimensional surface representing the number of distinct sites
visited by 500 random walkers. The different colors represent different numbers of steps. The
surface of the disk exhibits roughening as time progresses (Courtesy of P. Trunfio)

number of instances for analyzing flow in porous media. Only in the past twenty years has engineering empiricism given way to more systematic analytic techniques in the study of the kinetics of transport in disordered media. These have been mainly based on different versions of the theory of random walks.

The most elementary definition of the random walk is that it is a sum of random variables. This, as will be seen, is much too simple for many of the applications that we have in mind. In order to formulate a background for generalizations found useful in the study of transport in a disordered medium it will be necessary for us to discuss a number of the basic ideas, techniques, and results useful in the study of properties of random walks in strictly homogeneous media. The definition of a strictly homogeneous medium will be given later in this section.

The terminology "random walk" is fairly recent [5.2], dating from a query by K. Pearson published in *Nature* that posed a question phrased as follows: "A man starts from a point O and walks a distance a in a straight line. He repeats this process n times. I require the probability that after these n stretches he is at a distance between r and $r+dr$ from his starting point at O." While the terminology relating to random walks may have dated from 1905, the ideas of the random walk date back to the initial formulation and application to gambling games of what we now know as mathematical probability [5.3]. Indeed, many important results and a considerable amount of formalism useful for elucidating the theory of random walks had been developed prior to the publication of Pearson's query. Useful mathematical references to the general area of random walks are to be found in [5.4,5], and four reviews written with the physicist in mind are given in [5.6–9].

The random walk is useful because it allows one to study motion which has random features. An important distinction between random walks on a homogenous substrate and those on a disordered one is that disorder in the underlying substrate introduces a source of fluctuations additional to those inherent in the random walk itself. We are interested in studying the interplay between both types of fluctuations, but in order to do so we must understand a number of properties of random walks in a homogeneous medium. The following two sections contain an introduction to the formalism and properties of a homogeneous random walk which should be found useful for understanding material in the remainder of these chapters. Section 5.4 discusses a small, and by no means exhaustive, number of models for transport in disordered media.

5.2 Basic Formalism

5.2.1 Jump Probabilities

Let us start by considering the most elementary version of the random walk, which is a random walk on a (d-dimensional) translationally invariant lattice. In our initial version of the problem steps occur at uniform intervals in time

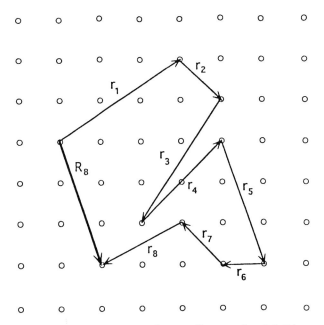

Fig. 5.1. An illustration of 8 steps of a random walk on a $d = 2$ lattice

thereby giving rise to what will be termed a random walk in discrete time. The lattice points will be denoted by d-dimensional vectors of integers, and the random walker is initially located at the origin of coordinates. Thus, the position of the random walker after n steps is $\boldsymbol{R}_n$ which can be written as the sum

$$\boldsymbol{R}_n = \boldsymbol{r}_1 + \boldsymbol{r}_2 + \cdots + \boldsymbol{r}_n. \tag{5.1}$$

where $\boldsymbol{r}_j$ is the random displacement on the jth step of the walk (see Fig. 5.1).

As an example, we show how a game called the gambler's ruin can be described in these terms. In the gambler's ruin problem, two gamblers play a game in which a coin is flipped at periodic intervals, heads resulting in the transfer of one unit of money from A to B and tails leading to a similar transfer from B to A. This game can be regarded as being a random walk in one dimension, in which r_j is the amount of money won or lost by A at toss j and R_n is the cumulative amount of money won or lost by A after n tosses of the coin. When the coin is unbiased, that is, the probability that a head appears is equal to $1/2$, the random variable r_j is characterized by the probability distribution

$$\text{Prob}\{r_j = 1\} = \text{Prob}\{r_j = -1\} = 1/2. \tag{5.2}$$

where one assumes that the result on any given toss has no influence on any other, which is to say that successive tosses of the coin give rise to random variables that are independent.

Let the $\boldsymbol{r}_j$ be random variables whose properties do not change with step number. The properties of such variables can be described in terms of a set of single-step transition probabilities, $\{p(\boldsymbol{j})\}$, in which $p(\boldsymbol{j})$ is the probability that the displacement of the random walker in a single step is equal to $\boldsymbol{j}$. In this chapter we restrict ourselves only to random walks for which the transition probabilities are independent of time. In the absence of any features of the model which can remove random walkers from the system (exemplified by traps, which we discuss in Sect. 5.4) the $p(\boldsymbol{j})$ are required to satisfy the normalization condition

$$\sum_{\boldsymbol{j}} p(\boldsymbol{j}) = 1. \tag{5.3}$$

If, instead of having a lattice random walk, we think about a random walk taking place in a continuum, say the Pearson random walk mentioned earlier, the equivalent descriptive functions will be denoted by a function written $p(\boldsymbol{r})$, which is now a probability density. That is to say, the probability that the displacement in a single step lies in an infinitesimal volume $(\boldsymbol{r}, \boldsymbol{r} + d^d\boldsymbol{r})$ is equal to $p(\boldsymbol{r})d^d\boldsymbol{r}$, where d is the number of dimensions. The normalization condition equivalent to (5.3) is

$$\int p(\boldsymbol{r})d^d\boldsymbol{r} = 1. \tag{5.4}$$

The formal mathematics required to solve for the properties of discrete and continuous random walks are quite similar. Indeed, lattice random walks can be subsumed under the category of continuous random walks through the artifice of introducing delta functions, that is, writing

$$p(\boldsymbol{r}) = \sum_{\boldsymbol{j}} p(\boldsymbol{j})\delta(\boldsymbol{r} - \boldsymbol{j}). \tag{5.5}$$

Because of this possibility we will use the terminology specific to continua, pointing out any features that might be unique to lattice random walks when these are relevant.

Let $p_n(\boldsymbol{r})$ be the probability density for the location of an (ordinary, rather than a persistent) random walker after n steps, and let $p(\boldsymbol{r}|\boldsymbol{r}')$ be the probability density for the displacement in a single step of the walk, conditional on the random walker being at the point $\boldsymbol{r}'$. This function can be found as the solution of what we will term an evolution equation which, in the most general case of spatially varying transition densities, can be written

$$p_{n+1}(\boldsymbol{r}) = \int p_n(\boldsymbol{r}')p(\boldsymbol{r} - \boldsymbol{r}'|\boldsymbol{r}')d^d\boldsymbol{r}. \tag{5.6}$$

A strictly homogeneous medium is one for which $p(\boldsymbol{r}|\boldsymbol{r}') = p(\boldsymbol{r})$, in which case the integral on the right-hand side of (5.6) is a convolution. This will not be the case in a disordered medium. Most models for disordered media have a property that we term statistical homogeneity, which means that there are local fluctuations in properties of the medium but the probability distribution of these fluctuations does not vary with location.

The principal tool for analyzing a number of properties of the simplest version of the random walk is that of the characteristic function, which is the Fourier series whose coefficients are the $p(\boldsymbol{j})$ for lattice random variables, and the Fourier transform of the $p(\boldsymbol{r})$ for continuum random variables. Why this can be regarded as the key tool for analyzing random walks is the subject of the following section.

5.2.2 Characteristic Functions

A common, although not the only, problem in the application of random walks, is that of finding the probability density of $\boldsymbol{R}_n$ given the transition for single steps of the walk. This function will be denoted by $p_n(\boldsymbol{r})$ and termed the n-step transition density. We will show that the use of characteristic functions greatly simplifies the analysis required for finding an expression for $p_n(\boldsymbol{r})$ for a strictly homogeneous medium.

Let $\langle f(\boldsymbol{r}) \rangle$ denote the average of the function $f(\boldsymbol{r})$ with respect to $p(\boldsymbol{r})$, which is to say that

$$\langle f(\boldsymbol{r}) \rangle \equiv \int f(\boldsymbol{r})p(\boldsymbol{r})d^d\boldsymbol{r}. \tag{5.7}$$

The assumption that random variables are independent is equivalent to a statement that the joint probability density of the n variables $\boldsymbol{r}_1, \boldsymbol{r}_2, \cdots, \boldsymbol{r}_n$, $p(\boldsymbol{r}_1, \boldsymbol{r}_2, \cdots, \boldsymbol{r}_n)$ can be factorized as

$$p(\boldsymbol{r}_1, \boldsymbol{r}_2, \cdots, \boldsymbol{r}_n) = \prod_{j=1}^{n} p(\boldsymbol{r}_j), \tag{5.8}$$

which is equivalent to the statement that there are no correlations between the $\boldsymbol{r}_j$. In more colloquial terms a knowledge of any one of the $\boldsymbol{r}_j$ cannot influence the value of any other one of them.

The characteristic function, or Fourier transform of $\boldsymbol{R}_n$ is defined in terms of a vector of transform variables $\boldsymbol{\omega} = (\omega_1, \omega_2, \cdots, \omega_d)$ and will be denoted by $C_n(\boldsymbol{\omega})$. This function can be expressed as

$$C_n(\boldsymbol{\omega}) = \langle \exp(i\boldsymbol{\omega} \cdot \boldsymbol{R}_n) \rangle = \langle \exp\{(i\boldsymbol{\omega} \cdot (\boldsymbol{r}_1 + \boldsymbol{r}_2 + \cdots + \boldsymbol{r}_n)\} \rangle. \tag{5.9}$$

Because of the factorization property in (5.8) this can be further simplified to

$$C_n(\boldsymbol{\omega}) = \prod_{j=1}^{n} \langle \exp(i\boldsymbol{\omega} \cdot \boldsymbol{r}_j) \rangle = \langle \exp(i\boldsymbol{\omega} \cdot \boldsymbol{r}) \rangle^n = \hat{p}^n(\boldsymbol{\omega}). \tag{5.10}$$

where $\hat{p}(\boldsymbol{\omega})$ is defined as the characteristic function for a single step, that is, it is equal to $\langle \exp(i\boldsymbol{\omega} \cdot \boldsymbol{r}) \rangle$. As a result $p_n(\boldsymbol{r})$ can be represented in terms of the inverse Fourier transform as

$$p_n(\boldsymbol{r}) = \frac{1}{(2\pi)^d} \int_{-\infty}^{\infty} \cdots \int_{-\infty}^{\infty} \hat{p}^n(\boldsymbol{\omega}) \exp(-i\boldsymbol{\omega} \cdot \boldsymbol{r}) d^d\boldsymbol{\omega}. \tag{5.11}$$

for the continuum random walk in d dimensions.

Although the same formalism can be used to treat both the continuum and the periodic lattice, it is customary to define the characteristic function for a lattice in terms of a Fourier series. That is to say, in the lattice case one defines the characteristic function by

$$\hat{p}(\boldsymbol{\theta}) = \sum_j p(\boldsymbol{j}) e^{i\boldsymbol{\theta} \cdot \boldsymbol{j}}. \tag{5.12}$$

in which case the analog of (5.11) is

$$p_n(\boldsymbol{j}) = \frac{1}{(2\pi)^d} \int_{-\pi}^{\pi} \cdots \int_{-\pi}^{\pi} \hat{p}^n(\boldsymbol{\theta}) \exp(-i\boldsymbol{\theta} \cdot \boldsymbol{j}) d^d\boldsymbol{\theta}. \tag{5.13}$$

The difference between the expressions for the n-step densities for the continuum and the lattice cases appears only in the limits of integration as can be seen by comparing (5.11) and (5.13). We remark, parenthetically, that while the characteristic function is generally defined in terms of a Fourier transform, it is sometimes convenient to define it in terms of other transforms. For example, when the random variables are allowed to assume non-negative values only, a definition in terms of Laplace transforms generally proves to be convenient in further mathematical analysis. Another way to introduce the characteristic function for a random walk in a homogeneous medium is to solve (5.6) by means of Fourier transforms. Let $\mathcal{F}\{p_n(\boldsymbol{r})\} = \hat{p}_n(\boldsymbol{\omega})$. The transform of (5.6) in a strictly homogeneous space yields the relation $\hat{p}_{n+1}(\boldsymbol{\omega}) = \hat{p}(\boldsymbol{\omega})\hat{p}_n(\boldsymbol{\omega})$ or $\hat{p}_n(\boldsymbol{\omega}) = [\hat{p}(\boldsymbol{\omega})]^n$, which is equivalent to (5.10).

It will later be shown that the characteristic function is a useful tool not only for generating the probability density of the end-to-end distance of the random walk, but also for providing asymptotic approximations for this function in the limit of large numbers of steps. As will also be seen, a knowledge of analytic properties of this function furnishes the starting point in the calculation of many useful properties of different functionals of lattice walks. Because of the general utility of the characteristic function it is useful to understand some of its more elementary properties. Examples of these will be given, mainly for one-dimensional systems, but we emphasize that these can be generalized to higher dimensions quite straightforwardly.

A most important property of the characteristic function is that $C_n(\boldsymbol{0}) = 1$, since $p(\boldsymbol{r})$ is normalized, its integral over all of accessible space being equal to 1. The characteristic function is also bounded by this value in the sense that

$$|\hat{p}(\boldsymbol{\omega})| \leq 1 = \hat{p}(\boldsymbol{0}), \tag{5.14}$$

which is a consequence of the inequality

$$|\hat{p}(\boldsymbol{\omega})| = \left| \int e^{i\boldsymbol{\omega}\cdot\boldsymbol{r}} p(\boldsymbol{r}) d^d\boldsymbol{r} \right| \leq \int |e^{i\boldsymbol{\omega}\cdot\boldsymbol{r}}| p(\boldsymbol{r}) d^d\boldsymbol{r} = \int p(\boldsymbol{r}) d^d\boldsymbol{r} = 1. \tag{5.15}$$

We will often make use of an assumption that the only solution of the equation $\hat{p}(\boldsymbol{\omega}) = 1$ is $\boldsymbol{\omega} = \boldsymbol{0}$. This can only be violated when the displacement $\boldsymbol{r}$ is a lattice random variable, and the random walk is constrained to take place on a sublattice [5.11]. In the analysis that follows we will assume that the inequality in (5.14) is a strict one whenever $\boldsymbol{\omega} \neq \boldsymbol{0}$.

The final property to be mentioned is that the characteristic function can be regarded as a moment-generating function. Consider a random variable in one dimension, and expand the exponential that appears in the definition of the characteristic function. When all of the moments are finite we have the formal expansion

$$\hat{p}(\omega) = \int_{-\infty}^{\infty} e^{i\omega x} p(x) dx = \int_{-\infty}^{\infty} \sum_{n=0}^{\infty} \frac{i^n \omega^n x^n}{n!} p(x) dx$$

$$= \sum_{n=0}^{\infty} \frac{(i\omega)^n}{n!} \int_{-\infty}^{\infty} x^n p(x) dx = \sum_{n=0}^{\infty} \frac{(i\omega)^n}{n!} \langle x^n \rangle, \qquad (5.16)$$

so that the integer moments, when they exist, can be expressed in terms of derivatives of $\hat{p}(\omega)$ as

$$\langle x^n \rangle = (-1)^n \frac{d^n \hat{p}}{d\omega^n} \bigg|_{\omega=0}, \quad n = 0, 1, 2, \ldots . \qquad (5.17)$$

This formula with $\hat{p}(\omega)$ replaced by $C_n(\omega)$ from (5.9) can, for example, be used to generate moments of the displacement in an n-step walk in terms of the moments of single steps. Let $\langle x^k(n) \rangle$ be the kth moment of displacement in an n-step walk. Using (5.17) we find for the mean and variance of the n-step walk

$$\langle x(n) \rangle = \langle x \rangle n, \quad \sigma^2(n) \equiv \langle x^2(n) \rangle - \langle x(n) \rangle^2 = \sigma^2(1)n, \qquad (5.18)$$

and so forth.

What can be said about the characteristic function when only a finite number of moments exist? In that case nothing can be said which has a comparable level of generality to that of the power series in (5.16). When additional information is available about the form of $p(x)$ it is sometimes possible to derive information about the analytical behavior of $\hat{p}(\omega)$ in the neighborhood of $\omega = 0$. This will be shown to be of some significance later when asymptotic properties of the one-dimensional $p_n(x)$ will be discussed. Suppose, for example, that $p(x)$ has the property

$$p(x) \sim X^\alpha / |x|^{\alpha+1}, \qquad (5.19)$$

for $|x/X| \gg 1$, where X is a constant and $0 < \alpha < 1$. Such a probability density has no integer moments except for the zero'th. The characteristic function can be expressed as

$$\hat{p}(\omega) = \int_{-\infty}^{\infty} p(x) dx - \int_{-\infty}^{\infty} (1 - e^{i\omega x}) p(x) dx$$

$$= 1 - \int_{-\infty}^{\infty} (1 - e^{i\omega x}) p(x) dx = 1 - \frac{1}{\omega} \int_{-\infty}^{\infty} p\left(\frac{\rho}{\omega}\right) (1 - e^{i\rho}) d\rho. \qquad (5.20)$$

We conclude from the last integral on the right-hand side that the behavior of $\hat{p}(\omega)$ near $\omega = 0$ depends mainly on the behavior of $p(x)$ in the tails of the curve, that is its behavior at large values of $|x/X|$. On substituting (5.19) into this integral, we find that for $\omega \to 0$

$$\hat{p}(\omega) \sim 1 - (X|\omega|)^\alpha \int_{-\infty}^{\infty} \frac{1 - \cos \rho}{|\rho|^{\alpha+1}} d\rho, \qquad (5.21)$$

so that the asymptotic property of $p(x)$ in (5.19) implies the appearance of a noninteger power of $|\omega|$ in the expansion of $\hat{p}(\omega)$ in the neighborhood of $\omega = 0$.

A somewhat less familiar result is that noninteger and negative moments can also be calculated in terms of the characteristic function. Consider, for example, a one-dimensional case in which a random variable, x, is restricted to lie on the positive x axis (this can, for example, refer to the time when it is a random variable), and one wishes to calculate the noninteger moment $\langle x^\beta \rangle$ where $0 < \beta < 1$. The power x^β can be written as $x/x^{1-\beta}$. The term $1/x^{1-\beta}$ will be represented in terms of an integral as follows:

$$\frac{1}{x^{1-\beta}} = \frac{1}{\Gamma(1-\beta)} \int_0^\infty \xi^{-\beta} e^{-\xi x} dx. \tag{5.22}$$

If $p(x)$ is the probability density characterizing the random variable we can, by making use of our representation of x^β, write

$$\langle x^\beta \rangle \equiv \int_0^\infty x^\beta p(x) dx = \frac{1}{\Gamma(1-\beta)} \int_0^\infty \xi^{-\beta} d\xi \int_0^\infty x p(x) e^{-\xi x} dx$$

$$= -\frac{1}{\Gamma(1-\beta)} \int_0^\infty \xi^{-\beta} \frac{d\hat{p}(\xi)}{d\xi} d\xi, \tag{5.23}$$

in which, since the range of x is $(0, \infty)$, the characteristic function is defined in terms of a Laplace transform.

There is an important distinction to be made between the expressions for the integer moments and those for noninteger or negative moments. The former depend only on local properties of the characteristic function at the origin, while the latter, exemplified by (5.23), depend on values taken by the characteristic function over the entire range of the transform parameter. Thus, a calculation of these moments is intrinsically more difficult than that for the integer moments.

A function closely related to the characteristic function is termed the cumulant-generating function, $K(\omega)$. This is defined as the logarithm of the characteristic function. When all of the moments exist the one-dimensional $K(\omega)$ can be expanded as

$$K(\omega) = \ln \hat{p}(\omega) = \sum_{n=1}^\infty \frac{i^n \omega^n}{n!} \kappa_n, \tag{5.24}$$

where the κ_n are known as cumulants. Since, when all of the moments are finite, $\hat{p}(\omega)$ can be written as a series in terms of the moments as shown in (5.16), the cumulants are also expressible in terms of moments. On carrying out in detail calculations based on this fact one finds that the first three in the hierarchy of cumulants are $\kappa_1 = \langle x \rangle$, $\kappa_2 = \langle x^2 \rangle - \langle x \rangle^2$, and $\kappa_3 = \langle x^3 \rangle - 3\langle x^2 \rangle \langle x \rangle + \langle x \rangle^3$. Exact, but rather complicated, expressions are available for converting moments into cumulants and *vice versa*. These relations have been tabulated up to order 10 [5.12]. When the moments do not exist the expansion in integer powers of ω must generally be replaced by a more complex representation. For example, when $p(x)$ has the stable-law property indicated in (5.19) the first term in the expansion of $K(\omega)$ will be replaced by a term proportional to $|\omega|^\alpha$.

5.2.3 The Continuous-Time Random Walk (CTRW)

One of the earliest approaches to a purely phenomenological theory of transport properties of a random medium is based on the continuous-time random walk, or CTRW as it is almost universally termed [5.13]. This variation of the basic random-walk model is based on the notion that the time between successive steps of the random walk is itself a random variable. Hence at an arbitrary time one cannot say how many steps have been taken. Let τ_i be the time at which the ith step is made, where $i = 1, 2, \ldots$. The CTRW is generally defined through the requirement that the $\{t_i\}$ (where $t_i = \tau_{i+1} - \tau_i$) are identically distributed independent random variables (see Fig. 5.2).

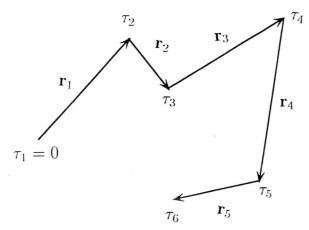

Fig. 5.2. An illustration of the CTRW in $d = 2$. The time at which the ith step is made is τ_i. The time differences $t_i = \tau_{i+1} - \tau_i$ are chosen randomly from the same distribution

The probability density for the time interval between two successive steps will be denoted by $\psi(t)$, that is, the probability that $t \le t_i \le t + dt$ is equal to $\psi(t)dt$. The probability that a given interstep interval is $\ge t$ will be denoted by $\Psi(t)$, which is defined in terms of $\psi(t)$ by $\Psi(t) = \int_t^\infty \psi(\tau)d\tau$. Let $\psi_n(t)$ be the probability density for the time at which the nth step occurs. This function satisfies the recurrence relation

$$\psi_1(t) = \psi(t); \quad \psi_n(t) = \int_0^t \psi(\tau)\psi_{n-1}(t - \tau)\, d\tau, \; n > 1. \qquad (5.25)$$

Because the relation between successive $\psi_n(t)$ is written in terms of a convolution integral in time, it is natural to switch to the equivalent relation in terms of Laplace transforms in order to transform this into a purely algebraic form. Let $\hat{\psi}(s)$ denote the Laplace transform of $\psi(t)$. Then (5.25) can be translated to the recursion relation $\hat{\psi}_n(s) = \hat{\psi}(s)\hat{\psi}_{n-1}(s)$, or

$$\hat{\psi}_n(s) = [\hat{\psi}(s)]^n. \qquad (5.26)$$

Later calculations are simplified by assuming that $\psi_0(t) = \delta(t)$ which, by convention, means that a step is always taken at $t = 0$.

Let us calculate the probability density $p(r, t)$, i.e., the probability that the position of a random walker is r at time t. Since the number of steps is discrete we may use the integral representation in (5.11) to define the probability density for the position of the random walker after n steps. Remembering that $\Psi(t)$ is the probability that the time between two successive steps is $\geq t$ we may write

$$p(r, t) = \sum_{n=0}^{\infty} p_n(r) \int_0^t \psi_n(\tau) \Psi(t - \tau) \, d\tau, \qquad (5.27)$$

where $p_0(r) = \delta(r)$ since the random walker always starts from the origin. The nth term of this expansion says that the nth step occurred at a time τ, the random walker then being at r. The time to the immediately following step must then be greater than $t - \tau$, to ensure that the random walker remains at that position until at least time t.

Equation (5.27) can be simplified by taking its Laplace transform. Note first that the transform of $\Psi(t)$ is

$$\int_0^{\infty} e^{-st} dt \int_t^{\infty} \psi(\tau) \, d\tau = \frac{1 - \hat{\psi}(s)}{s}. \qquad (5.28)$$

The Laplace transform of $p(r, t)$ will be denoted by $\hat{p}(r, s)$ which, by combining (5.27) and (5.28), can be expressed in the form

$$\hat{p}(r; s) = \frac{1 - \hat{\psi}(s)}{s} \sum_{n=0}^{\infty} p_n(r) \hat{\psi}^n(s), \qquad (5.29)$$

which is seen to have the form of a generating function, i.e., a power series, the expansion variable being $\hat{\psi}(s)$. This equation is further simplified by substituting the integral representation for $p_n(r)$ from (5.11) into it. Since the condition $\text{Re}\{s\} > 0$ implies that $|\hat{\psi}(s)| < 1$, (5.29) is converted to a geometric series which can be summed exactly, allowing us to write

$$\hat{p}(r; s) = \frac{1 - \hat{\psi}(s)}{(2\pi)^d s} \int_{-\infty}^{\infty} \cdots \int_{-\infty}^{\infty} \frac{\exp(-i\boldsymbol{\omega} \cdot r)}{1 - \hat{p}(\boldsymbol{\omega})\hat{\psi}(s)} d^d\boldsymbol{\omega}. \qquad (5.30)$$

This type of representation, (5.30), will later be used to generate asymptotic results that are independent of the detailed form of $p(r)$, but depend only on the moments, which are very coarse properties of the distribution. Before showing how this comes about we point out that (5.30) can be used to furnish exact expressions for Laplace transforms of the moments of the cumulative displacement at time t. For example, let $\langle x^n \rangle$ denote the nth moment of a single jump in a one-dimensional random walk, and let $\langle \hat{x}^n(s) \rangle$ denote the Laplace transform of the nth moment of the position of the random walker at time t. In our derivation we make use of a representation of the delta function, i.e., $\delta(\omega) = (2\pi)^{-1} \int_{-\infty}^{\infty} \exp(-i\omega x) \, dx$, and form the averages

$$\langle \hat{x}^n(s) \rangle \equiv \int_{-\infty}^{\infty} x^n \hat{p}(x,s) dx. \tag{5.31}$$

After multiplying (5.30) by x and x^2 successively, one finds for the Laplace transforms of the first two moments

$$\langle \hat{x}(s) \rangle = \frac{\langle x \rangle \hat{\psi}(s)}{s[1 - \hat{\psi}(s)]},$$

$$\langle \hat{x}^2(s) \rangle = \frac{2\langle x \rangle^2 \hat{\psi}^2(s)}{s[1 - \hat{\psi}(s)]^2} + \frac{\langle x^2 \rangle \hat{\psi}(s)}{s[1 - \hat{\psi}(s)]}, \tag{5.32}$$

and so forth. The moments in higher dimensions can be derived by a straightforward extension of the one-dimensional calculation.

5.2.4 The Characteristic Function and Properties of the Lattice Random Walk

Up to this point we have illustrated a number of ways in which the characteristic function can be used to generate information related to random walks on strictly homogeneous spaces. In this section we show that results of the calculation of a number of properties of the lattice random walk, exemplified by the average number of distinct sites visited by an n-step walk, are also expressible in terms of the characteristic function. The relations to be derived will form the basis of our derivation of asymptotic properties of these random variables.

A function that will prove of some importance in all of the following calculations is the first passage time probability $f_n(\boldsymbol{j})$, defined as the probability that the random walker reaches site $\boldsymbol{j}$ for the first time at step n. This differs from the function $p_n(\boldsymbol{j})$ which allows the random walker to be at $\boldsymbol{j}$ an arbitrary number of times before reaching that site at step n. The relation between these two functions is

$$p_n(\boldsymbol{j}) = \delta_{n,0}\delta_{\boldsymbol{j},\boldsymbol{0}} + \sum_{k=1}^{n} f_k(\boldsymbol{j}) p_{n-k}(\boldsymbol{0}). \tag{5.33}$$

The first term on the right-hand side accounts for the initial step and the summation term accounts for the possibility that the random walker, reached $\boldsymbol{j}$ for the first time at step k, thereafter returning to that point after $n - k$ further steps. A numerical solution for $f_k(x)$ is shown in Fig. 5.3.

To produce a more useful relation between these two sets of probabilities we introduce generating functions with respect to n. Specifically we define

$$p(\boldsymbol{j}; z) = \sum_{n=0}^{\infty} p_n(\boldsymbol{j}) z^n, \quad f(\boldsymbol{j}; z) = \sum_{n=0}^{\infty} f_n(\boldsymbol{j}) z^n. \tag{5.34}$$

A relation between these generating functions is found by multiplying both sides of (5.33) by z^n and summing over all n. In this way we find that the generating function $f(\boldsymbol{j}; z)$ is related to the $p(\boldsymbol{j}; z)$ by

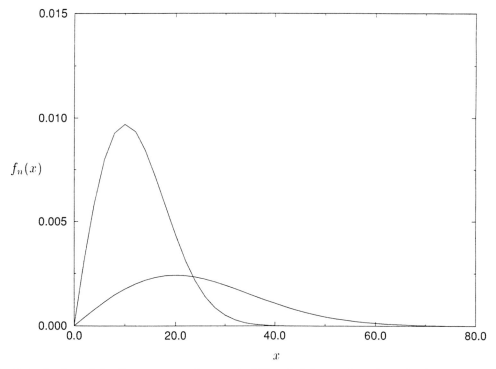

Fig. 5.3. A plot of the first passage time probability $f_n(x)$ as a function of x for $d = 1$ and $n = 20$ and 50 single-step random walks, (5.2). Note that in contrast to $p_n(x)$, shown in Fig. 5.4, the maximum shifts to $|x| \gg 1$ when n is increased

$$f(\boldsymbol{j}; z) = \frac{p(\boldsymbol{j}; z)}{p(\mathbf{0}; z)}, \ \boldsymbol{j} \neq \mathbf{0}; \quad f(\mathbf{0}; z) = 1 - \frac{1}{p(\mathbf{0}; z)}. \tag{5.35}$$

These relations are useful since we already have integral representations for the $p_n(\boldsymbol{r})$. This enables us to obtain integral representation of the $p(\boldsymbol{j}; z)$ by inserting the expression for $p_n(\boldsymbol{j})$ from (5.13) into the definition of the generating function in (5.34). By doing so we find that

$$p(\boldsymbol{j}; z) = \frac{1}{(2\pi)^d} \int_{-\pi}^{\pi} \cdots \int_{\pi}^{\pi} \frac{\exp(-i\boldsymbol{\theta} \cdot \boldsymbol{j})}{1 - z\hat{p}(\boldsymbol{\theta})} d^d \boldsymbol{\theta}. \tag{5.36}$$

An n-step random walk is likely to visit fewer than n distinct sites, since it may double back on itself. The number of distinct sites visited by an n-step random walk will be denoted by D_n. The problem of calculating the distribution of D_n proves to be an extremely difficult one except for one dimensional random walks which are constrained to make steps to nearest neighbors only. An easier version of this problem is to find the first moment of this variable, or the expected number of distinct sites visited in n steps. This will be shown to be an important parameter in the study of the trapping problem, as discussed in our description of that problem in Sect. 5.4.3. However, we will show that it is not too difficult to calculate the generating function associated with the set of expectations $\langle D_n \rangle$, which will suffice for finding its asymptotic behavior.

Our derivation is based on the observation that the probability that a new site is visited at step n, Δ_n, can be written in terms of the $f_n(\boldsymbol{j})$ as

$$\Delta_n = \sum_j f_n(\boldsymbol{j}), \tag{5.37}$$

where the sum is over all lattice sites.

The value of $\langle D_n \rangle$, where the average is over all n-step random walks, is found by remarking that either one or no new site is visited on step k, the probabilities for these events being equal to Δ_k and $[1 - \Delta_j]$ respectively. Consequently the contribution of step k to $\langle D_n \rangle$ is equal to $1 \cdot \Delta_k + 0 \cdot [1 - \Delta_k] = \Delta_k$ This gives us an expression for $\langle D_n \rangle$:

$$\langle D_n \rangle = \sum_{k=1}^{n} \Delta_k. \tag{5.38}$$

Let $\Delta(z)$ be the generating function of the Δ_n and $D(z)$ be that of the $\langle D_n \rangle$. Since the generating function of the $f_n(\boldsymbol{j})$ is given in (5.35) we can multiply both sides of (5.37) by z^n, sum, and then substitute the expressions in (5.35) into the result. The result of doing this is an expression for $\Delta(z)$ in terms of the $p(\boldsymbol{j}; z)$:

$$\Delta(z) = \frac{\sum_j p(\boldsymbol{j}; z) - 1}{p(\boldsymbol{0}; z)}. \tag{5.39}$$

But since the $p_n(\boldsymbol{j})$ are normalized in the sense that

$$\sum_j p_n(\boldsymbol{j}) = 1, \tag{5.40}$$

we have

$$\sum_j p(\boldsymbol{j}; z) = \frac{1}{1 - z}, \tag{5.41}$$

which is found by taking the generating functions of both sides of (5.40). Hence

$$\Delta(z) = \frac{z}{(1 - z)p(\boldsymbol{0}; z)}. \tag{5.42}$$

Finally we note that since $\Delta_n = \langle D_n \rangle - \langle D_{n-1} \rangle$, the generating function of the $\langle D_n \rangle$ is

$$D(z) = \frac{z}{(1 - z)^2 p(\boldsymbol{0}; z)}. \tag{5.43}$$

Recently, this last equation has been generalized by deriving the generating function of the expected number of sites visited by $N > 1$ independently moving random walkers each of which is initially located at the origin [5.14]. Let $\Gamma_n(\boldsymbol{r})$ be the probability (for the case $N = 1$) that $\boldsymbol{r}$ has not been visited by the random walker by step n. This is expressed in an obvious way in terms of the $f_n(\boldsymbol{j})$. The average number of distinct sites visited by N random walkers in n steps is then

$$\langle D_N(n) \rangle = \sum_{\boldsymbol{j}} [1 - \Gamma_n^N(\boldsymbol{j})]. \tag{5.44}$$

In this formulation of the problem it is convenient to take the generating function with respect to N rather than with respect to n. Since this involves only a geometric sum, as seen from the last equation, one finds that

$$D(z; n) = \frac{z}{1-z} \sum_{\boldsymbol{j}} \frac{1 - \Gamma_n(\boldsymbol{j})}{1 - z\Gamma_n(\boldsymbol{j})}, \tag{5.45}$$

from which the asymptotic properties of the $D_N(n)$ can be deduced. The generating functions of other kinds of random parameters exemplified by the average occupancy, i.e., the average number of times an n-step random walk visits a given site or given set of sites can also be exactly expressed in terms of the $p(\boldsymbol{j}; z)$ [5.13].

If we suppose that we are dealing with a CTRW rather than a discrete-time random walk, we can still use the formalism leading to (5.43) with only one slight difference. Let $\langle D_{\text{CTRW}}(t) \rangle$ be the expected number of distinct sites visited by a CTRW in time t. The analog of $D(z)$ in continuous time requires the replacement of the discrete generating function by a Laplace transform. This will be denoted by $\langle \hat{D}_{\text{CTRW}}(s) \rangle$. Recall that our derivation of $D(z)$ in discrete time depended on our having an expression for the generating function of first-passage times to site $\boldsymbol{j}$, i.e., the function $f(\boldsymbol{j}; z)$, which in turn can be related to the $p(\boldsymbol{j}; z)$. Since the number of steps taken by a CTRW at time t is a random variable, the Laplace transform $\hat{p}(\boldsymbol{j}; s)$ can be expressed as shown in (5.30). The point to be emphasized is that $\hat{p}(\boldsymbol{j}; s)$ is proportional to a generating function, the transform parameter z being replaced by $\hat{\psi}(s)$. Hence $\langle \hat{D}_{\text{CTRW}}(s) \rangle$ can be found from (5.43) by simply replacing the generating function variable z by the Laplace transform $\hat{\psi}(s)$.

5.3 Asymptotic Properties

5.3.1 The Central-Limit Theorem and Some Generalizations

Up to this point all that we have shown is that the use of characteristic functions is useful in translating exact results into a somewhat simpler, but equivalent, mathematical form. This may be cosmetically pleasing, but the characteristic function has much more powerful properties. In this section we show that certain universal asymptotic properties can be calculated from the analytical behavior of the characteristic function in the neighborhood of the origin in the transform parameter space. A peripheral remark is that if no universal properties existed random-walk theory would be fragmented into a large number of special examples, which would make it very much less interesting than it is in practice. There are many properties of the random walk that are asymptotically independent of the detailed form of the underlying probability density for single-step displacements but depend only on the moments of this density. The first important example of how an asymptotic property is described is the so-called central-limit theorem, which gives an asymptotic expression for the probability density for the end-to-end distance $p_n(\mathbf{r})$ for large n. We give a brief and somewhat heuristic, derivation of this result for a random walk in one dimension. Our derivation contains the essence of fully rigorous derivations, details of which can be found, for example, in the monograph by Gnedenko and Kolmogoroff [5.11].

A natural starting point for such a derivation is one of the exact expressions found in (5.11) or (5.13), depending on whether the random walk occurs on a lattice or in a continuum. To keep the appearance of our results simple only the case of the one-dimensional continuum walk will be discussed. In order for the central limit theorem to be valid it is necessary for the first two moments, $\langle x \rangle$ and $\langle x^2 \rangle$, to be finite. The central-limit form for $p_n(x)$ is based on an approximate evaluation of the integral representation in (5.13). The intuitive reason why it is possible find an approximation to it is based on the inequality for $\hat{p}(\omega)$ given in (5.14). This, together with the assumption that $\omega = 0$ is the only root of the equation $\hat{p}(\omega) = 1$, implies that in the limit $n \to \infty$ the major contribution to the integral is due to the behavior of $C_n(\omega)$ in the neighborhood of $\omega = 0$. Hence one strategy for generating an approximation to $p_n(x)$ is to accurately approximate to $\hat{p}^n(\omega)$ near $\omega = 0$ and essentially ignore contributions from other values of ω. Write $\hat{p}^n(\omega) = \exp[nK(\omega)]$ in terms of the cumulant generating function $K(\omega)$. Because of our assumption that the first two moments of the displacement are finite we may expand the expression for $\hat{p}^n(\omega)$ around $\omega = 0$ up to second order

$$\hat{p}^n(\omega) \sim \exp\left\{ n \left[i\langle x \rangle \omega - \frac{\sigma^2}{2}\omega^2 \right] \right\}. \tag{5.46}$$

On substituting this expression into the integral in (5.11) we find that

$$p_n(x) \sim \frac{1}{\sigma\sqrt{2\pi n}} \exp\left[-\frac{(x - n\langle x \rangle)^2}{2\sigma^2 n} \right]. \tag{5.47}$$

That is, the lowest order of approximation to $p_n(x)$ furnished by the central-limit theorem is a Gaussian whose mean is $n\langle x \rangle$ and whose variance is equal to $\sigma^2 n$. To illustrate the accuracy of (5.47) see the comparison of results in Fig. 5.4.

Our derivation has been given in terms of a random walk along a line, but the result for the lattice random walk is essentially identical to that for the random walk in a continuum. The reason for this is that the expression for $p_n(\mathbf{j})$ in (5.11) is identical to that in (5.13) except for the limits of integration. Since, when $n \to \infty$, the major contribution to the integral comes from the neighborhood of $\omega = 0$ the contribution from larger values of $|\omega|$ will be much smaller, and indeed can be shown to go to zero in the limit $\omega \to 0$, generally as a power of ω higher than the second, in compared to the Gaussian function of ω in (5.46).

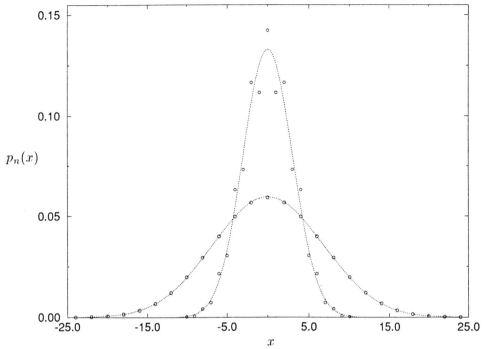

Fig. 5.4. Comparison of the approximate expression $p_n(x)$ from (5.47) with the exact numerical solution from the binomial distribution at $n = 10$ and 50 steps

In our exposition we have assumed that successive steps are identically distributed random variables. However, the central-limit theorem provides an approximation that may also be useful when the jth step is characterized by a probability density for single-step displacements that is a function of j. This is an important consideration in crystallographic applications of random walk theory. In such a case, let $\hat{p}_j(\omega) = \langle \exp(i\omega x_j) \rangle$. Then the characteristic function for the sum $x_1 + x_2 + \cdots + x_n$ is

$$C_n(\omega) = \prod_j \hat{p}_j(\omega). \tag{5.48}$$

The central-limit approximation to $p_n(x)$ is found by expanding each of the $\hat{p}_j(\omega)$ around $\omega = 0$, and also leads to an approximation that can be expressed as a Gaussian. How accurate this approximation will be depends on the degree of similarity between the $p_j(x)$.

At least four comments are in order about the Gaussian approximation resulting from the central-limit theorem in the last paragraph. Firstly, these results are trivially extended to higher dimensions, at least at the level of rigor of our demonstration in one dimension. The result of such a calculation is expressible in terms of the variance-covariance matrix. Let d be the number of dimensions. The vector of mean displacements will then be denoted by $\langle \boldsymbol{r} \rangle$ of which the jth component is the average displacement along the jth coordinate, $\langle x_j \rangle$, and the associated covariances by $\langle x_j x_k \rangle$, $j, k = 1, 2, \ldots, d$, which is defined by

$$\langle x_j x_k \rangle = \int_{-\infty}^{\infty} \cdots \int_{-\infty}^{\infty} x_j x_k p(\boldsymbol{r}) d^d \boldsymbol{r}. \tag{5.49}$$

Let $\boldsymbol{M}$ denote the $d \times d$ variance-covariance matrix whose jkth element is equal to $\langle x_j x_k \rangle - \langle x_j \rangle \langle x_k \rangle$, and $|\boldsymbol{M}|$ the determinant of $\boldsymbol{M}$. The Gaussian approximation to $p_n(\boldsymbol{r})$ to second order in the k's can then be written

$$p_n(\boldsymbol{r}) = \frac{1}{(2\pi)^{d/2}|\boldsymbol{M}|^{1/2}} \exp \left\{ -\frac{(\boldsymbol{r} - \langle \boldsymbol{r} \rangle)' \cdot \boldsymbol{M}^{-1} \cdot (\boldsymbol{r} - \langle \boldsymbol{r} \rangle)}{2n} \right\}. \qquad (5.50)$$

Our second comment is that the Gaussian approximation is generally found to be accurate in the neighborhood of the peak, the accuracy falling off rather drastically in the tails of the distribution (See Fig. 5.4). Other techniques are required to deal with the problem of approximating to the tail of the density, one of which, the saddlepoint approximation, will be mentioned below.

Thirdly, it is possible to calculate corrections to the basic Gaussian approximation provided that moments higher than the second are finite. A calculation of the next correction to the Gaussian requires that the third moment be finite and of order $n^{-1/2}$ relative to 1. Further corrections in higher powers of $n^{-1/2}$ can be found provided that the appropriate number of moments of $p(x)$ are finite.

A fourth comment is that one can achieve a considerable improvement over the Gaussian approximation by choosing a different strategy for approximating the integral representation of $p_n(x)$ in (5.13). One excellent technique available is the so-called saddlepoint approximation, which makes use of the method of steepest descents to evaluate the integral [5.15,16]. Without going into great detail about this technique we point out two considerable virtues of the approximation. The first is that the accuracy of the approximation tends to be nearly uniform across the range of the curve rather than falling off in the tails of $p_n(x)$, and the second is that it is intrinsically more accurate than the Gaussian approximation (to which it reduces in the immediate vicinity of the peak) as a function of n. In fact it can be shown that the lowest-order correction to the saddlepoint approximation goes like $n^{-3/2}$ rather than $n^{-1/2}$, and higher order corrections fall off in powers of n^{-1} rather than $n^{-1/2}$ [5.17]. On the debit side is the fact that the equation(s) determining the saddlepoint is (are) usually transcendental which means that it (they) must be solved numerically. Applications of this technique to the analysis of simple random walks in one dimension have been made by Domb and Offenbacher [5.18] and by Weiss and Kiefer [5.20].

Separate consideration must be given to the case in which moments of the single-step displacements are infinite, but the general method of deriving asymptotic results in such cases is still based on the characteristic function. Let us, for example, suppose that $p(x)$ has a stable-law form. That is, for large values of $|x|$ it is proportional to

$$p(x) \sim L^\alpha / |x|^{\alpha+1}, \qquad (5.51)$$

where L is a scaling parameter and α satisfies $0 < \alpha \le 2$. Such walks have been called Levý walks. For more results generalizations and application of the Levý walk model see [5.19,20]. The nth power of the characteristic function $\hat{p}(\omega)$ in such a case can be expanded in the neighborhood of $\omega = 0$ as

$$\hat{p}^n(\omega) \sim (1 - |L\omega|^\alpha)^n \sim \exp(-n|L\omega|^\alpha), \qquad (5.52)$$

with the result that

$$
\begin{aligned}
p_n(x) &\sim \frac{1}{2\pi} \int_{-\infty}^{\infty} \exp(-n|L\omega|^\alpha) \cos(\omega x)\, d\omega \\
&= \frac{1}{2\pi|x|} \int_{-\infty}^{\infty} \exp\left(-\frac{nL^\alpha |v|^\alpha}{|x|^\alpha}\right) \cos(v)\, dv,
\end{aligned} \qquad (5.53)
$$

which reduces to a Gaussian when $\alpha = 2$. This integral can be evaluated in closed form for only a few values of α, but extensive numerical tables exist for the range $(0, 2)$ [5.21]. It can be shown that if a limit law exists for the sum of independent random variables then that law can only be a Gaussian or a stable law [5.11]. An important observation is that the approximation to $p_n(x)$ in (5.53) has a scaling form, which is to say that the integral depends on n and x only in the combination $|x|^\alpha/n$. We can find the behavior of $p_n(x)$ in the limit $|x/L| \gg n^{1/\alpha}$ by expanding the exponential in (5.53). The first two terms of such an expansion are

$$p_n(x) \sim \frac{1}{2\pi|x|} \int_{-\infty}^{\infty} \left(1 - \frac{nL^\alpha |v|^\alpha}{|x|^\alpha}\right) \cos(v)\, dv. \qquad (5.54)$$

The first term in brackets just gives rise to a delta function at $x = 0$, which cannot contribute to the tail behavior of $p_n(x)$. The second term, which does contribute to this behavior, gives

$$
\begin{aligned}
p_n(x) &\sim -\frac{nL^\alpha}{2\pi|x|^{\alpha+1}} \int_{-\infty}^{\infty} |v|^\alpha \cos(v)\, dv \\
&= -\frac{nL^\alpha}{2\pi|x|^{\alpha+1}} \lim_{\epsilon \to 0} \int_{-\infty}^{\infty} e^{-\epsilon v} |v|^\alpha \cos(v)\, dv \\
&= \Gamma(1 + \alpha) \sin\left(\frac{\pi\alpha}{2}\right) \frac{nL^\alpha}{2\pi|x|^{\alpha+1}}.
\end{aligned} \qquad (5.55)
$$

A comparison of (5.51) with this equation indicates that the large $|x|$ dependence of $p_n(x)$ is the same as that for $p(x)$. The same technique can be used to generate an asymptotically convergent infinite series [5.21]. A simple example which can be solved in closed form is that of the Cauchy density

$$p(x) = \frac{L}{\pi(L^2 + x^2)}, \qquad (5.56)$$

whose characteristic function is $\hat{p}(\omega) = \exp(-L|\omega|)$. The nth power of $\hat{p}(\omega)$ has the same functional form except that L is replaced by nL, which means that $p_n(x)$ is also a Cauchy density with the same replacement of L by nL.

5.3.2 The Diffusion Approximation

The reader will notice that the Gaussian approximation in (5.50) can also be regarded as the solution to a diffusion equation provided one replaces the factor $2\sigma^2 n$ by $4Dt$, where D is a diffusion constant. It is instructive to derive the diffusion equation directly from the master equation in (5.6) to see which scaling assumptions ensure its validity. A heuristic derivation will be given here, but a more general, and mathematically more rigorous, treatment of this topic is to be found in [5.22]. We discuss the process in one dimension only, in order to keep the analysis simple. The diffusion equation can only be expected to hold for random motion on a strictly homogeneous line, which is to say that without further validation we cannot expect the diffusion equation to be a correct description of the motion along a line which has disorder built into it. Hence we restrict our remarks at this point to the case of a random walk on a homogeneous line for which the evolution equation is that given in (5.6).

The basic idea is that the diffusion equation is a correct approximation when jumps are not too large in a sense to be specified shortly. The time between successive steps is constant and will be set equal to Δt, which will be taken to zero in the derivation. We will then write $t = n\Delta t$, which implies that since all values of t are considered, the parameter n will be large. In the one-dimensional version of (5.6) we have

$$p_{n+1}(x) = \int_{-\infty}^{\infty} p_n(x-y)\, p(y)\, dy. \tag{5.57}$$

An expansion of the left-hand side of this equation to lowest order with respect to Δt then reads

$$p_{n+1}(x) \sim p(x,t) + \Delta t \frac{\partial p(x,t)}{\partial t}. \tag{5.58}$$

On the right-hand side of (5.57), remembering that jumps are required to be small so that the position is localized, we expand $p_n(x-y)$ to second order:

$$p_n(x-y) \sim p(x,t) - y\frac{\partial p(x,t)}{\partial x} + \frac{1}{2}y^2\frac{\partial^2 p(x,t)}{\partial x^2}. \tag{5.59}$$

Since the time between successive steps is taken to zero we also require that the displacement in any step should also be small. A more precise way of stating this requirement is that

$$\lim_{\Delta t \to 0} \frac{1}{\Delta t}\int_{-\infty}^{\infty} \xi p(\xi)d\xi = v, \qquad \lim_{\Delta t \to 0} \frac{1}{2\Delta t}\int_{-\infty}^{\infty} \xi^2 p(\xi)d\xi = D,$$

$$\lim_{\Delta t \to 0} \frac{1}{\Delta t}\int_{-\infty}^{\infty} \xi^n p(\xi)d\xi = 0, \; n = 3,4,\ldots, \tag{5.60}$$

where v and D are constants. The combination of (5.57), (5.59), and (5.60) then leads to a diffusion equation of the form

$$\frac{\partial p}{\partial t} = D\frac{\partial^2 p}{\partial x^2} - v\frac{\partial p}{\partial x}. \tag{5.61}$$

It is necessary to retain terms up to second order in (5.59) because when the process is symmetric $v = 0$. The flip side of our derivation is that when a second moment does not exist in the sense of (5.60) a diffusion equation will not provide an accurate characterization of the transport process.

A point which can sometimes be a cause of confusion in the simulation of diffusion processes is that the diffusion equation is a mean-field equation, which is to say that if one looks at a single realization of a symmetric diffusion process it is quite likely to be asymmetric. It is highly probable, for example, that a single diffusing particle initially at the origin will visit the half-line on only one side of the origin rather than spending an equal amount of time on both sides of the origin. This was originally remarked on in the context of an analysis of polymer configurations [5.23,24] and a simpler model of this phenomenon was given in [5.25].

5.3.3 A Mathematical Excursion: Abelian and Tauberian Theorems

Random-walk properties that may be termed universal are generally asymptotic ones. In order to find asymptotic properties of quantities exemplified by $\langle D_n \rangle$, we must be able to invert the generating function $D(z)$ given in (5.43). However, only very large values of n are interesting as since when $n = O(1)$ random properties of physical interest generally depend on the specific choice of the jump probabilities rather than on the moments, which depend only weakly on these probabilities. Similar problems arise in trying to find many asymptotic properties of the CTRW. The functions describing these properties are usually representable in terms of Laplace transforms. While there are inversion formulae for both generating functions and Laplace transforms, it is quite difficult to make use of these for computation in the asymptotic range. Fortunately, many of the resulting problems can be handled by methods based on what are known as Abelian and Tauberian theorems which apply to both power series and different kinds of transforms [5.26]. These theorems, in the context of power series, relate singular behavior of a power series to the asymptotic behavior of the coefficients, and, in the context of Laplace transforms, relate the singular behavior of the transform to asymptotic behavior of the original function.

Abelian theorems relate to series in which the asymptotic behavior of coefficients of a power series is given. The knowledge of this behavior enables one to predict the singular behavior of either the generating function or Laplace transform in the neighborhood of the origin in transform parameter space. Tauberian theorems, which are generally harder to prove, proceed in the opposite direction, which is to say that a knowledge of the transform in any neighborhood in which it is singular allows the inference of asymptotic behavior of the function in the original space. Both types of theorem have been used extensively in the theory of random walks. We will not prove these theorems but try to make plausible the motivation behind them and then list the most important ones for our purposes. Observe that the Laplace transform can be regarded as a continuum analog of a power series, which is to say that if one takes the series

$$f(z) = \sum_{n=0}^{\infty} f_n z^n, \tag{5.62}$$

replaces the discrete parameter n by a continuous variable t, the summation by an integration, and the parameter z by e^{-s}, the result will be the Laplace transform

$$\hat{f}(s) = \int_0^{\infty} f(t) e^{-st} \, dt. \tag{5.63}$$

Henceforth we will use the parameter z to refer to the generating function in (5.62) and the parameter s to refer to the Laplace transform as in this last equation.

We now present some of the more useful results of Abelian and Tauberian theorems. Define a slowly varying function, $L(t)$, to be one which has the property

$$\lim_{t \to \infty} L(ct)/L(t) = 1$$

for every $c > 0$. A typical example of such a function is $L(t) = \ln(t)$. Suppose that we know that a function $f(t)$ behaves asymptotically as $f(t) \sim At^{\beta}L(t)$, where A is a constant. An Abelian theorem relates this behavior to that of $\hat{f}(s)$ in the limit $s \to 0$. To see how this comes about heuristically, change variables in (5.63) to $\rho = st$, in which case we have

$$\hat{f}(s) = \frac{1}{s} \int_0^{\infty} f\left(\frac{\rho}{s}\right) L\left(\frac{\rho}{s}\right) e^{-\rho} d\rho$$

$$\sim \frac{A}{s^{1+\beta}} L\left(\frac{1}{s}\right) \int_0^{\infty} \rho^{\beta} e^{-\rho} d\rho = A \frac{\Gamma(1+\beta)}{s^{1+\beta}} L\left(\frac{1}{s}\right), \tag{5.64}$$

where we have made use of the property $L(\rho/s) \sim L(1/s)$ as $s \to 0$. The corresponding Abelian theorem for power series follows from the observation that the limit $s \to 0$ corresponds to $z \to 1$. Consequently if $f_n \sim An^{\beta}L(n)$ in (5.62), then we can assert that

$$f(z) \sim A \frac{\Gamma(1+\beta)}{(1-z)^{1+\beta}} L\left(\frac{1}{1-z}\right) \tag{5.65}$$

in the limit $z \to 1$. Both (5.64) and (5.65) show that the behavior of f_n or $f(t)$ for large n or t implies specific singular behavior of the transforms. As an application of (5.65) let $f_n = \exp(-1/n)/\ln(n)$. The logarithmic term is a slowly varying function and the exponential approaches 1 in the limit $n \to \infty$. Hence (5.65) permits the conclusion that $f(z) \sim 1/[(1-z)\ln\{1/(1-z)\}]$ in the limit $z \to 1$.

As mentioned, Tauberian theorems allow one to go from a knowledge of singular behavior in the transform domain to asymptotic behavior in the independent variable. Here, some caution is required because the inversion requires the use of assumptions regarding the behavior of the function in real space additional to a knowledge of the nature of the singularity in transform space. The most useful form of a Tauberian theorem for power series that has been used in random walk applications is the following. Let $U(z)$ be defined by

$$U(y) = \sum_{n=0}^{\infty} a_n e^{-ny} \tag{5.66}$$

where $a_n > 0$. Let $U(y)$ have the specific singularity in the limit $y \to 0$,

$$U(y) \sim y^{-\gamma} L(y^{-1}), \tag{5.67}$$

where $L(z)$ is a slowly varying function. From this we can conclude that the a_m have the asymptotic property [5.26]

$$a_1 + a_2 + \cdots + a_n \sim \frac{n^{\gamma} L(n)}{\Gamma(1+\gamma)} \tag{5.68}$$

as $n \to \infty$. If it is also known that the a_n are monotonic for $n > N$ one can formally differentiate this formula to find

$$a_n \sim \frac{\gamma n^{\gamma-1} L(n) + n^{\gamma} L'(n)}{\Gamma(1+\gamma)}. \tag{5.69}$$

The tools provided by Abelian and Tauberian theorems allow us to exploit exact results expressible in terms of either generating functions or transforms for the CTRW and for lattice random walks.

5.3.4 Asymptotic Properties of the CTRW in an Unbounded Space

A number of phenomenological techniques for analyzing properties of transport in disordered media are based on the CTRW [5.13,24]. We therefore return to this type of random walk to see how the relevant asymptotic properties can be derived through the application of techniques discussed in the last subsection. Consider, for example, the expression for the Laplace transform of the average displacement of a one-dimensional CTRW given in (5.32), and suppose that $\psi(t)$ has the fractal time property $\psi(t) \sim T^{\alpha}/t^{\alpha+1}$, where $0 < \alpha < 1$ and $(t/T) \gg 1$. As we have already shown, when the average displacement in the course of a single step is finite this property of the probability density $\psi(t)$ in the time domain implies the property $\hat{\psi}(s) \sim 1 - (Ts)^{\alpha}$ as $s \to 0$ with the result that

$$\langle \hat{x}(s) \rangle \sim \frac{\langle x \rangle}{T^{\alpha} s^{\alpha+1}}, \quad s \to 0. \tag{5.70}$$

The continuum analog of the Tauberian theorem summarized in (5.69) can be invoked to predict the long-time behavior of $\langle x(t) \rangle$, which is

$$\langle x(t) \rangle \sim \frac{\langle x \rangle}{\Gamma(1-\alpha)\Gamma(1+\alpha)} \left(\frac{t}{T}\right)^{\alpha}. \tag{5.71}$$

This should be contrasted to the asymptotic behavior of the mean displacement $\langle x(t) \rangle \sim \langle x \rangle (t/\langle t \rangle)$ which is valid provided that $\langle t \rangle$, the average time between successive steps, is finite. In similar fashion we can find the asymptotic behavior of the variance of the displacement for $\psi(t)$ having the stable-law property. The results of the analysis depend on whether the average displacement in a single step equals zero or not. The two results are

$$
\begin{aligned}
\sigma^2(x(t)) &\sim \frac{\sigma^2}{\Gamma(1-\alpha)\Gamma(1+\alpha)} \left(\frac{t}{T}\right)^{\alpha}, \quad \langle x \rangle = 0, \\
\sigma^2(x(t)) &\sim \frac{\langle x \rangle^2}{\Gamma^2(1-\alpha)} \left[\frac{2}{\Gamma(1+2\alpha)} - \frac{1}{\Gamma^2(1+\alpha)} \right], \quad \langle x \rangle \neq 0,
\end{aligned}
\tag{5.72}
$$

as first derived by Shlesinger [5.28].

Essentially the same techniques can be exploited to find asymptotic properties of $p(\boldsymbol{r},t)$. We illustrate how such a calculation goes in the case of the CTRW in one dimension, starting from the integral representation in (5.30). Recall that $\hat{p}(0) = 1$ and $|\hat{p}(\omega)| < 1$ for $\omega \neq 0$. Since $\hat{\psi}(s)$ is also a characteristic function it can be shown that when $\mathrm{Re}\{s\} > 0$, $|\hat{\psi}(s)| < 1$. The asymptotic properties of $p(x,t)$ result from the singular behavior of the integrand of (5.30), which is defined by the condition $\hat{\psi}(s)\hat{p}(\omega) = 1$. By our earlier remark, this can only occur when $s = \omega = 0$. Hence the strategy in developing the asymptotic expansion will be to expand the denominator in (5.30) around the singular point to lowest order enabling us to evaluate the resulting approximation in closed form. Before examining any more exotic variety of CTRW we show that the central-limit approximation can be found by this technique provided that both the mean time between jumps, $\langle t \rangle$, and the second moment of the displacement in a single step are both finite. Without loss of generality we can restrict ourselves to the case of the symmetric one-dimensional walk. This allows us to utilize the approximations $\hat{\psi}(s) \sim 1 - s\langle t \rangle$ and $\hat{p}(\omega) \sim 1 - \sigma^2\omega^2/2$ in (5.30). If we retain only the lowest order terms in the denominator we find that

$$
\hat{p}(x,s) \sim \frac{\langle t \rangle}{2\pi} \int_{-\infty}^{\infty} \frac{e^{-ix\omega}}{\frac{\sigma^2\omega^2}{2} + s\langle t \rangle} \, d\omega.
\tag{5.73}
$$

The inverse Laplace transform of the integrand is readily found, which allows us to write

$$
p(x,t) \sim \frac{1}{2\pi} \int_{-\infty}^{\infty} \exp\left(-ix\omega - \frac{\sigma^2\omega^2 t}{2\langle t \rangle} \right) d\omega = \sqrt{\frac{\langle t \rangle}{2\pi\sigma^2 t}} \exp\left(-\frac{\langle t \rangle x^2}{2\sigma^2 t} \right).
\tag{5.74}
$$

This result is identical to that in (5.50), provided that one replaces the number of steps, n, appearing in that equation by $t/\langle t \rangle$, i.e., by setting $t = n\langle t \rangle$.

The same strategy can be used to find the form of $\hat{p}(x,s)$ in cases in which the first moment of $\psi(t)$ fails to exist. Because there are an infinite number of forms of $\psi(t)$ whose first moment is infinite we cannot expect to find results quite as general as that shown in (5.74). They rather depend on the asymptotic properties of $\psi(t)$. We will choose the so-called fractal time form for $\psi(t)$, which is defined by the property $\psi(t) \sim T^\alpha/t^{\alpha+1}$ for large t/T and $0 < \alpha < 1$, and which has been used in a number of investigations of transport in disordered media [5.29,30]. Without losing generality we can put $T = 1$ and work in terms of a dimensionless time $\tau = t/T$. We have shown that the Laplace transform can be expanded around $s = 0$ as $\hat{\psi}(s) \sim 1 - s^\alpha$. It will be convenient for the following calculation to replace this form by

$$\hat{\psi}(s) \sim (1 + s^\alpha)^{-1}. \tag{5.75}$$

For simplicity we consider only the case of a one-dimensional symmetric random walk for which $\sigma^2 < \infty$, although more general cases can also be analyzed [5.31]. To illustrate a number of points we consider a lattice random walk and calculate the asymptotic time dependence of the probability of being found at the origin, $p(0, \tau)$. To this end we go to the Laplace domain, again expanding (5.38) around the singularity at $s = \omega = 0$. In this way we find that

$$\hat{p}(0, s) \sim \frac{s^{\alpha-1}}{2\pi} \int_{-\pi}^{\pi} \frac{d\theta}{s^\alpha + \sigma^2\theta^2/2}. \tag{5.76}$$

One further simplification can be made in finding the final form of the result. Since the integral converges over the entire interval $(-\infty, \infty)$ but the principal determinant of asymptotic behavior is the analytic behavior near the origin, we can, in fact, extend the interval of integration to the entire line. The integral is then trivially evaluated and leads to the result

$$\hat{p}(0, s) \sim \frac{s^{\alpha-1}}{2\pi} \int_{-\infty}^{\infty} \frac{d\omega}{s^\alpha + \sigma^2\omega^2/2} = \frac{s^{\alpha/2-1}}{\sigma\sqrt{2}} \tag{5.77}$$

in the limit $s \to 0$. This implies that as $\tau \to \infty$

$$p(0, \tau) \sim \frac{1}{\sigma\sqrt{2}\Gamma(1 - \alpha/2)\tau^{\alpha/2}}. \tag{5.78}$$

Notice that the power of τ that appears is $\alpha/2$. When $\alpha > 1$ the fractal time form of $\psi(t)$ does have a finite first moment and one finds that $p(0, \tau) \sim \tau^{-1/2}$ so that there is a transition in the behavior of the exponent at $\alpha = 1$. When α decreases from 1, $p(0, \tau)$ will go to zero more slowly than when $\alpha = 1$. This is understandable on the consideration that the first jump, on the average, takes much longer to execute, so that the chance that a random walker remains at the origin without ever leaving before time τ is considerably increased.

Essentially the same technique can be used to find an approximation to $p(x, \tau)$ as a function of x in the neighborhood of the origin. For this purpose one leaves the exponential in (5.30) and evaluates the resulting integral for $\hat{p}(\omega, s)$, the final result being

$$\hat{p}(x, s) \sim \frac{s^{\alpha/2-1}}{\sigma\sqrt{2}} \exp\left(-\frac{|x|\sqrt{2}}{\sigma}s^{\alpha/2}\right). \tag{5.79}$$

While the indicated transform can be expressed as an integral of the function $f_\alpha(t) = \mathcal{L}^{-1}\{\exp(-s^{\alpha/2})\}$, for present purposes it is simpler to expand the exponential since we are interested in the small-$|s|$ regime. The transforms that appear after this step are readily evaluated, which then gives

$$p(x, \tau) \sim p(0, \tau) \left[1 - \frac{\Gamma(1 - \alpha/2)}{\sigma\Gamma(1 - \alpha)} \frac{|x|}{\tau^{\alpha/2}} + \cdots \right]$$

$$\sim p(0, \tau) \exp \left[-\frac{\Gamma(1 - \alpha/2)}{\sigma\Gamma(1 - \alpha)} \frac{|x|}{\tau^{\alpha/2}} \right]. \tag{5.80}$$

The derivative of this function is discontinuous at $x = 0$, in contrast to the behavior of the Gaussian that follows from the central-limit theorem. Notice that the case $\alpha = 1$ cannot be recovered from the expansion in this last equation because of the factor $\Gamma(1-\alpha)$ that occurs in the denominator. Rather, one must set $\alpha = 1$ in (5.79) and evaluate the inverse directly, which leads to the Gaussian density in (5.74). The first and the second moments scale as $\langle x(t) \rangle \sim t^{\alpha/2}$ and $\langle x^2(t) \rangle \sim t^\alpha$ respectively. The diffusion exponent d_w, defined by the relation $\langle x^2(t) \rangle \sim t^{2/d_w}$, is therefore given by $d_w = 2/\alpha$. For a detailed discussion of anomalous diffusion on fractals see Chap. 3 in [5.8,9a]. The extension of this analysis to symmetric models in a greater number of dimensions is described in [5.31], as is the approximation suitable for describing the behavior of $p(x, \tau)$ in the tail of the curve defined by $[x^2/(\sigma^2\tau^\alpha)] \gg 1$.

A slightly different strategy, based on the combined utilization of both Abelian and Tauberian theorems, is required to find results analogous to those of the last few paragraphs when the underlying random walk is biased. As an illustration of the formalism, let us show how it may be used to find the small-$|s|$ behavior of $\hat{p}(0, s)$ in one dimension. For this purpose we write

$$\hat{p}(0, s) \sim \frac{s^{\alpha-1}}{2\pi} \int_{-\infty}^{\infty} \frac{d\omega}{s^\alpha + 1 - \hat{p}(\omega)}$$

$$= \frac{s^{\alpha-1}}{2\pi} \int_0^\infty e^{-s^\alpha \xi} d\xi \int_{-\infty}^{\infty} e^{-\xi[1 - \hat{p}(\omega)]} d\omega. \tag{5.81}$$

The idea behind the calculation is that the integral with respect to ξ can be regarded as a Laplace transform. Hence the small-$|s|$ behavior is determined by the behavior of the ω-integral at large ξ. However, when ξ is large, the major contribution to the ω-integral comes from the neighborhood in which the exponent vanishes, i.e., from the neighborhood of $\hat{p}(\omega) = 1$ or $\omega = 0$. In that neighborhood we expand $\hat{p}(\omega)$ to the lowest non-vanishing term in ω, i.e., $\hat{p}(\omega) \sim 1 + i\langle x \rangle k$, and evaluate the resulting integral with respect to ω. Since this is a delta function we find the small-$|s|$ approximation $\hat{p}(0, s) \sim s^{\alpha-1}/(2\langle x \rangle)$, equivalent to the asymptotic τ result

$$p(0, \tau) \sim \frac{1}{2\langle x \rangle \Gamma(1 - \alpha)\tau^\alpha}. \tag{5.82}$$

This result goes to zero faster than the corresponding result for the symmetric random walk (which goes like $\tau^{-1/2}$) because the effect of the field is to diminish the probability that the random walker will return to the origin after it leaves it for the first time. For a more complete development of the theory, including a derivation of the asymptotic expression for $p(r, \tau)$ in d dimensions, the reader is again referred to [5.31].

5.3.5 Asymptotic Properties of Random Walks on a Lattice: Recurrent and Transient Behavior

Up to this point we have shown how to calculate properties of the probability density for the displacement of a random walk at time τ. There are, however, a number of properties of lattice random walks that are indirectly related to this function. In this subsection we present a sampling of asymptotic results for lattice walks, not only because these are significant in and of themselves but also because in the course of our derivation of them we will be able to illustrate mathematical techniques based on Abelian and Tauberian theorems discussed earlier.

The first property illustrates a very important transition in certain properties of random walks attributable to the dimensionality of the walk. The basis of this distinction was originally suggested in a paper by Polyá in 1921 [5.32], who posed the question of finding the probability that a random walker initially found at the origin will, at some point during the course of the walk, return to its starting point. A natural generalization is to find the probability that the random walk eventually reaches a specified site j. The answers to both of these questions are related and can be determined using essentially the same analytical tools. Let us first address the Polyá problem in its simplest form. Let the random walker initially be at the origin of a d-dimensional uniform lattice. The probability that the random walker returns to the origin for the first time at step n is $f_n(0)$ (with the convention that $f_0(0) = 0$) so that the probability of eventual return is

$$P_{\text{ret}} = \sum_{n=0}^{\infty} f_n(0). \tag{5.83}$$

But the sum can be regarded as the value of the generating function $f(0; z)$ defined in (5.34) evaluated at $z = 1$. Consequently one has from (5.35) that

$$P_{\text{ret}} = 1 - \frac{1}{p(0; 1)}, \tag{5.84}$$

where an integral representation for $p(0; z)$ can be found from (5.36). Equation (5.84) states that the probability of return to the origin is equal to 1 when $p(0; 1) = \infty$ and is less than 1 otherwise. One is therefore directed towards determining the convergence or divergence of the integrals in (5.36) with z set equal to 1.

We will examine the behavior of these integrals under the specific assumptions that

$$\sum_{\boldsymbol{j}} j_k p(\boldsymbol{j}) = 0;$$

$$\sum_{\boldsymbol{j}} j_k j_m p(\boldsymbol{j}) = \sigma^2 \delta_{k,m} < \infty, \qquad k, m = 1, 2, \cdots, d. \tag{5.85}$$

The first of these says that the average displacement in a single step is equal to zero and the second says that the variances associated with the transition probability along each of the axes are equal, and, furthermore, that there are no correlations between the components of displacement. Each of these assumptions can easily be relaxed but the resulting expressions are then somewhat more complicated in appearance.

If the integral representation of $p(\mathbf{0}; 1)$ is singular it can only be so because the denominator of the integrand in (5.36) vanishes. By our earlier assumptions this can only occur at $\boldsymbol{\theta} = \mathbf{0}$. Consequently, to determine the behavior of $p(\mathbf{0}; z)$ in the neighborhood of $z = 1$, we expand the denominator around the origin in $\boldsymbol{\theta}$-space, with the assurance that there can be no other source of singular behavior of the integral. Doing so leads to the approximation

$$p(\mathbf{0}; z) \sim \frac{1}{(2\pi)^d} \int_{-\pi}^{\pi} \cdots \int_{-\pi}^{\pi} \frac{d^d \boldsymbol{\theta}}{1 - z + \frac{\sigma^2}{2} \theta^2}. \tag{5.86}$$

Since we eventually want to set $z = 1$ we see that the convergence or divergence of this integral can only be due to the behavior of the integrand in the neighborhood of $\boldsymbol{\theta} = \mathbf{0}$. With this as justification we can extend the limits of integration to $(-\infty, \infty)$ and examine singular behavior only in the neighborhood of $\boldsymbol{\theta} = \mathbf{0}$.

Because the integrand is a function of the magnitude θ only, we can introduce d-dimensional spherical coordinates and integrate over all angles, which contributes only a numerical factor multiplying the integral. Hence we have

$$p(\mathbf{0}; z) \sim K \int_{-\infty}^{\infty} \frac{\theta^{d-1} d\theta}{1 - z + \frac{\sigma^2}{2} \theta^2}, \tag{5.87}$$

where K is a constant which has no effect on the convergence properties being sought. An examination of this integral shows than when $d = 1$ and 2 the integral will diverge when z is set equal to 1 (the behavior of the integrand at $\theta = \pm\infty$ being of no consequence), and when $d \geq 3$ the integral will converge. Hence in $d = 1$ and 2 the random walker returns to the origin with a probability equal to 1, while in higher dimensions the probability of return to the origin is less than 1, which is to say that in higher dimensions the random walker can escape without ever returning to the origin. In mathematical terminology one says that one and two-dimensional random walks are *recurrent* while those in higher dimensions are *transient*. It is easy to show that the probability of reaching a lattice site other than the origin is certain when the random walk is recurrent and occurs with a probability < 1 when the random walk is transient.

An interesting observation is that even when the random walk is recurrent, the average number of steps required for the first return to the origin is infinite. A derivation of this is similar to that for the probability of return to the origin.

Although the results stated so far apply to lattice random walks there are also continuum analogs. It would be incorrect to try to calculate the probability of return of a diffusion process to a point in dimensions greater than 1, since that probability is identically equal to 0. This is because a point has no dimensions. However, suppose that a hypersphere is centered at the origin, the surface of the sphere consisting entirely of traps. The trapping surface is defined by saying that a particle that reaches the surface of the sphere immediately disappears. In this system the probability that a diffusing particle is eventually trapped is equal to 1 in $d = 1$ and 2 dimensions and is strickly less than 1 in $d \geq 3$ dimensions which is the analog to the original Polyá property for lattice walks. A further point is that when the jump lengths of the random walk do not have finite second moments these results will be modified. In one dimension, for example, when $p(j)$ is a stable law having the form shown in (5.19) it can be shown that the random walk is transient.

5.3.6 The Expected Number of Distinct Sites Visited by an n-Step Random Walk

The generating function for the expected number of distinct sites visited by an n-step random walk, $D(z)$, is given in (5.43). In this section we make use of Abelian and Tauberian techniques to calculate the leading terms in the asymptotic form of $\langle D_n \rangle$ in different numbers of dimensions. This again depends on, and can be calculated from, a knowledge of the analytic behavior of $p(\mathbf{0}; z)$ in the neighborhood of $z = 1$. We do this for the case of one dimension. Since the singular behavior comes from the behavior of the integrand of (5.36) in the neighborhood of $\theta = 0$ we may extend the limits of integration to $(-\infty, \infty)$, thereby finding that

$$p(\mathbf{0}; z) \sim \frac{1}{2\pi} \int_{-\infty}^{\infty} \frac{d\theta}{1 - z + \frac{\sigma^2}{2}\theta^2} = \frac{1}{\sigma\sqrt{2(1 - z)}}. \tag{5.88}$$

On substituting this estimate into (5.43) we find that

$$D(z) \sim \frac{\sigma\sqrt{2}}{(1 - z)^{3/2}}. \tag{5.89}$$

This approximation to $D(z)$ implies according to the Tauberian theorem for power series discussed early, that for large n

$$\langle D_n \rangle \sim \sigma\sqrt{\frac{8n}{\pi}}. \tag{5.90}$$

In one dimension one can also calculate the probability distribution of D_n for large n, while in higher dimensions a calculation of the analogous results requires the use of much more complicated mathematical tools [5.33]. The asymptotic form for $\langle D_n \rangle$ is readily found for random walks in higher dimensions by applying techniques similar to those we have used in the case of one dimension.

Details of the computation are to be found in [5.13]. When properties of the random walk are completely isotropic and the variances of individual steps are finite, one finds the results

$$\langle D_n \rangle \sim \frac{\pi \sigma^2 n}{\ln(n)}, \quad d = 2, \qquad \langle D_n \rangle \sim \frac{n}{p(\mathbf{0}; 1)}, \quad d \geq 3, \qquad (5.91)$$

where in $d \geq 3$ dimensions it can be shown that $p(\mathbf{0}; 1)$ is finite. One can observe that in one and two dimensions the asymptotic form for $\langle D_n \rangle$ depends only on σ^2 and on no other property of the transition probabilities, while in higher dimensions it depends on the detailed form of these probabilities since these are used in the evaluation of $p(\mathbf{0}; 1)$. In Fig. 5.5 we compare exact numerical results of $\langle D_n \rangle$ for $d = 1$ and 2 with the asymptotic results [5.46]. Asymptotic forms for $\langle D_n \rangle$ similar to those just summarized for the case $\sigma^2 < \infty$ have been derived for classes of random walks with $\sigma^2 = \infty$ by Gillis and Weiss [5.34].

As mentioned, a generalization of the problem of finding $\langle D_n \rangle$ for large n has recently appeared, in which one is interested in the analogous parameter for N independent random walkers is sought [5.14]. Let $\langle D_n(N) \rangle$ denote this function. The analysis in this case is based on the generating function given in (5.45). We summarize the results obtained in one dimension since these are also representative of results obtained in higher dimensions. We will suppose that the transition probabilities are such that the random walker can visit no more than a finite number of sites, say M, on a single step. In the limit in which n is held fixed and $N \to \infty$ it is easy to convince oneself that $\langle D_n(N) \rangle \sim Mn$. That is to say, all possible accessible sites will be visited. In contrast, when N is fixed and n increases indefinitely one finds that

$$\langle D_n(N) \rangle \sim \sigma \sqrt{n \ln N}, \qquad (5.92)$$

so that the effect of having N, rather than a single, random walkers is contained in the factor $\sqrt{\ln N}$ while the dependence on n has the square root form found for the asymptotic behavior of $\langle D_n \rangle$ for a single random walker. It can be shown that the crossover time between these two behaviors is of the order of $\ln N$. Numerical results in support of (5.92) are shown in Fig. 5.6. Higher dimensional results are also characterized by a number of regimes that depend on n and N. Figure 5.0 shows some contours of the surface obtained from snapshots at successive times of the territory covered by 500 random walkers in two dimensions as a sequence of times. As might be expected, the surface of the disk roughens as the number of steps increases since the density (random walkers/area) decreases. The random-walk approach was generalized to include long-range correlations between different steps. This has been applied to generate one and two-dimensional surfaces in which the roughness can be varied. For more details on rough surfaces see Chaps. 1, 4 and [5.10,35].

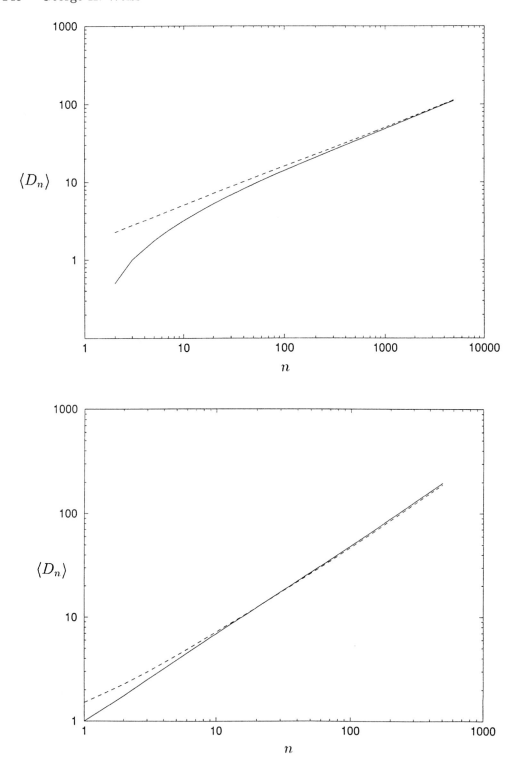

Fig. 5.5. Comparison of the asymptotic results (5.90) and (5.91) (*dashed line*) with the exact numerical results (*continuous line*). The upper figure is for $d = 1$, the lower figure is for $d = 2$

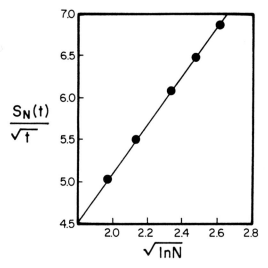

Fig. 5.6. The number of distinct sites visited by N random walkers in $d = 1$. The numerical results agree with the prediction of (5.92)

5.4 Random Walks in Disordered Media

5.4.1 Introductory Remarks

Since this book is primarily dedicated to properties of fractals and disordered media we briefly discuss some approaches to transport in disordered media that are based on the theory of random walks. Classical approaches to describing the transport of matter in homogeneous materials are generally based on or related to diffusion theory. However, when properties of the material are random such an approach is no longer adequate. A number of analogs of the diffusion equation have been proposed (cf., for example [5.36,37]), but these are purely phenomenological models. Hence very different approaches to those based on the diffusion equation are required, and these, by and large, have been phrased in terms of random walks. Many physical and mathematical examples of disordered media that have found application in the natural sciences are described in [5.10] as well as in the review articles of Haus and Kehr [5.7], Havlin and ben-Avraham [5.8], and Bouchard and Georges [5.9]. We describe here just a small number of models to give the reader a feel for the flavor of the anomalous properties to be expected in transport in disordered media, and to present a sampling of some of the techniques which have been used to analyze anomalous diffusion.

One most common manifestation of transport in a disordered medium is that the linear time (or step number, in discrete time) dependence of the mean-squared displacement of a particle in a homogeneous medium, $\langle r^2(t) \rangle$, i.e., the relation

$$\lim_{t \to \infty} \langle r^2(t) \rangle \propto t, \tag{5.93}$$

is replaced by a dependence which is commonly expressed as

$$\lim_{t\to\infty} \langle r^2(t) \rangle \propto t^{2/d_w}. \tag{5.94}$$

In most, but not all, models that have been studied, the fractal dimension d_w exceeds 2. The implication of this is that disorder tends, on average, to slow down the motion of a particle moving in such a medium. In one of the most widely studied models of this phenomenon originally proposed by Sinai [5.36], the mean-squared displacement exhibits the even more drastic asymptotic dependence

$$\langle r^2(n) \rangle \sim \ln^4 n \tag{5.95}$$

in discrete time. We will see that properties other than the mean-squared displacement are also generally changed when a particle moves in a disordered medium. Equations (5.94) and (5.95) indicate that the order of the mean-squared displacement is diminished by disorder. This can be understood in terms of the random walker being trapped in regions in space in which, roughly speaking, it tends to oscillate for a long period before resuming motion outside of the region. For example, the Sinai model is a one-dimensional model in which site i has attached to it a probability p_i which is a random variable. At site i, the random walker steps with probability p_i to site $i + 1$ and with probability $1 - p_i$ to site $i - 1$. In one version of this class of models $p_i = p$ with probability $1/2$ and $p_i \equiv q = 1 - p$ with probability $1/2$. Suppose that $p = 0.95$, which, in an almost deterministic manner, biases the motion of the random walker to move in the direction of increasing site number. Then a quasi-trapping region would be one in which the sequence of transition probabilities is $\cdots pppqqq \cdots$ over some wide range of contiguous sites (see Fig. 5.7).

The slow behavior exemplified by that in (5.95) simply means that the random walker inevitably finds itself in such a region. While most of the commonly studied models of disordered media lead to the slowing of net displacement of the random walker [5.8,9,9a] this is by no means inevitable, and there are other models in which $\langle r^2(t) \rangle \propto t^\beta$ where $\beta > 1$ [5.38–40].

What I mean to suggest in these preliminary remarks is that the possible variety of behavior that occurs in random walks in a disordered medium is much richer than that for transport in a strictly homogeneous medium. Furthermore there are no theorems that have the nearly universal scope of the central-limit theorem for strictly homogeneous media; hence the study of transport

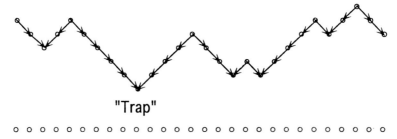

"Trap"

Fig. 5.7. A sketch of the Sinai model. At each site, the prefered direction of the random walker is designated by *downhill* arrow, pointing to the *right* or to the *left*

in a disordered medium often consists of the analysis of disparate models, the strategy being to discover phenomena which one might hope to see in physical systems. We consider a few of these in the following sections, with a view towards illustrating some of the effects caused by local fluctuations in properties of the medium as well as some of the analytical tools used in this field.

5.4.2 The Trapping Model

One of the earliest models of a disordered system is the so-called trapping model as originally suggested in the context of models for annealing in metals by Beeler and Delaney [5.41] and Beeler [5.42], and in relation to solid state models by Rosenstock [5.43]. Many related models have been analyzed in relation to models for chemical reaction rates [5.44]. The most common version of the trapping problem is defined in terms of a lattice random walk. Each site of the lattice is designated as a trap with probability c or as a conducting site with probability $1 - c$. A random walker is injected at a random site on the lattice. When the walker moves onto a trapping site it immediately vanishes. The problem most often posed about such a system relates to the survival probability of the walker, which is the probability that the walker survives for at least n steps before being trapped. A second parameter whose dependence on n has also been analyzed is the mean-squared displacement of random walks conditional on survival till step n.

We indicate some of the typical results for such a system in one dimension where exact results are available. An exact expression may be given for the survival probability, S_n, in terms of the number of distinct sites visited by an n-step random walk, D_n. This is valid in any number of dimensions, but of course an evaluation in detail of such an expression poses some extremely tricky mathematical problems. The survival probability may be written as

$$S_n = \langle (1 - c)^{D_n} \rangle \tag{5.96}$$

where the average is taken with respect to the placement of the traps and to all realizations of the random walk. Equation (5.96) is derived by noting that in order for the random walker to survive till step n, each of the distinct sites visited must have been a conducting rather than a trapping site. It has not proved possible to evaluate (5.96) exactly, but it can be used to generate heuristic approximations [5.45].

The one example of the trapping problem that can be solved exactly and that sheds considerable light on the nature of what might be expected more generally is the one-dimensional random walk. The solution of this case is simple because of the restricted geometry possible on a line. That is to say, a line with traps consists of segments of trap-free sites separated by the trapping sites. We briefly outline the solution of the continuum version of this problem since a number of asymptotic results that appear in higher dimensions occur in a rather transparent form in the one-dimensional case. Because only the large-n limit will be discussed we can replace the random walk in discrete time by a diffusion process.

The analog of having a constant probability of a site being a trap on a lattice is that the density of traps is constant on the continuous line. Hence for the diffusion process the probability that a trap occurs in the interval $(x, x + dx)$ is equal to cdx, where c is a constant. The probability density for the spacing between two adjacent traps to be equal to L is therefore

$$\varphi(L) = ce^{-cL}. \tag{5.97}$$

The survival probability in this model will be found by considering a single trap-free interval of length L with a diffusing particle randomly inserted along the interval at $t = 0$, and two traps at the ends of the interval at $x = 0$ and $x = L$. At the final step one performs an average over all L to find the survival probability for the entire system. The mathematical statement of the problem requires a solution of the diffusion equation

$$\frac{\partial p}{\partial t} = D\frac{\partial^2 p}{\partial x^2}, \tag{5.98}$$

where the solution is to be found subject to the boundary conditions

$$p(0, t) = p(L, t) = 0 \tag{5.99}$$

together with the initial condition $p(x, 0) = \delta(x - x_0)$, the initial position, x_0, being uniformly distributed over the interval $(0, L)$. An equivalent statement is that $p(x, 0) = 1/L$. The solution to this problem can be found by standard methods for the solution of the diffusion equation [5.46], and, with the fixed boundary condition, is given by

$$p(x, t) = \frac{2}{L}\sum_{n=0}^{\infty} \exp\left(-\frac{n^2\pi^2 D}{L^2}t\right) \sin\left(\frac{n\pi x_0}{L}\right) \sin\left(\frac{n\pi x}{L}\right). \tag{5.100}$$

The survival probability for a particle on an interval of length L, $S_L(t)$, is related to $p(x, t)$ by

$$S_L(t) = \frac{1}{L}\int_0^L dx_0 \int_0^L p(x, t)\,dx = \frac{8}{\pi^2}\sum_{n=0}^{\infty} \frac{\exp\left[-\frac{\pi^2(2n+1)^2 Dt}{L^2}\right]}{(2n+1)^2}. \tag{5.101}$$

The final step in our calculation requires an average of $S_L(t)$ with respect to L. One might naively think that this average is to be taken with respect to the function $\varphi(L)$ defined in (5.97), but this would be incorrect because it does not take into account the fact that the diffusing particle is more likely to be found initially in a large rather than a small interval. The relative probability of finding it in an interval of length L' rather than one of size L is equal to L'/L. It therefore follows that the probability density for the size of interval that surrounds the initial position of an arbitrary random walker is given by a new probability density $\rho(L)$, which must be proportional to L. This function is therefore found from $\varphi(L)$ by multiplying it by L and by a constant chosen to ensure that $\rho(L)$ is normalized to 1. Thus, we have the relation

$$\rho(L) = \frac{L\varphi(L)}{\int_0^\infty \ell\varphi(\ell)d\ell} = c^2 L\, e^{-cL}, \qquad (5.102)$$

with which we can calculate the survival probability of an arbitrary diffusing particle as the average value

$$S(t) = \int_0^L S_L(t)\rho(L)dL. \qquad (5.103)$$

The result of inserting the exact expression for $S_L(t)$ given in (5.101) into this integral will be a series of integrals each of which must be integrated numerically. However, if we restrict ourselves to the long-time regime defined by the condition $c^2 Dt \gg 1$, then we can derive an asymptotic expression for $S(t)$. Because of the restriction to large times we need only consider the $n = 0$ term in (5.101), finding in this way that

$$S(t) \sim \frac{8}{\pi^2}c^2 \int_0^\infty L\exp\left(-cL - \frac{\pi^2 Dt}{L^2}\right)dL$$

$$= 8\left(\frac{c^5 Dt}{\pi^4}\right)^{1/3} \int_0^\infty \ell\exp\left[-(\pi^2 c^2 Dt)^{1/3}\left(\ell + \frac{1}{\ell^2}\right)\right]d\ell. \quad (5.104)$$

Since the coefficient of the ℓ-dependent term in the exponent is large by assumption, the dominant contribution to the integral comes from the neighborhood of the value of ℓ that minimizes $\ell + 1/\ell^2$. An expansion of the integrand in the neighborhood of this value of ℓ to second order in ℓ and evaluation of the resulting integral (this is essentially Laplace's method for the asymptotic evaluation of integrals [5.15]) leads to the approximation

$$S(t) \sim A(c^2 Dt)^{1/2}\exp\left[-\frac{3}{2^{2/3}}(\pi^2 c^2 Dt)^{1/3}\right], \qquad (5.105)$$

where A is a constant of no significance to our point.

The important feature of this formula is the form of the exponential, which is proportional to $t^{1/3}$ rather than t as one might naively guess. The origin of this seemingly curious behavior is readily explained in more intuitive terms since the behavior of the survival probability at long times is mainly due to particles initially found in large trap-free intervals. One can take advantage of this general notion to derive, by means of a heuristic argument, the asymptotic form of S_n in d dimensions [5.47–49] (when $d > 1$ we must return to the lattice picture). This is found to have the form

$$\ln S_n \sim -K_d(c)n^{d/(d+2)}, \qquad (5.106)$$

in which $K_d(c)$ is a constant. A related result is that for the long-time limit of the mean-squared displacement of a particle surviving until time n, $\langle r^2(n)\rangle$. This is found to be proportional to $n^{2/(d+2)}$, which, since the exponent differs from 1, is indicative of anomalous diffusion, as can be verified in detail in the case of the one-dimensional trapping problem. While the argument suggesting the validity of (5.106) is definitely not a rigorous proof, this deficiency has been remedied by Donsker and Varadhan [5.50], whose analysis relies on rather complicated mathematical reasoning not readily summarizable in this article. For other variants of the trapping problem see also [5.51].

Even though the formula shown in (5.106) is well known in both the mathematical and physical literature, there remains the issue of how large t in continuous time, or the step number n in discrete time, has to be in order for it to yield a useful approximation. This problem can be approached by calculating corrections to (5.103), which has only been done in one dimension in which the correction term has been found to be $O(t^{-1/3})$ [5.52]. Another approach is through a numerical calculation using the so-called method of exact enumeration, which is basically a numerically exact solution of the evolution equation [5.8,53]. Such a calculation has been used to suggest that (5.106) is useful only at times for which S_n is of the order of approximately 10^{-13}, the exact probability depending on the concentration, in two and three dimensions. This very small trapping probability is in a range unlikely to be measureable in any physical realization of a trapping system.

Approximations in the opposite limit of small and therefore physically useful values of n can be derived by returning to the exact representation of the survival probability in (5.96) and expanding it in a series of powers of the parameter $\lambda = -\ln(1 - c)$. This cumulant expansion leads to a series of approximations in terms of the moments of D_n, of which the second is

$$S_n \sim \exp\left[-\lambda\langle D_n\rangle + \frac{\lambda^2}{2}\left(\langle D_n^2\rangle - \langle D_n\rangle^2\right)\right]. \qquad (5.107)$$

The lowest-order approximation in this hierarchy was originally suggested by Rosenstock [5.43], and the validity of higher-order approximation based on this expansion was explored by Zumofen and Blumen [5.45], who showed that (5.107) is quite accurate at very low concentrations and over a sufficiently wide range of survival probabilities to be physically interesting. The only problem with a more systematic exploitation of (5.107) is that while one has asymptotic approximations to $\langle D_n\rangle$ as we have already seen, the same cannot be said for the second moment, $\langle D_n^2\rangle$, although there are some results for the asymptotic behavior of this quantity in two and three dimensions in the mathematical literature [5.33]. The dependence on n of higher moments of D_n has not been explored so far.

Results obtained in the trapping problem have a number of properties found in a number of other versions of models for transport in disordered media. That is, one can explicitly demonstrate the property of anomalous diffusion in the sense of (5.94), and asymptotic behavior tends to set in at very large step numbers. Because of the many chemical and physical applications the trapping problem has been studied quite extensively. A recent summary of research on this subject is to be found in [5.54].

5.4.3 Some Models Based on the CTRW

One of the early approaches taken to develop phenomenological models for transport in disordered media was based on the CTRW [5.55,56]. The major assumption in these models is that the disorder of the underlying medium can be incorporated into a CTRW on a regular lattice, the disorder being incorporated into the pausing-time density $\psi(t)$. In particular, a number of investigations have been based on CTRW's having an anomalous probability density for interstep time intervals [5.27]. The most widely applied models are based on fractal-time forms for $\psi(t)$, i.e., $\psi(t) \sim T^{\alpha}/t^{\alpha+1}$ for $t \gg T$, where $0 < \alpha \leq 1$. With this $\psi(t)$ the average time between successive steps is infinite. It is clear that a model based on the CTRW can never be rigorously correct because it does properly account for the quenching property by which we mean that once a transition probability has been assigned to a site by some random mechanism, that site retains the particular probability for all time. In spite of this obvious deficiency, models based on the CTRW have been applied quite successfully in solid state physics [5.57] and are therefore worthy of mention. They tend to work better as the number of dimensions is increased. Applications to one-dimensional models must always be checked carefully.

The first such model was discussed by Scher and Lax who were interested in developing a phenomenological model for frequency-dependent conductivity in disordered solids [5.55]. Conductivity is implicitly related to transport properties since it is proportional to the frequency-dependent diffusion constant. We will describe a second model due to Scher and Montroll for transport in amorphous semiconductors [5.56], because it predicts an interesting transition in behavior which is experimentally observable. The experiment to which the theory applies is schematized in Fig. 5.8.

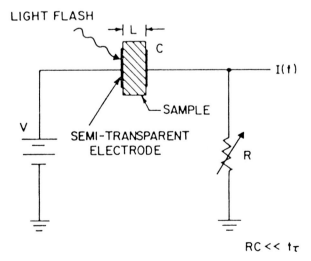

Fig. 5.8. A schematic diagram of the physical experiment in which transient photoconductivity is induced by a laser pulse impinging on one side of a semiconducting slab and is measured as a current on the second side of the slab

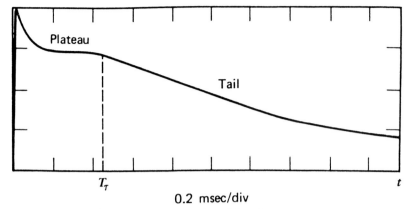

Fig. 5.9. A plot of a typical current trace produced by the experiment performed on $a - As_2$ by G. Pfister [5.58]

The left-hand face of the slab is illuminated by a pulse of light which creates electron–hole pairs in the sample. An imposed electrical field causes the holes to migrate to the opposite side of the sample where it produces an output current $I(t)$. A measured current in such an experiment is shown in Fig. 5.9, where one sees that an initial spike is followed by a plateau followed by a long-tailed decay. This suggests some sort of two-phase behavior, in contrast to the qualitatively different single-phase behavior predicted for a model in which the holes migrate by ordinary diffusion.

The two regimes in Fig. 5.9 can be described in terms of the behavior of a CTRW with a fractal time form for $\psi(t)$. A physical description of the two regimes attributes them, respectively, to holes in the neighborhood of the illumination point, $x = 0$, and those in the neighborhood of the second face at $x = L$. The current is proportional to the velocity, $v(t)$, of the holes, which can be related to the mean displacement $\langle x(t) \rangle$ by $v(t) = d\langle x \rangle / dt$. Since holes near $x = 0$ tend to be unaffected by the presence of the second face at L we can approximate $\langle x(t) \rangle$ from (5.32) as $\langle x(t) \rangle \sim \langle x \rangle (t/T)^\alpha$ so that the current from this set of holes is proportional to $(t/T)^{\alpha-1}$.

A calculation of the second component of the current starts by finding an expression for the probability density for the displacement of a particle at time t in the presence of an absorbing boundary at $x = L$. This will be denoted by $p(x, t)$. It is found in terms of the corresponding density in the absence of the boundary, $p_0(x, t)$, by subtracting from this quantity the contribution of all of those random walks which have reached $x = L$ before time t. Let $F(L, t)$ be the probability density for a random walker to reach L for the first time at t. The relation between $p(x, t)$ and $p_0(x, t)$ can be written in terms of this quantity as

$$p(x,t) = p_0(x,t) - \int_0^t F(L,\tau) p_0(L-x, t-\tau)\, d\tau. \qquad (5.108)$$

The function $F(L, t)$ can also be found in terms of $p(x, t)$ by noting that

$$p_0(L,t) = \int_0^t F(L,\tau) p_0(0, t-\tau)\, d\tau. \qquad (5.109)$$

Both Eqs (5.108) and (5.109) are simplified by taking their Laplace transforms. Denote the Laplace transform of $p(x,t)$ by $\hat{p}(x,s)$, with a similar notation for other transforms. Then (5.109) implies that

$$\hat{F}(L,s) = \hat{p}_0(L,s)/\hat{p}_0(0,s), \tag{5.110}$$

which, together with (5.108), yields

$$\hat{p}(x,s) = \hat{p}_0(x,s) - \frac{\hat{p}_0(L,s)}{\hat{p}_0(0,s)}\hat{p}_0(x-L,s). \tag{5.111}$$

This relation can be used to find an expression for the Laplace transform of the average displacement, by multiplying both sides by x and integrating with respect to x in the interval $(0,L)$. This leads to the relation

$$\langle\hat{x}(s)\rangle = \langle\hat{x}_0(s)\rangle[1-\hat{F}(L,s)] + L\hat{F}(L,s)\int_0^L \hat{p}_0(x,s)dx. \tag{5.112}$$

When the fractal form for $\psi(t)$ is assumed, the small-s expansion of $\langle\hat{x}_0(s)\rangle$ can be shown, by a Tauberian theorem, to imply that

$$\langle\hat{x}_0(s)\rangle \sim \frac{\langle x\rangle}{T^\alpha s^{\alpha+1}} \tag{5.113}$$

so that

$$\langle\hat{x}(s)\rangle \sim \frac{\langle x\rangle}{T^\alpha}\left(\frac{a}{s} + \frac{b}{s^{1-\alpha}}\right) + L\int_0^L \hat{p}_0(x,s)\,dx. \tag{5.114}$$

The transform of the average velocity is $\langle\hat{v}(s)\rangle = s\langle\hat{x}(s)\rangle$, from which we infer that the leading term in the singular behavior of $\langle\hat{v}(s)\rangle$ is proportional to $s^{-\alpha}$. This is equivalent to the statement that the asymptotic behavior of $\langle v(t)\rangle$ in the second, or boundary-determined, regime is

$$\langle v(t)\rangle \sim K_2 t^{-(1+\alpha)}, \tag{5.115}$$

where K_2 is a constant. This is to be contrasted to our earlier result in which $\langle v(t)\rangle$ at earlier times was found to be asymptotically proportional to $t^{-(1-\alpha)}$. The significant feature emerging from this comparison is that the sum of the two exponents is equal to -2. This prediction of the theory is borne out by the data shown in Fig. 5.10. Other varieties of models for disordered media based on the CTRW include those for transport in chromatographic columns [5.59] and comb models [5.60,61].

An abbreviated introduction to the somewhat related subject of the effective-medium approximation will be the subject of the following section. As our final comment we point out that the class of CTRW models, developed on purely phenomenological grounds, has proved to be effective and insightful in a number of analyses of problems involving disordered media.

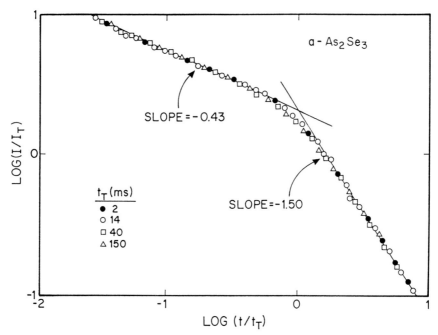

Fig. 5.10. A plot of experimental data that illustrates the prediction of CTRW theory that the sum of exponents should be equal to -2. The parameter t_T is a standard transit time, which is proportional to slab thickness. The data is taken from [5.58]

5.4.4 The Effective-Medium Approximation

The effective-medium approximation was first introduced for the approximate solution of problems in multiple-scattering problems by Lax [5.62] and is now a very commonly used tool for the study of random walks in disordered media. Rather than discuss the theory in great generality we present the basic idea in the context of a solvable example without carrying out all the steps in the analysis. Consider a one-dimensional lattice random walk which is allowed to move in one direction only. Let the jump rate for the transition $j \rightarrow j + 1$ be k_j so that the set of equations describing the evolution of the state probability is

$$\dot{p}_n(t) = k_{n-1}p_{n-1} - k_n p_n, \quad n \geq 0, \tag{5.116}$$

where $p_{-1}(t) \equiv 0$, which reflects the random walker's initial position at $n = 0$. The basic idea in this type of approximation is that the disordered medium is homogeneous and its properties are summarized by the single non-local rate function $K(t)$, which is initially unknown. This function characterizes the effective medium. Thus, the somewhat complicated set of equations in (5.116) is replaced by another approximate set for state probabilities $\{q_n(t)\}$ which take the form

$$\dot{q}_n(t) = \int_0^t K(t - \tau)[q_{n-1}(\tau) - q_n(\tau)] \, d\tau \tag{5.117}$$

where, as before, $q_{-1}(t) \equiv 0$. The Laplace transform for this set of equations is readily solved. The crucial step in this technique is that of determining the function $K(t)$.

In the lowest order of a hierarchy of approximations one replaces the rate function $K(t)$ at a single site by a random rate constant k and solves the resulting equations for a homogeneous medium in which is embedded a single random rate constant. Let the solution to this set of equations at site n be denoted by $q_{eff,n}(t;k)$. The final step of this analysis requires that the solution of the modified equations be self-consistent with (5.116), which translates into the requirement that

$$\langle q_{eff,n}(t;k)\rangle_k = q_n(t), \tag{5.118}$$

where the average is taken with respect to the random variable k. This step suffices to determine the function $K(t)$ and therefore, to lowest order, the nature of the effective medium. Higher-order approximations are equivalent to solving the equations for the random walk in a homogeneous medium with $2, 3, \ldots$, rate functions being replaced by random rate constants, afterwards enforcing the requirement of self-consistency in (5.118). A more complete discussion of some of the results obtained using this type of approximation is to be found in the review by Haus and Kehr [5.7]. Variations on this general theme in which a disordered medium is replaced by a homogeneous one, the link between the two being based on the requirement that some property of the solution is required to be self-consistent, can be found in articles by Lax and Odagaki [5.63], Movaghar *et al* [5.64], and many others.

A number of methods have been developed for analyzing properties of random walks in random media based on the exact evolution equations from which one derives systematic approximations to the exact solution based on such techniques as a scaling *ansatz* [5.65], renormalization group methodology [5.66], and perturbation theory [5.67,68]. Several of these are restricted in applicability because they depend strongly on the underlying model being one-dimensional.

The variety of techniques brought to bear on the class of problems involving disordered media is a rich one, suggesting both the difficulties inherent in the analysis of random walks in random media and the amount of work still remaining in exploring qualitative and quantitative properties of such systems. This article has hopefully provided a general introduction to random-walk theory as well as to the challenges posed by a class of problems involving transport in random media, which might be said to be understood in outline but not in detail.

References

5.1 A.E. Scheidegger: *The Physics of Flow through Porous Media* (University of Toronto Press, Toronto 1974)
5.2 K. Pearson: Nature **72**, 294 (1905)
5.3 A. Hald: *A History of Probability and Statistics and their Applications before 1750* (John Wiley, New York 1990)
5.4 F. Spitzer: *Principles of Random Walk*, 2nd ed. (Springer-Verlag, Berlin 1976)
5.5 G.F. Lawler: *Intersections of Random Walks* (Birkhäuser, Boston 1991)
5.6 G.H. Weiss, R.J. Rubin: Adv. Chem. Phys. **52**, 363 (1983)

5.7   J.W. Haus, K.W. Kehr: Phys. Rep. **150**, 263 (1987)

5.8   S. Havlin, D. ben-Avraham: Adv. Phys. **36**, 695 (1987)

5.9   J.P. Bouchaud, A. Georges: Phys. Rep. **195**, 127 (1990)

5.10   A. Bunde, S. Havlin, eds.: *Fractals and Disordered Systems* (Springer-Verlag, Heidelberg 1991)

5.11   B.V. Gnedenko, A.N. Kolmogorov: *Limit Distributions for Sums of Independent Random Variables*, (Addison-Wesley, Cambridge 1954)

5.12   A. Stuart, J.K. Ord: *Kendall's Advanced Theory of Statistics*, vol. 1 (Oxford University Press, Oxford 1987)

5.13   E.W. Montroll, G.H. Weiss: J. Math. Phys. **6**, 167 (1965)

5.14   H. Larralde, P. Trunfio, S. Havlin, H.E. Stanley, G.H. Weiss: Phys. Rev. A **45**, 7128 (1992); Nature **356**, 168 (1992)

5.15   N. Bleistein, R.A. Handelsman: *Asymptotic Expansions of Integrals* (Dover reprint, New York 1986)

5.16   H.E. Daniels: Ann. Math. Stat. **25**, 631 (1954)

5.17   H.E. Daniels: Biometrika **43**, 169 (1956)

5.18   C. Domb, E. Offenbacher: Am. J. Phys. **46**, 49 (1978)

5.19   M.F. Shlesinger: Ann. Rev. Phys. Chem. **39**, 269 (1988); M.F. Shlesinger, I. Klafter, in: *On Growth and Form: Fractal and non-Fractal patterns in Physics*, ed. by H E. Stanley and N. Ostrovsky (Nijhoff, Dordrecht 1985)

5.20   J.E. Kiefer, G.H. Weiss: AIP Conference Proceedings **109**, 11 (1984)

5.21   M. Dishon, J.T. Bendler, G.H. Weiss: J. Res. Nat. Inst. Stand. Techn. **95**, 433 (1990)

5.22   A. Khintchin: *Asymptotische Gesetze der Wahrscheinlichkeitsrechnung* (Chelsea, New York 1948)

5.23   W. Kuhn: Koll. Zeit, **68**, 2 (1934)

5.24   K. Solč, W. Stockmayer: J. Chem. Phys. **71**, 2756 (1971)

5.25   G.H. Weiss, H. Weissman: J. Stat. Phys. **52**, 287 (1988)

5.26   G.H. Hardy: *Divergent Series* (Clarendon, Oxford 1949)

5.27   M.F. Shlesinger: Ann. Rev. Phys. Chem. **39**, 269 (1988)

5.28   M.F. Shlesinger: J. Stat. Phys. **10**, 421 (1974)

5.29   H. Scher, M. Lax: Phys. Rev. B **7**, 4502 (1973)

5.30   H. Scher, E.W. Montroll: Phys. Rev. B **12**, 2455 (1975)

5.31   H. Weissman, G.H. Weiss, S. Havlin: J. Stat. Phys. **57**, 301 (1989)

5.32   G. Polyá: Math. Ann. **84**, 149 (1921)

5.33   N.C. Jain, W.E. Pruitt, in: *Proceedings of the Sixth Berkeley Symposium*, vol. III (University of California Press, Berkeley 1971), p. 31

5.34   J.E. Gillis, G.H. Weiss: J. Math. Phys. **11**, 1308 (1970)

5.35   J. Feder: *Fractals* (Plenum, New York 1988)

5.36   Ya.G. Sinai: Theory Prob. Appl. **27**, 256 (1982)

5.37   H. Kesten: Physica A **138**, 299 (1986);
       B. Derrida, Y. Pomeau: Phys. Rev. Lett., **48**, 627 (1982);
       A. Bunde, S. Havlin, H.E. Stanley, B.L. Trus, G.H. Weiss: Phys. Rev. B **34**, 8129 (1986);
       E. Koscielny-Bunde, A. Bunde, S. Havlin, H.E. Stanley: Phys. Rev. A **37**, 1821 (1988)

5.38   G. Mathéron, G. de Marsily: Water Resources Res. **16**, 901 (1980)

5.39   S. Redner: Physica D **38**, 287 (1989)

5.40   G. Zumofen, J.K. Klafter, A. Blumen: Phys. Rev. A **42**, 4601 (1990);
       M. Araujo, S. Havlin, H.E. Stanley: Phys. Rev. A **44**, 6913 (1991);
       S. Havlin, M. Araujo, H. Larralde, H.E. Stanley, P. Trunfio: Physica A **191**, 143 (1992);
       S. Havlin, M. Araujo, H. Larralde, A. Shehter, H.E. Stanley: Fractals **1**, 405 (1993)

5.41   R.J. Beeler, J.A. Delaney: Phys. Rev. **130**, 962 (1963)

5.42   R.J. Beeler: Phys. Rev. A. **134**, 1396 (1964)

5.43   H.B. Rosenstock: Phys. Rev. **187**, 1166 (1969); J. Math. Phys. **11**, 487 (1970)

5.44   D.F. Calef, J.M. Deutch: Ann. Rev. Phys. Chem. **34**, 493 (1983)

5.45   G. Zumofen, A. Blumen: Chem. Phys. Lett. **83**, 372 (1981); **88**, 63 (1982)

5.46   H.S. Carslaw, J.C. Jaeger: *Conduction of Heat in Solids* (Clarendon, Oxford 1959)

5.47   B.Ya. Balagurov, V.G. Vaks: Sov. Phys. JETP **38**, 968 (1974)

5.48   P. Grassberger, I. Procaccia: J. Chem. Phys. **77**, 6281 (1982)

5.49   R.F. Kayser, J.B. Hubbard: Phys. Rev. Lett. **51**, 79 (1983); J. Chem. Phys. **80**, 1127 (1984)

5.50 M.D. Donsker, S.R.S. Varadhan: Comm. Pure and Appl. Math. **32**, 721 (1979)
5.51 A. Bunde, L. Mosely, H.E. Stanley, D. ben-Avraham, S.Havlin: Phys. Rev. A **34**, 2575 (1986);
      M. Araujo, S. Havlin, H. Larralde, H.E. Stanley: Phys. Rev. Lett. **68**, 1791 (1992)
5.52 J.K. Anlauf: Phys. Rev. Lett. **52**, 1845 (1984)
5.53 S. Havlin, M. Dishon, J.E. Kiefer, G.H. Weiss: Phys. Rev. Lett. **53**, 407 (1984)
5.54 *Proceedings of the NIH Meeting on Diffusion Controlled Reactions*, J. Stat. Phys. **65**, 837 (1991)
5.55 H. Scher, M. Lax: Phys. Rev. B **7**, 4491 (1973)
5.56 H. Scher, E.W. Montroll: Phys. Rev. B **12**, 2455 (1975)
5.57 G. Pfister, H. Scher: Phys. Rev. B **15**, 2062 (1977)
5.58 G. Pfister: Phys. Rev. Lett. **33**, 1474 (1976)
5.59 G.H. Weiss, in: *Transport and Relaxation in Random Materials*, ed. by J. Klafter, R.J. Rubin, M.F. Shlesinger (World Scientific, Singapore 1986), p. 394
5.60 I. Goldhirsch, Y. Gefen: Phys. Rev. A **33**, 2583 (1986)
5.61 G.H. Weiss, S. Havlin: Physica A **134**, 474 (1986); Phil. Mag. B **56**, 941 (1987)
5.62 M. Lax: Rev. Mod. Phys. **23**, 287 (1951); Phys. Rev **85**, 621 (1962)
5.63 M. Lax, T. Odagaki: AIP Conf. Proc. **109**, ed. by M. F. Shlesinger, B. J. West (American Institute of Phys. New York 1984), p.133
5.64 B. Movaghar, M. Grünewald, B. Pohlmann, D. Würtz, W. Schirmacher: J. Stat. Phys. **30**, 315 (1983)
5.65 S. Alexander, J. Bernasconi, W.R. Schneider, R. Orbach: Rev. Mod. Phys. **53**, 175 (1981)
5.66 J. Machta: J. Stat. Phys. **30**, 305 (1983)
5.67 R. Zwanzig: J. Stat. Phys. **28**, 218 (1982)
5.68 K. Kundu, P. Phillips: Phys. Rev. A **135**, 857 (1987)

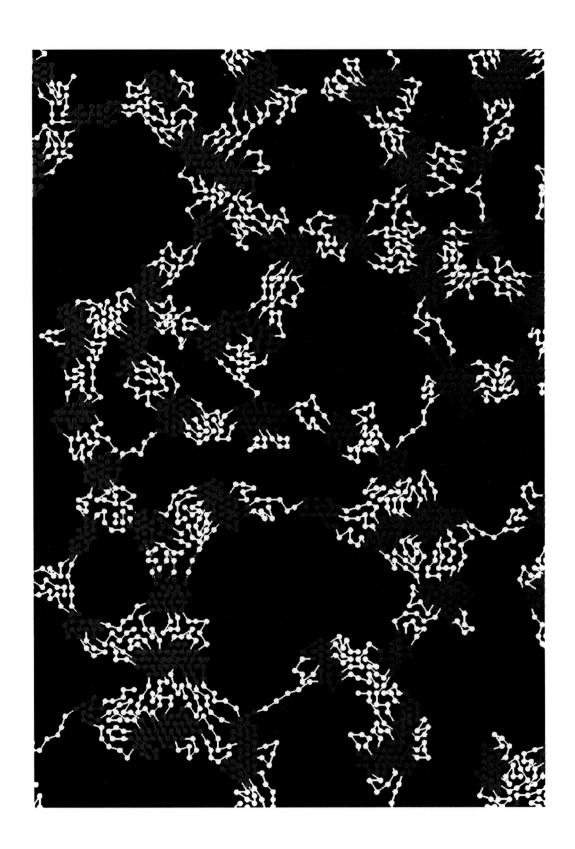

6 Polymers

Mohamed Daoud

6.1 Introduction

Even people living in a remote part of the jungle have probably heard about nylon, polystyrene, or plexiglas. If they have not, they may be forgiven; they are good (and wise) savages. Macromolecules or polymers are essential products in everyday life and tend to become more and more so, because they are essentially synthetic materials that substitute for natural materials that are becoming scarce and expensive. Plastics is the best-known name to designate them, but they are much more than that: rubbers, paints, adhesives, fibers, and glasses are made of polymers, and probably very soon batteries, will be too. They have in common the fact that they are huge macromolecules, made of millions of atoms. They are usually the repetition of a unit called a monomer. But they may also be made of two different monomers A and B in either regular or random succession. They are then called copolymers. They may be neutral or electrically charged. In the latter case they are named polyelectrolytes. All biological molecules are copolymers of this type. They may be linear or branched. In the latter case, they may eventually form a solid, a gel. Finally, they may be either amorphous or crystalline. In the former case, their structure is random, whereas in the latter, they form a regular structure which is never completely ordered: an amorphous fraction is most usually present between organized regions.

Because of the diversity of the possible structures, it is easy to realize that these polymers will exhibit a wide variety of properties. Interestingly, those that seemed to be the most difficult to understand a priori turn out to be the ones that we master most these days. The reason for this is that they are fractals [6.1] (for an introduction to fractals see Chap. 1), and it is their features we are going to discuss in this chapter. The discussion will be limited to those cases that we understand best, and to only some of their properties, because of space limitation. The first case will be the conformation of amorphous, linear, neutral

---

◄ **Fig. 6.0.** A model of a cross-linking polymer system quenched into the two-phase region. Note that this system has the potential to be a "filter" of a mesh size given by the characteristic spacing between the polymer strands. Courtesy of S. Glotzer

chains [6.2-4]. Actual chains are dissolved in a solvent, and this case will deal mostly with what is called a good solvent, which will be defined below. As we will see, the simplest way to model a polymer chain is to assume that it is a random walk (see Chap. 5). Actual chains, however, have a steric interaction that prevents monomers from lying on top of each other. This is the excluded-volume interaction and leads to a swelling of the chains, which we will discuss in detail in Sect. 6.2.

The second example we will consider concerns surfaces. Solutions such as those considered above always have surfaces: those of the container, and possibly the interface between air and the solution. Most usually, polymers interact very strongly with these surfaces. They may be either attracted or repelled. This corresponds respectively to adsorption and depletion of the polymer by the surface. We will consider mainly the case of the adsorption of linear chains by an attractive surface (Sect. 6.3). Self-similarity comes naturally into the problem both in the resulting concentration profile that builds up from the surface because of the adsorption and in the loop structure of the adsorbed polymers.

In Sect. 6.4, we will consider the structure of branched macromolecules, and the sol-gel transition [6.5]. A very important class of the latter which is characterized by a very broad fractal distribution of molecular weights for the polymers (called polydispersity) may be described by percolation [6.6] (see also Chaps. 2 and 3 in [6.7]). Two consequences will be discussed: in a dilute solution where every branched polymer is a fractal, the observed "fractal" dimension is an effective dimension that depends on the Euclidean dimension, and also on the distribution of masses; the dynamics is also characterized by a fractal distribution of relaxation times that leads to unusual anomalous relaxations.

6.2 Linear Chains and Excluded Volume

In this section, we will consider the conformation of random linear chains. As mentioned above, this means that we consider a solution of a polymer in a solvent. Typical examples are polystyrene in benzene or polydimethylsiloxane in toluene or cyclohexane. We assume in the following that the macromolecules are made of N statistical units which are randomly oriented with respect to each other. Because the actual monomers have to respect chemical bond angles, independent units can be regarded as being made of some monomers. It is possible to define such independent units, which will be used in all cases. This procedure was first presented by Kuhn, who defined the concept of local rigidity of a polymer [6.4]. Here, we consider the chains to be completely flexible, and we do not distinguish between actual monomers and statistical independent units. In general, however, a statistical unit is made of some monomers. We will first consider the single-polymer case, corresponding to very dilute solutions where one may neglect the interactions between macromolecules in a first approximation. Then we will study the effect of finite concentration. Static properties will be considered first.

6.2.1 The Random Walk

The simplest model to describe the structure of a linear chain made of N units each of length ℓ is the random walk (see Chap. 5). This is an ideal chain where no interactions are present between monomers. As the statistics are well known in this case, we merely recall some results. The distribution function $P(r, N)$, which is the probability that a chain made of N steps starts at the origin and ends at a distance r is a Gaussian. In three-dimensional space it is

$$P(r, N) = (3/2\pi N\ell^2)^{3/2} \exp\{-3r^2/2N\ell^2\}. \tag{6.1}$$

From the second moment we define the fractal dimension d_f of the walk by $\langle r^2 \rangle^{d_f/2} \sim N$ (see Chap. 1). It is easy to show that, for any dimension d, the second moment $R_0^2 \equiv \langle r^2 \rangle$ of $P(r, N)$ is

$$R_0^2 \sim N\ell^2. \tag{6.2}$$

Thus the fractal dimension is $d_f = 2$ for any d. It is important to stress that any definition of a characteristic length for the random walk leads to this result.

For a polymer chain, it is possible to use the mean square end-to-end distance, as we did above. It is also possible to define the average radius of gyration. One finds that all the lengths are proportional to each other, and that the fractal dimension is 2 (for a discussion, see Chap. 1 in [6.7]). This is the reason for using the proportionality sign $\sim$ in all the relations below. It is important that the precise way the length is defined will change the prefactor, but not the exponents. In this sense, we can say that there is only one characteristic length, and we will not be interested in the differences between the constants.

The fractal dimension may be observed experimentally by light or neutron scattering [6.8]. The scattered intensity $S(q)$ is the Fourier transform of the pair correlation function

$$S(q) = \sum_{i,j=1}^{N} \langle \exp[i\boldsymbol{q}(\boldsymbol{r}_i - \boldsymbol{r}_j)] \rangle, \tag{6.3}$$

where the brackets $\langle ... \rangle$ represent an average over all configurations, and $\boldsymbol{q}$ is the momentum transfer in the scattering experiment: for a neutron with wavelength λ elastically scattered with an angle $\boldsymbol{\theta}$ (see Fig. 6.1), we have

$$q = \frac{4\pi}{\lambda} \sin\frac{\theta}{2}. \tag{6.4}$$

Because (6.1) is valid for any pair of units in a random walk, (6.3) may be calculated exactly. This was done by Debye [6.4] some years ago. He found that

$$S(q) = \frac{2}{X^2}\left(e^{-X} - 1 + X\right), \tag{6.5}$$

with

$$X = q^2 R_0^2 / 3. \tag{6.6}$$

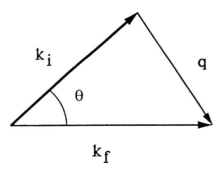

Fig. 6.1. The scattering vector as a function of the incident and final wavevectors and the scattering angle

Here, R_0 is the radius of gyration of the ideal chain. In the intermediate range, $\ell^{-1} \gg q \gg R_0^{-1}$, where the fractal nature of the walk is observed, relation (6.5) may be approximated by

$$S(q) \sim q^{-2}. \tag{6.7}$$

This relation provides a convenient way to measure the fractal dimension of a single polymer, whenever the intermediate range may be reached experimentally. Neutron scattering is an excellent technique for this: the available wavevector range is particularly well suited for polymers, since the typical unit size is around 10 Å, and the radius of gyration is several hundred Å. Linear chains actually behave as random walks in two cases: in a melt, when no solvent is present, and in the so-called theta solvent [6.9]. The latter is introduced in the next section, when we discuss the actual interactions between monomers.

6.2.2 The Self-Avoiding Walk

Random walks are ideal chains in the sense that there is no interaction between monomers. For actual polymers, there is an interaction between any two monomers that has the shape shown in Fig. 6.2.

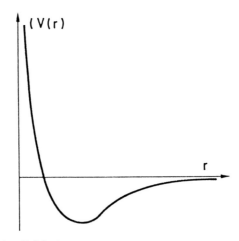

Fig. 6.2. The actual potential between monomers

The interaction consists of an attractive part for large distances, goes through a minimum at intermediate distances, and becomes a repulsive core at short distances. Because of this steric constraint, two monomers cannot be in the same location. At high temperatures, the repulsive core is dominant, and the local minimum may be neglected completely. This is the excluded-volume effect and corresponds to what is called a good solvent [6.10,11]. As the temperature is lowered, the presence of the minimum becomes more and more important. There exists a special temperature called the Flory theta temperature, where the excluded-volume effect and the attractive part compensate each other. Such solutions are said to be in a theta solvent [6.12-14]. For still lower temperatures, the attractive part of the potential becomes dominant, and although two monomers are not allowed to be in the same location, they tend to be in the vicinity of each other. As a consequence, the chain tends to collapse on itself [6.15-17]. Solvents in which this happens are known as poor solvents.

As mentioned above, at the theta temperature, because of the compensation between attractive and repulsive parts of the potential, the random-walk model gives an adequate description of a polymer chain in three-dimensional space. Actually, there are still logarithmic corrections, but they may be neglected. In two dimensions, a chain at theta temperature is still not equivalent to a random walk [6.18]. In what follows, we will be concerned with solutions in a good solvent. It was realized by Edwards [6.10] that the exact shape of the potential is not important and that it could be described by a parameter $v(T)$, where T is the temperature, called the excluded-volume parameter, defined as

$$v(T) = \int [1 - e^{-V(r)/kT}] dr. \tag{6.8}$$

This parameter is positive in a good solvent, vanishes at the theta temperature, and becomes negative in a poor solvent.

In the good solvent, steric interactions are dominant, as mentioned above, and the polymer is swollen compared to the ideal chain. This swelling, however, does not correspond to a change in the prefactor but to a change in the fractal dimension of the chain, which now becomes smaller than 2.

The fractal dimension was calculated by various renormalization-group techniques and by computer simulations [6.19,20]. Here, we describe the Flory approximation, which, although wrong, gives the fractal dimension to very high accuracy for all dimensions. In this approximation one assumes that the free energy can be written as

$$F = \frac{R^2}{R_0^2} + v \frac{N^2}{R^d}. \tag{6.9}$$

The first term is the elastic energy, in which one considers the chain to be a spring with spring constant $1/R_0^2$, where R_0 is the ideal radius from (6.2). The radius R is the actual radius of the chain and is to be determined. The second term is the interaction energy which can be estimated as follows. In a unit volume, the concentration of monomers is N/R^d, the number of pair interactions scales as $(N/R^d)^2$, and the interaction energy is therefore $v(N/R^d)^2$. Thus, the total interaction energy in the volume R^d scales as $R^d v(N/R^d)^2$, which is the second term in (6.9). Minimizing F with respect to R gives the fractal dimension d_f of a linear chain in the Flory approximation,

$$N \sim R^{d_f}, \tag{6.10a}$$

$$d_f = \frac{d+2}{3}. \tag{6.10b}$$

Note that we recover the ideal chain dimension for $d = 4$. This is the upper critical dimension above which the excluded-volume interaction becomes irrelevant, and the chain is ideal. For higher dimensions, the interaction with itself is negligible, because space is sufficiently large that the polymer almost does not cross itself. Therefore, for $d \geq d_c$ chains with or without interactions are equivalent.

Equation (6.10) was checked directly, using polymers with different masses. It was also tested using scattering experiments, by measuring the Fourier transform $S(q)$ of the pair correlation function. As above (6.7), one can show that the scattered intensity is related to the wavevector q by the fractal dimension. In $d = 3$ one finds, using (6.10), that

$$S(q) \sim q^{-\frac{5}{3}} \quad (\ell^{-1} \gg q \gg R_0^{-1}). \tag{6.11}$$

Relations (6.10) and (6.11) were tested experimentally by small-angle neutron scattering. Figure 6.3 shows the scattered intensity in the intermediate range for a dilute solution of polystyrene in CS_2. In the same plot are also shown the scattered intensities for a dilute solution of polystyrene (PS) in deuterated cyclohexane, which is a theta solvent for PS, and for a deuterated chain of PS in a matrix of hydrogenated PS. The latter two curves show that the fractal dimension of a chain is 2 for the theta solvent, as discussed above.

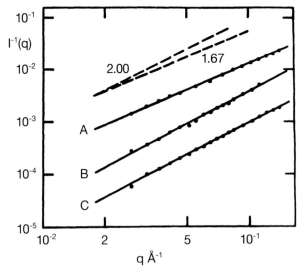

Fig. 6.3. The scattered intensities in a small angle neutron experiment by a chain in a good solvent (A), a theta solvent (B), and a melt (C)

6.2.3 Dilute Solutions

So far, we have considered only a single polymer chain. Actual solutions contain several of them. We expect the above results to hold as long as the various polymers are far from each other. This is the case for dilute solutions, where we expect the concentration effects to be only perturbations to the various laws that we have found.

Let C be the monomer concentration. It is common to define the overlap concentration C^*, where the distance between centers of masses of the chains is of the order of the radius of the macromolecules. If we assume that the polymers are randomly distributed, the average distance δ between their centers of masses is

$$\delta \sim (C/N)^{-1/d}. \tag{6.12}$$

Equating (6.12) to the radius of gyration, and using (6.10), we get

$$C^* \sim N^{1-d/d_f} \approx N^{-4/5} \qquad (d=3). \tag{6.13}$$

Relation (6.13) exhibits the fractal character of the chains; because they are fractals, their volume grows faster than their mass. Therefore, the overlap concentration decreases as the polymers become larger.

As mentioned above, we expect two concentration regimes, with C/C^* smaller or larger than unity. Therefore, we do not expect N and C to act as independent variables for all the properties but to appear only through the ratio C/C^*. This scaling behavior occurs in many properties, but we will consider here only the radius of gyration R and the osmotic pressure π. In both cases, one may write a scaling relation deduced from the definition of the fractal dimension (6.10) (for a general discussion of scaling behavior see Sect. 2.5 in [6.7]):

$$R(N,C) \sim N^{1/d_f} f(C/C^*) \tag{6.14a}$$

and

$$\pi(N,C) = \frac{C}{N} g(C/C^*), \tag{6.14b}$$

where the prefactor $C_p \equiv C/N$ in (6.14b) is merely the pressure of an ideal gas that is obtained for very low concentrations when the chains are very far from each other.

The unknown functions $f(x)$ and $g(x)$ may be expanded for small x in the dilute regime but have singular behavior for large x in the semi-dilute regime (see Sect. 6.4.2). Therefore, in the dilute-concentration regime, one expects corrections for both the radius and the osmotic pressure. In the latter case, we may write

$$\pi(N,C) = \frac{C}{N}\{1 + \alpha C/C^* + \beta(C/C^*)^2 + ...\}, \tag{6.15a}$$

where $\alpha, \beta, ...$ are constants. This may be identified with a virial expansion,

$$\pi(N,C) = \frac{C}{N} + A_2 C^2 + ... \tag{6.15b}$$

Comparing (6.15a) and (6.15b) and using (6.13) and (6.10) leads to the following expression for the second virial coefficient:

$$A_2 \sim (NC^*)^{-1} \sim N^{3/d_f}. \tag{6.16}$$

A similar expansion can be obtained for the scattering intensity $S(q, C)$. For very low q, in the Guinier regime $qR \ll 1$, this expansion is the basis for the so-called Zimm plots that are commonly used to determine the radius of a chain and the second virial coefficient of a solution.

6.2.4 Semi-Dilute Solutions

When the concentration C is increased above the overlap concentration C^*, one reaches a different regime where the macromolecules interpenetrate each other, and we expect the concentration effects to become dramatic. In dilute solutions, the concentration effects are represented by corrections to the power laws. Because the chains are fractals, the volume they occupy grows much faster than their mass. As indicated by (6.13), the larger the macromolecule, the smaller C^* is. For a typical polymer of 10^5 units, C^* is of the order of $10^{-2} g/cm^3$.

For infinite chains, the overlap concentration vanishes, and one is left with only the semi-dilute range. In this range, because the chains are flexible, they overlap each other, and we expect the simple laws we discussed above to break down. Still, the scaling laws (6.14) are valid, but one has to look for other limits. The basic idea for understanding the behavior of a polymer in this regime was given in the limit of a melt by Flory [6.4] and was later generalized to semi-dilute solutions by Edwards [6.21].

In a melt, the average interaction should cancel, since each monomer is surrounded by other monomers, and therefore the polymer should behave as an ideal chain.

We will see below that although this argument is valid for linear chains, it turns out to be wrong for branched polymers. The reason for this is related to the interpenetration of the various chains. For linear chains, this interpenetration effect was studied by Edwards [6.10], who introduced the concept of a screening length ξ. The idea here is that if we consider two monomers on a given chain, their total interaction is the sum of the direct excluded-volume interaction and all the contributions coming from indirect interactions between them via other monomers belonging to other chains. This is equivalent to the Debye-Hückel screening in an electrolyte solution.

a) **Distances.** In the following, we will assume the existence of a screening length and evaluate it using the above scaling laws. First note that this screening length coincides with the radius of gyration at C^*, since in dilute solutions the polymers are characterized by a single length. In the semi-dilute regime, we expect the solution to be characterized both by the radius of the chains and by the screening length ξ. Therefore, the scaling relation (6.14a) should lead to both lengths for $C \gg C^*$. To obtain them we assume that the scaling functions

$f_R(x)$ and $f_\xi(x)$ behave as power laws. The exponents are determined by the following conditions on R and ξ. As discussed above, the screening length is only due to concentration effects and thus should be independent of the mass N of the polymers in the semi-dilute regime. Using this condition and (6.14a), one finds [6.22] that

$$\xi \sim C^{1/(d_f - d)} \approx C^{-3/4}, \quad d = 3. \tag{6.17}$$

Similarly, the radius of gyration is obtained by assuming that the excluded-volume interaction between monomers is screened. Therefore, for distances larger than ξ, one should recover the ideal chain behavior. The condition for R is that its mass dependence is the same as for the ideal chain. This yields

$$R \sim N^{1/2}C^{(2-d_f)/2(d_f - d)} \approx N^{1/2}C^{-1/8}, \quad d = 3, \tag{6.18}$$

where we have used Flory's result (6.10) in the last equality.

Both relations (6.17) and (6.18) were tested by small-angle neutron-scattering experiments with polystyrene dissolved in good solvents. The radius was measured by labelling several chains in a semi-dilute solution in order to create a contrast between those and the rest of the matrix. The labeling is performed by replacing hydrogen by deuterium atoms. This is deuteration, which does not change the chemical nature of the labeled polymers and avoids, to some extent, the phase separation that occurs commonly between different polymers. From this point of view, neutron scattering is an invaluable tool because it allows the direct observation of a chain in the presence of others without perturbing the system. The screening length is measured without any labeling since it is a property of the solution rather than of the individual chains. The experimental data are shown in Figs. 6.4 and 6.5 and are in very good agreement with the above considerations.

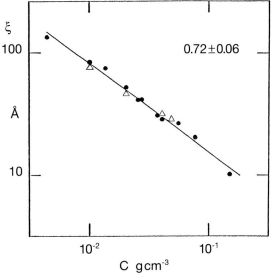

Fig. 6.4. Variation of the screening length with concentration in a semi-dilute solution in a good solvent

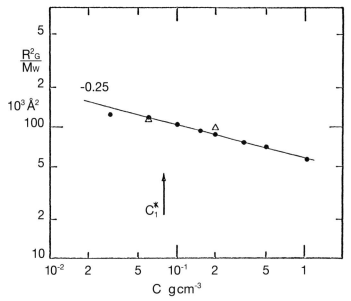

Fig. 6.5. Variation with concentration of the squared radius of a labeled chain in a semi-dilute solution in a good solvent

b) Osmotic pressure. The last quantity we will consider is the osmotic pressure. We may use the same arguments as above to determine its dependence on C, starting with (6.14b). In the semi-dilute regime, we do not expect the expansion (6.15) to be valid, since the variable $x = C/C^*$ is larger than unity. Instead, we assume that $g(x)$ behaves as a power law. Its exponent is determined by the following condition. In this concentration range, we expect the osmotic pressure to be given by the density of contacts between polymers. This is again a collective property of the solution that should depend only on concentration and not on the mass of the individual chains. Using this condition, we find that

$$\pi \sim C^{-d/(d_f-d)} \approx C^{9/4}, \qquad (6.19a)$$

a relation that was found first by des Cloizeaux [6.23]. A more direct argument is as follows. The screening length may be considered the average distance between successive contacts of a chain. Therefore, the osmotic pressure, which is proportional to the density of contacts, scales as

$$\pi \sim \xi^{-d}. \qquad (6.19b)$$

Using (6.19b) and (6.17), one recovers (6.19a). Equation (6.19a) was tested experimentally by Noda *et al.* [6.24] and found to be in very good agreement.

Many points are remarkable in (6.17-19). The first is that these results differ greatly from what one would expect in a mean-field approach. The latter would give the classical $C^{-1/2}$ variation for the screening length, as in the Debye-Hückel theory for instance. The second, and most remarkable, result is that the fractal dimension controls the thermodynamic properties of the solution. This is extremely interesting because the fractal dimension was introduced to describe the properties of a single chain, where only small concentrations and distances in the order of several hundreds of Ångströms were considered. We are now discussing thermodynamic, macroscopic properties of a solution that is semi-dilute and where the polymers strongly interact. Thus what was introduced to describe a local property of a single chain controls a solution that may be quite concentrated: even a 20% solution may be in this concentration range!

6.2.5 Dynamics

So far, we have considered the average conformation of polymers and have shown that fractal geometry is very important for the description of dilute and semi-dilute solutions. The dynamics [6.2,25] of the chains in the dilute regime is classic in the sense that hydrodynamic interactions are present. Thus the motion of a monomer at position r is dependent on the motion of the solvent in the surroundings. The dynamics is more interesting in the concentrated regime when hydrodynamic interactions are screened. This occurs for distances comparable to the screening length ξ.

In order to avoid unnecessary difficulties, we will consider only the case of polymer melts, in the absence of solvent. For the intermediate cases, one has to generalize the considerations that we had in semi-dilute solutions and distinguish between various behaviors depending on the distance scales. In a melt, two regimes may be found, depending on the mass of the chains. For short chains, the viscosity is proportional to the mass of the polymers. For large macromolecules, it becomes proportional to $N^{3.4}$. Simple models describe these two regimes, namely the Rouse [6.2,4,25] model for low masses, and the reptation model for large ones [6.2,25-29]. Both models lead to interesting behavior at intermediate times smaller than the characteristic time required by a polymer to diffuse on a length of the order of its radius of gyration. In both cases, one gets anomalously slow diffusion owing to the polymeric nature of the chains.

a) The Rouse model. For short chains, because the hydrodynamic interactions are screened, the monomers can be regarded as independent. This implies that the friction coefficient of a chain is N times that of a monomer. Therefore the diffusion coefficient D_R is

$$D_R \sim N^{-1}. \tag{6.20}$$

The characteristic time T_R for a chain to diffuse over a distance of the order of its radius R is

$$T_R \sim \frac{R^2}{D_R} \sim N^2, \tag{6.21}$$

where we have used the fact that a chain has an ideal behavior in a melt. For times larger than T_R, we expect normal diffusion to occur. The average distance the chain will have traveled during a time t is

$$\langle r^2(t) \rangle \sim D_R t, \quad t \gg T_R. \tag{6.22a}$$

For a time t smaller than T_R, only a portion of the length $N(t)$ of the chain will have diffused with a diffusion coefficient $N^{-1}(t)$. The time dependence of $N(t)$ can be found by generalizing relation (6.21). This implies that

$$N(t) \sim t^{1/2}, \tag{6.23}$$

which yields

$$\langle r^2(t) \rangle \sim t^{1/2} \quad t \ll T_R. \tag{6.22b}$$

b) Reptation. For larger masses, a chain in a melt moves along itself. This motion, called reptation by de Gennes, is due to the fact that the other chains are exerting a topological constraint because a probe cannot cross the other macromolecules. This constraint was also modeled by Doi and Edwards [6.25] who introduced a tube to represent the average effect of the matrix, inside which the diffusion of the polymer takes place. Therefore, the motion of sufficiently long chains is a one-dimensional diffusive motion along a tube that is roughly centered around their initial position.

The tube model is also interesting because it gives an idea about the relaxation of a polymer: during its motion, the chain ends escape from the initial tube. As time goes on, the initial tube decreases in time where the central part of the chain is still located, as shown in Fig. 6.6. The outer parts have reptated outside and lost the memory of the initial location of the chain.

There is a characteristic time, T_{rep}, at which the chain has diffused completely out of its initial tube. This corresponds to a complete renewal of the configuration. Therefore, any constraint that was initially applied to the polymer is relaxed after such a time. It is possible to evaluate this relaxation time in a simple way. We assume that the radius of the tube is $r_0 \sim N_e^{1/2} a$, where N_e is the contour length along the chain characteristic of the topological constraints mentioned above. The contour length N_e is also called the *entangle-*

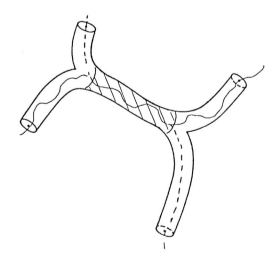

Fig. 6.6. The relaxation of a reptating chain is controlled by the remaining part *(dashed region)* of the initial tube as time goes on. *Full line:* the original configuration of the chain at time $t = 0$. *Dashed line:* the configuration at time t

ment length. The motion of the probe can be divided into two successive steps: (a) a local motion, and (b) a reptation motion.

The local motion is within the tube, that is, for distances shorter than r_0. This motion is the Rouse motion as discussed above and corresponds to times smaller than

$$T_R \sim N_e^2 t_1, \tag{6.24}$$

where t_1 is the typical diffusion time for a monomer. For larger distances or times, the motion is reptation. We are led to consider the reptation of a chain that is made of N/N_e units. Every unit has a Rouse dynamics. This leads us to define a diffusion coefficient D_t for the motion of the chain inside its tube. As this is a one-dimensional motion, we have

$$D_t \sim N^{-1} D_1, \tag{6.25}$$

where D_1 is the diffusion coefficient for a monomer and we have used the fact that there is a local Rouse motion inside the units. The longest time corresponds to the one-dimensional diffusion out of the original tube, that is, along a distance $L = (N/N_e) N_e^{1/2}$,

$$T_R \sim \frac{L^2}{D_t} \sim \frac{N^3}{N_e} t_1. \tag{6.26}$$

Finally, the diffusion coefficient D_c of the center of mass [6.30] in real space is evaluated in a direct way:

$$D_c \sim \frac{R^2}{T_R} \sim \frac{N_e}{N^2} D_1. \tag{6.27}$$

Note that when N is of the order of N_e, both results cross over to the Rouse equations (6.21) and (6.20) respectively. It is interesting to consider the diffusion of a monomer as a function of time. For times t larger than T_R, it simply follows the diffusion of the whole chain, and we expect the conventional result

$$\langle r^2(t) \rangle \sim D_c t. \tag{6.28a}$$

c) Local motions. For times smaller than the reptation time, but larger than a time T_d to be discussed below, (6.29) we expect the chain to have a motion that is a local reptation inside a part of the tube with length $L(t)$. One may consider that the chain diffuses in the tube with a diffusion coefficient D_t. The contour length $s(t)$ that it has moved along is

$$\langle s^2(t) \rangle \sim D_t t \sim \frac{D_1}{N} t;$$

the distance it has diffused in real space is therefore

$$\langle r^2(t) \rangle \sim s(t) N_e^{1/2} a \sim \left(\frac{N_e}{N} \right)^{1/2} t^{1/2} \quad (T_{rmrep} \gg t \gg T_d) \tag{6.28b}$$

The motion of the polymer is made by the diffusion of defects along it. There is an equilibration time T_d needed by these defects to equilibrate along the chain. This corresponds to the time that is necessary for a defect to diffuse all along the chain. We find that

$$T_d \sim N^2 t_1. \tag{6.29}$$

For times smaller than the equilibration time, the defects are not equilibrated along the chain. But one may find a portion of the macromolecule with length $n(t)$ along which local equilibrium is reached. Generalizing (6.29) for times shorter than T_d, we find that

$$n(t) \sim (t/t_1)^{1/2}. \tag{6.30}$$

We are then led to consider tube portions made of $n(t)$ units instead of the entire tube as before. The mobility of any portion is $\mu(n) = n\mu_1$. The diffusion coefficient inside the portion is $D_n \sim 1/\mu_n = D_1/n$. Generalizing the approach that we used above, we find for the diffusion of a monomer that

$$s^2(t) \sim D_1 \frac{t}{n} \sim a^2 \left(\frac{t}{t_1}\right)^{1/2}$$

and

$$\langle r^2(t) \rangle \sim N_e^{1/2} t^{1/4} \quad (T_d \gg t \gg T_R). \tag{6.28c}$$

Finally, for times shorter than T_R, one gets back to a Rouse motion:

$$\langle r^2(t) \rangle \sim t^{1/2} \quad (t \ll T_R). \tag{6.28d}$$

The various regimes (6.28) were tested by computer simulations [6.31-33] and by neutron spin echo [6.34,35].

6.3 Adsorption

The last section dealt with solutions. The presence of surfaces was neglected completely. Polymers, however, interact very strongly with surfaces, which may be either the walls of the container or the solution-air interface. The reason for such a strong interaction lies again in the fact that a macromolecule is made of a large number of units. We will see that because of this, a large number of monomers stay on the surface. Therefore, even when the interaction between the surface and a monomer is small, the total energy gained by the polymer may be large, so that it remains adsorbed. This has many practical applications, ranging from surface protection to the stabilization of colloids or water purification [6.36]. In what follows, we will see that in practical cases, the concentration profile due to the higher surface concentration is self-similar. To this profile one may associate the presence of a broad distribution of loops in the polymer structure, which is self-similar. We will first discuss the case of a single-chain adsorption where the conformation of the chain is a pancake.

Then we will consider the case of the so-called plateau, when the surface is saturated with polymers. This is the most interesting case, since the polymers have a fractal distribution of loop sizes.

6.3.1 The Single Chain

a) Monomers on the surface. Let us consider a polymer solution with a flat interface. Let δ be the energy gain per monomer on the surface. Because of the polymeric nature of the macromolecule, whenever a monomer is on the surface, there is an increased probability for a monomer of the same chain also to be on the surface, even when the chain is not adsorbed. Because of this, the number N_s of monomers that are located on the surface when the polymer is still isotropic is

$$N_s \sim N^\phi. \tag{6.31}$$

This may be understood by considering a Gaussian chain with one point on a plane. Then, it may be shown that such a chain cuts the plane $N^{1/2}$ times. For an actual polymer, where excluded-volume interactions are present, the number of monomers is also a power of N that depends on the space dimension and the nature of the solvent and the surface. In the following, we will consider only a good solvent, with a two-dimensional surface. Even with these characteristics given, one still has to define the surface more carefully. Two possibilities [6.37,38] are present, as shown in Fig. 6.7.

In the first one, the surface is impenetrable, and the polymer may be on one side only. In the second one, the surface may be crossed by the chain. In the latter case, the surface is called a surface of defects, whereas in the first one, it is called an impenetrable surface. The exponents ϕ are different for both cases. It was shown that for surfaces of defects [6.39], the cross-over exponent, denoted by ϕ_d, is related to the fractal dimension d_f of the chain,

$$\phi_d \sim 1 - d_f^{-1}. \tag{6.32}$$

Such a simple relation does not exist for impenetrable surfaces. However, com-

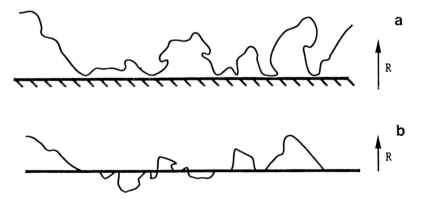

Fig. 6.7. (a) Adsorption of a linear chain on an impenetrable surface and (b) on a surface of defects

puter simulations show that for a single chain and a three-dimensional solution, ϕ is very close to d_f^{-1}. This is the only case that will be considered below, and we will assume [6.40] that ϕ has the Flory value 3/5. For simplicity, we will consider a chain with one end fixed on the attractive surface. The energy gain is therefore,

$$F_a \sim \delta N^{3/5}. \tag{6.33}$$

b) Conformation of the chain. The entropy loss corresponding to the confinement of the center of mass close to the surface when the chain is adsorbed is of the order of 1 in units of kT, where k is Boltzmann's constant and T the temperature. Therefore, there is a cross-over from a non-adsorbed, isotropic state as long as $\delta N^{3/5} \ll 1$ to an adsorbed state where the chain has an anisotropic pancake shape along the wall when the inequality is reversed [6.41,42]. Note that in the adsorbed state, the number of monomers on the surface becomes proportional to N. Using (6.31) and (6.33), one may write a scaling relation for N_s,

$$N_s \sim N^{3/5} f(\delta N^{3/5}). \tag{6.34a}$$

Assuming that $f(x)$ behaves as a power law for large values of x, one finds that

$$N_s \sim N\delta^{2/3} \quad (\delta N^{3/5} \gg 1). \tag{6.34b}$$

It is also possible to study the width D of the pancake by using similar scaling arguments. The condition here is that the size normal to the surface is independent of N and depends on the interaction δ only. We find by assuming $\nu \approx \phi \approx 3/5$ that

$$D \sim \delta^{-1}. \tag{6.35}$$

It is also possible to show that the structure of the chain is a two-dimensional array of units [6.42], called adsorption blobs, each of these being isotropic. Let g be the number of monomers per blob; one obtains, using (6.33),

$$g \sim \delta^{-5/3}. \tag{6.36a}$$

The width of the adsorbed chain is the size of a blob,

$$D \sim g^{3/5}, \tag{6.36b}$$

whereas its extension $R_{\parallel}$ along the surface is

$$R_{\parallel} \sim (N/g)^{3/4} D. \tag{6.36c}$$

Because of space limitations we will not give more details about the structure of a single chain, and we will consider the plateau regime, when the surface is saturated with polymers.

6.3.2 The Plateau Regime

a) The concentration profile. In the previous section, we considered the single-chain case. In a dilute solution, there is a partition of the polymers between the bulk and the surface. As usual, there is a Boltzmann factor between the fraction of the polymers on the surface and on the wall. As the interaction δ or the bulk concentration C_b increases, various surface-concentration regimes may be obtained, which will not be discussed here. Our discussion will be restricted to the case where the surface is saturated with polymers. Saturation here does not mean that the surface is totally covered with monomers, except when δ is of the order of unity or higher. When the interaction is weak, the surface is covered with adsorption blobs of size D mentioned above. Thus the surface concentration ϕ_s is

$$\phi_s \sim \frac{g_s}{D^2} \sim \delta, \tag{6.37}$$

where $g_s \sim g^{3/5}$ is the number of monomers per adsorption blob on the surface. Thus, although the surface is covered with blobs, its concentration may be small, except when δ is large. Note that although the surface is saturated, the bulk may be either in the dilute or the semi-dilute regime.

Three distance regions were defined by de Gennes [6.43,44]. The proximal range is within a blob, for distances smaller than D. When $\delta \approx 1$, this regime reduces to the size of a monomer. The central range is the self-similar regime and extends to the radius of a chain or to the bulk screening length, depending on the bulk concentration. Finally, in the distal range, the concentration, which was higher at the surface than in the bulk, decays exponentially back to its bulk value. We will consider only the first two regimes, where characteristic power laws appear. The most commonly found one is the central regime, because the interaction between the surface and the monomers is usually large. Let us assume that it is of the order of unity. Then, as mentioned above, the proximal range reduces to the size of a monomer, and the surface concentration is unity.

The concentration profile $\phi(z)$ falls off as a power law, where z is the distance from the surface. Assuming power-law behavior, the exponent may be found in two different ways. The first way, due to de Gennes, is to realize that the profile is self-similar, so that no distance scale should be present, as shown in Fig. 6.8.

However, two length scales may be defined. The first is z. The second is a screening length $\xi(z)$. This is related to the fact that for any distance from the wall, the concentration is in the semi-dilute regime, so that there is a local

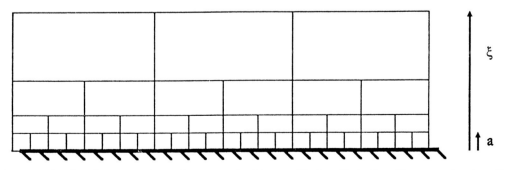

Fig. 6.8. In the plateau regime, the concentration profile is self-similar: the screening length at depth z is proportional to z

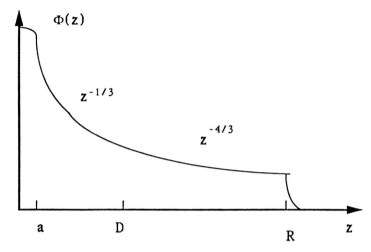

Fig. 6.9. The concentration profile as a function of distance z from the adsorbing surface

screening length. This is related to the local concentration by the same law as we saw in the previous subsection,

$$\xi(z) \sim \phi(z)^{-3/4}.$$

Because of self-similarity, these two lengths have to be identified with each other, that is, $\xi(z) \sim z$. Using the above relation between ξ and ϕ, we get

$$\phi(z) \sim z^{-4/3} \quad (D \ll z \ll R, \xi). \tag{6.38}$$

Another way to derive $\phi(z)$ is to assume a power law and to determine the exponent by the requirement that for $z \sim R$, ϕ should be of the order of the overlap concentration. For a bulk semi-dilute solution, for $z \sim \xi_b \sim C_b^{-3/4}$, ϕ should be of the order of the bulk concentration C_b. Using these conditions, one recovers (6.38).

The proximal range may be discussed by merely assuming continuity at both D and a: for both distances, one has to recover the value in the central range and the surface concentration ϕ_s. We find that

$$\phi(z) \sim \phi_s z^{-1/3} \quad (a \ll z \ll D). \tag{6.39}$$

The resulting concentration profile is shown in Fig. 6.9.

b) The loop structure. It is interesting to relate these power laws to the structure of the polymers. We will do this for the case $\delta \approx 1$ and for the central range, but a similar approach may be applied to the adsorption cross-over for a single chain or in the proximal range when $\delta \ll 1$. Because the macromolecules are adsorbed, they are made of a succession of loops of different sizes. Let $P(n)$ be the probability per unit surface of finding a loop with n monomers [6.45]. The distribution may be found by assuming that every loop behaves as a single polymer chain with excluded volume; its radius is

$$z(n) \sim n^{1/d_f} \approx n^{5/3}. \tag{6.40}$$

It is then possible to assume that the concentration at a distance z from the surface is mainly due to those loops that reach that size. Then we have

$$\phi(z)dz = nP(n)dn. \tag{6.41}$$

Using (6.40) and (6.41), we find that

$$P(n) \sim n^{-11/5}. \tag{6.42}$$

This distribution is also interesting because it allows us to consider the set of monomers that are located on the surface, or equivalently, the projection of an adsorbed polymer. Because the units are the projection of loops, it is easy to realize that they do not have a fixed length but a broad distribution [6.46]. Let $g(\ell)$ be the probability of finding a unit with length ℓ in this projection. This is related to $P(n)$:

$$g(\ell)d\ell = P(n)dn, \tag{6.43}$$

where ℓ is the end-to-end distance of a loop and is given by (6.40). Using (6.40) and (6.43), we find that

$$g(\ell) \sim \ell^{-3}. \tag{6.44}$$

Note that the distribution of step lengths, $g(\ell)$, decreases rapidly so that the projection is as in a random walk and not as in a Levy flight [6.47]. It would be interesting to look at the influence of roughness on this distribution and to see if there are cases where the distribution falls off to a lesser extent so that the projection of the polymer on the surface has exponents different from those of a self-avoiding walk.

The concentration profile was observed experimentally using various techniques. Although the power laws mentioned above are still in question, we discuss some experiments showing such behavior. The first and second moments of the profile were observed by ellipsometry and hydrodynamic thickness measurements [6.48,49]. All experiments were performed on systems with $\delta \approx 1$. It is interesting to note that because of the form that we found, the first moment is not proportional to the radius of gyration of the chains, whereas the second one is:

$$D_e \sim \int_a^R z\phi(z)dz \sim R^{2/3} \sim N^{2/5}. \tag{6.45}$$

This was observed by Takahashi, Kawaguchi, and co-workers [6.48]. Two sets of neutron experiments were performed. Small-angle neutron scattering by Auvray *et al.* [6.50] on polymers absorbed on porous silica, and neutron reflectometry by Sun *et al.* [6.51]. Both techniques led to reasonable agreement with the above results. Note, however, that some of these results were challenged by Cosgrove *et al.* [6.52], who could not reconcile their results with power laws. Further studies were performed by looking at rough and fractal surfaces [6.53]. As this is still being debated, we will not discuss their results but turn to a third example showing the fractal behavior of another class of macromolecules, namely branched polymers and gels.

6.4 Branched Polymers and Gels

So far we have considered polymers made of bifunctional units. These may react by means of two ends, or functionalities. When the monomers are more than bifunctional, polymerization leads to branched structures, and eventually to a solid called a gel [6.1,2]. In this section we will consider this case. As we will see, every polymer still has a fractal behavior. In addition to this, there is a very broad distribution of molecular weights, known as polydispersity. Because of this, what is observed is an effective dimension that also depends on the dimension of the distribution. This holds for many polydisperse systems, with restrictions that will be discussed below. We will first present the distribution of molecular weights that is naturally found in the reaction bath. We will turn to dilute solutions, where the fractal dimension is smaller because of swelling. We will discuss the effective dimension that is observable. Then we will turn to the semi-dilute solutions and to the swollen gels. Finally, we will discuss the dynamics of these systems in the reaction bath.

6.4.1 The Sol-Gel Transition

Let us consider multifunctional monomers in a vessel. These may react by f functionalities. We assume that the probability that any two neighboring functionalities react is a constant. As time proceeds, there is formation of dimers, trimers, ..., polymers; this is the sol. This process makes the solution more and more viscous, because of the presence of large macromolecules. The viscosity diverges, and this defines a threshold time t_c. For $t > t_c$, in addition to the sol, there is an infinite molecule, the gel. Thus, there appears an elastic modulus due to the presence of a solid phase.

Although there are probably other universality classes, this transition was successfully modeled by bond percolation [6.54]. Sites are the unreacted monomers, bonds are the reacted functionalities. Clusters are the polymers, and the infinite cluster is the gel. We recall very briefly some results of percolation that will be used below. For a general review on percolation see Chaps. 2 and 3 in [6.7]. The main result concerns the distribution of cluster sizes. This corresponds to what we called polydispersity. The distribution is very broad. If we call p the probability that a bond is reacted and p_c is its value at the threshold the probability $P(N, \varepsilon)$ of finding a polymer with size N at a distance $\varepsilon = p - p_c$ from the threshold is

$$P(N, \varepsilon) \sim N^{-\tau} f(\varepsilon N^\sigma), \tag{6.46}$$

where τ and σ are percolation exponents ([6.55], see also Chap. 2 in [6.7]) to be discussed below. The moments of the distribution have several interesting properties. The first moment is normalized below p_c. The higher moments diverge with different exponents

$$N_\omega \sim \frac{\int N^2 P(N, \varepsilon) dN}{\int N P(N, \varepsilon) dN} \sim \varepsilon^{-\gamma} \tag{6.47}$$

and

$$N_z \sim \frac{\int N^3 P(N,\varepsilon)dN}{\int N^2 P(N,\varepsilon)dN} \sim \varepsilon^{-1/\sigma}. \qquad (6.48)$$

Higher-order moments defined in the same way as above are proportional to N_z.

The exponent γ is the susceptibility exponent in percolation. Similarly, one may also define a characteristic length, corresponding to the size of the largest polymers in the sol or to the mesh size of the gel. This diverges as

$$\xi \sim \varepsilon^{-\nu}. \qquad (6.49)$$

Using (6.48) and (6.49), we find the fractal dimension d_f for percolation clusters,

$$d_f = \sigma\nu. \qquad (6.50)$$

Let us stress that this is the fractal dimension of the polymers in the reaction bath. We assume that all polymers that constitute the sol have this same fractal dimension. This was calculated by renormalization-group techniques and computer simulations [6.56-60]. We will give a simple Flory derivation [6.61] that is close to the former results for all space dimensions. The polydispersity exponent τ can be shown to be related to the fractal dimension:

$$\tau = 1 + d/d_f. \qquad (6.51)$$

This hyperscaling relation is valid for space dimensions $d \leq 6$. We will restrict ourself in the following to $d = 3$, although the two-dimensional case may also be of interest. Equation (6.51) implies that the distribution is very special; if one considers polymers with a given mass, they are in a C^* situation, that is, they are in a space-filling configuration. Since they are fractals, however, voids are left in the structure. These voids are filled by polymers with smaller masses, with the same requirement for every mass: each one is in a C^* situation. The total concentration is unity, since we are considering site percolation. Therefore, if one looks at the distribution, for any size considered, one always observes polymers at C^*. In this sense, the distribution is fractal [6.62,63]. Note that it is possible to relate the masses N_z and N_w defined above by eliminating ε,

$$N_z \sim N_w^{d_f/(2d_f-d)}. \qquad (6.52a)$$

Using (6.51), we get

$$N_z \sim N_w^{1/(3-\tau)}. \qquad (6.52b)$$

Note that both relations (6.52) hold only if $d_f < d$, or equivalently if $\tau > 2$. If $d_f = d$, or $\tau = 2$, both masses become proportional to each other, and in our definitions, there is only one mass present in the problem. This is the case in percolation, and it will prove to be important in the discussion for the scattered intensity for dilute solutions below. We will define two more exponents related to the viscosity and the modulus close to the gelation threshold when we consider the distribution of times below.

6.4.2 The Flory Approximation

Let us consider the large polymers, with mass N_z and radius ξ in the distribution. In the Flory approximation, one writes down a free energy made of two contributions

$$F = \frac{\xi^2}{\xi_0^2} + \frac{v}{N_w} \frac{N_z^2}{\xi^d}. \tag{6.53}$$

The first one is an entropic term where we assume that the polymer behaves like a spring with constant ξ_0^2, where ξ_0 is the radius of an ideal chain when no interactions are present. The second term is the interaction energy in which v is the excluded volume interaction, that was discussed for linear chains. Except for the presence of N_w this is very similar to what we considered for chains. The presence of this factor is due to the fact that the large polymers are penetrated by the small ones. Because of this, there is a screening of the interactions, as in the semi-dilute case for linear chains. The precise form for the energy was evaluated by Edwards [6.21] and de Gennes [6.64] in a Debye-Hückel approximation. The ideal chain radius ξ_0 was calculated on a Cayley tree [6.65] and was shown to be

$$N_z \sim \xi_0^4. \tag{6.54}$$

In the Flory approximation, all quantities, except the radius, which is to be calculated, are assumed to have a mean-field behavior. Therefore there is a relation between N_z and N_w,

$$N_z \sim N_w^2. \tag{6.55}$$

Minimizing the free energy with respect to ξ and using (6.54) and (6.55), we get the fractal dimension d_f of the large clusters,

$$d_f = \frac{1}{2}(d + 2). \tag{6.56}$$

This was tested indirectly by measurements of the mass distribution and the exponent τ (6.51). The results of the Kodak group [6.66] on polyesters are shown in Fig. 6.10. They were obtained by size exclusion chromatography coupled with light-scattering experiments.

6.4.3 Dilute Solutions

Once the distribution of polymer sizes is known, it is possible to dilute the sol, and to consider dilute solutions. Let us stress that the growth of the polymers is quenched before dilution and that the distribution function is given. Because of the excluded volume interactions the polymers swell and their fractal dimension changes from d_f to d_f^a. The new fractal dimension d_f^a may be obtained within a Flory approximation by considering a free energy similar to that in (6.53). The difference between a dilute solution and the reaction bath considered above is in the interaction term. We expect that the excluded volume interactions are present in the dilute case whereas they are fully screened in the previous case [6.67]. Therefore, this contribution has the same form as in (6.9) for linear chains. It is straightforward to minimize the free energy with respect to the radius, which yields

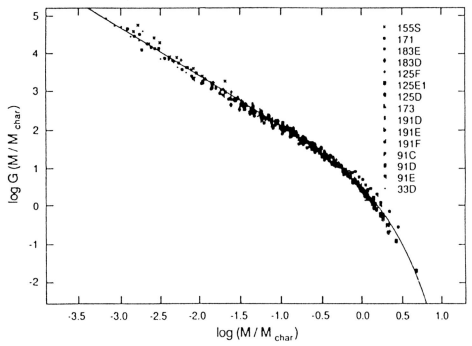

Fig. 6.10. The distribution of sizes in a sol close to the gelation threshold decreases as a power law with a cut-off for very large masses. The quantity M_{char} is the largest mass measured by light scattering and corresponds to N_z (after [6.66])

$$d_f^a = \frac{2}{5}(d+2). \tag{6.57}$$

The observation of this fractal dimension, however, is not easy, as we now discuss. Any experiment provides the average of the observed quantity over the whole distribution of masses. This averaging procedure leads to an effective dimension [6.68-71] that is different from the actual one. In order to see this, let us consider the scattered intensity. For a single mass, we have

$$S_1(q, N) = Nf[qR(N)], \tag{6.58a}$$

where the function $f(x)$ behaves as a power law in the fractal range, $qR > 1$ and $qa \ll 1$, in such a way that the mass dependence disappears. For a distribution of masses, and in the dilute regime where one may neglect correlations between monomers belonging to different polymers, the total scattered intensity [6.72] is

$$S_{\text{total}}(q) = \sum NP(N, \varepsilon)S_1(q, N). \tag{6.58b}$$

Using (6.58) and (6.46) and (6.47), we get

$$S_{\text{total}}(q) = CN_w f(qR_z), \tag{6.58c}$$

where C is the monomer concentration and R_z is the radius of the largest polymers,

$$R_z = \frac{\int N^2 R^2(N) P(N, \varepsilon) dN}{\int N^2 P(N, \varepsilon) dN}. \tag{6.59}$$

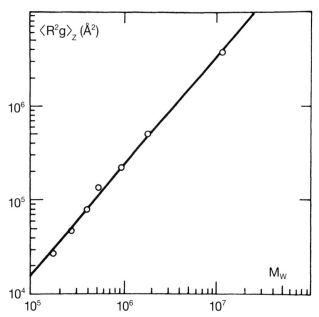

Fig. 6.11. The average radius of gyration in a diluted sol as a function of the average molecular weight (after [6.70])

The radius R_z can be related to the largest masses N_z, (6.48) through the fractal dimension

$$R_z \sim N_z^{d_f^a} \approx N_w^{5/8}. \qquad (6.60)$$

This relation was tested by light-scattering measurements and found to be in good agreement with experimental results (see Fig. 6.11). In the intermediate scattering range, $\ell^{-1} \gg q \gg R_z^{-1}$, the function $f(x)$ in (6.58c) behaves as a power law. The exponent of $S_{rmtotal}(q)$ is determined by the requirement that we are now in the fractal regime where no explicit mass dependence should appear. Using (6.52), we get

$$S_{\text{total}}(q) \sim q^{-d_f^a(3-\tau)} \approx q^{-8/5} \quad (\ell^{-1} \gg q \gg R_z^{-1}). \qquad (6.58d)$$

Therefore, an effective fractal dimension appears that describes the behavior of the polydisperse system. As can be seen from (6.58b), this effective dimension is related to the actual one but also to the exponent τ characterizing the distribution of masses. Note that this holds for percolation and for other distributions, as long as $\tau > 2$, as discussed above. The polydispersity effect disappears when $\tau = 2$. In this sense, we will say that such systems are not polydisperse. The difference between the scattered intensities [6.73] of a polydisperse solution and a monodisperse solution is shown in Fig. 6.12.

This has an important implication. Measuring an exponent in a scattering experiment does not necessarily imply that one gets the fractal dimension directly. First one has to check the polydispersity by independent measurements, either with viscosity or with second-virial-coefficient experiments. The latter may be calculated following the same steps as above and taking into account the interactions between the centers of masses of different polymers. We will

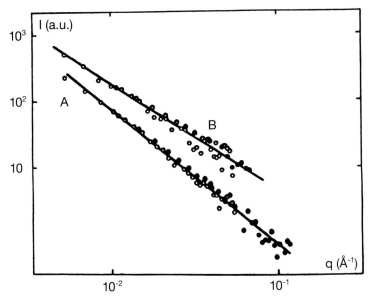

Fig. 6.12. Scattered intensities in a SANS experiment in the fractal regime. Curve (A) represents the polydisperse (natural) distribution. Curve (B) represents a monodisperse (fractionated) sample (after [6.73])

not give this derivation here because of lack of space. For a detailed derivation the reader is referred to [6.74]. The results are

$$I(q \to 0, C) = CN_w(1 - A_2 C), \tag{6.61a}$$

with

$$A_{2p} \sim N_w^{\frac{d}{(3-\tau)}\left(\frac{1}{d_f^a} - \frac{1}{d_f}\right)} \approx N_w^{3/8}. \tag{6.61b}$$

Here, the index p stands for polydispersity. Note that when no polydispersity is present the complicated relation (6.61b) simply becomes

$$A_{2m} \sim N^{d/d_f^a} \approx N^{3/2}. \tag{6.61c}$$

Here, the index m stands for monodisperse.

Similarly, it is possible to show that the measured [6.66,68] intrinsic viscosity $[\eta]$ is

$$\langle[\eta]\rangle \equiv \lim_{C \to 0} (\eta - \eta_s)/\eta_s C \sim A_2,$$

where η and η_s are respectively the viscosities of the solution and of the solvent.

Finally, the overlap concentration C^* that separates the dilute concentration regime from the semi-dilute range is also very different in both cases. It was found that $C^* \sim 1/A_2$. The overlap concentration C^* for a polydisperse system is very different from that of a monodisperse one, and it is quite straightforward to check whether the observed exponent is the actual dimension or an effective one due to polydispersity effects.

6.4.4 Semi-Dilute Solutions and Swollen Gels

In the semi-dilute concentration range, the polymers interpenetrate each other in a very special way. Whereas two macromolecules with similar sizes never overlap, smaller ones may penetrate large ones. The reason for this stems from the features of the reaction bath that were given above. When no solvent is left, the polymers interpenetrate as was discussed in Sect. 6.1. For $C^* \ll C \ll 1$, this can be partly effected. It is possible to divide the distribution into two parts [6.74]. One will include the large polymers that are penetrated by the smaller ones, and are screened. The other part will include those smaller ones that have the same behavior as in a dilute solution. Let ξ be the cross-over length between these two length scales. The behavior of a swollen gel is in principle identical to that of a semi-dilute solution. The only difference is the fact that one may dilute a solution made of the sol part to any extent. The gel, however, may be swollen only up to a finite size. We will assume that when the typical size becomes swollen, it corresponds to a C^* situation.

a) Volume change. It is now possible to use the fractal behavior of polymers to calculate the maximum and intermediate swelling of a gel. This is done as follows. Let us consider a cell of a gel of radius ξ_i containing N_z monomers in the reaction bath. As mentioned earlier, swelling corresponds to a change in the fractal dimension. This change is from the percolation value $d_f = 5/2$ in the reaction bath to a value which is assumed to be $d_f^A = 2$ in the swollen state. For the maximum swelling, we get

$$N_z \sim \xi_i^{5/2} \sim \xi_f^2, \tag{6.62a}$$

where ξ_f is the swollen length of the cell. From (6.62a), we get

$$\xi_f \sim \xi_i^{5/4}. \tag{6.62b}$$

The swelling ratio Q, which is the ratio between the final and initial volumes, is given by,

$$Q = \frac{V_f}{V_i} = \frac{C_i}{C_f} = \{\frac{\xi_f}{\xi_i}\}^3 \tag{6.62c}$$

$$\sim \xi_i^{3/4} \sim N_w^{3/8}, \tag{6.62d}$$

where we have used (6.52b) and (6.56). Because the initial concentration, C_i, in the reaction bath is of the order of unity, we get the overlap concentration,

$$C_f \sim C^* \sim N_w^{-3/8}. \tag{6.63}$$

This result is in agreement with (6.61b). For intermediate swelling concentration C, (6.62c) provides the mesh size

$$\xi_f \sim \xi_i C^{-1/3}. \tag{6.64}$$

Here C is the final concentration. We stress that the initial state is that of the reaction bath, without any solvent, but with the presence of the sol fraction.

Note that an implicit assumption in (6.62c) is that there is only one (or a finite number of) infinite clusters. We know that in the case of vulcanization, when linear chains are crosslinked in order to get a gel, this assumption does not hold since the critical region is very narrow [6.75-78]. Thus, we do not expect (6.63) and (6.64) to hold in this case. For vulcanization, it has been shown that in addition to the change in the fractal dimension there is also a full disinterpenetration of the various networks upon swelling. So far, it has been assumed that there is a full disinterpenetration [6.79]. Obviously, this assumption should break down for high crosslink densities [6.80-83], and more research work is needed to clarify this point.

Finally, the screening length ξ_f may also be calculated for the intermediate regime. Using (6.62b,c), we get

$$\xi_f \sim C^{-5/3}. \tag{6.65}$$

This characteristic length is also the screening length in a semi-dilute regime and corresponds to the radius of the separation between small and large polymers mentioned above.

b) Other properties. Among the other properties which may be calculated for semi-dilute solutions, we mention briefly the scattered intensity at zero angle, $S(0)$, and the osmotic pressure π. The first one may be estimated by simple scaling arguments. In a dilute solution, we have

$$S(0) \sim C N_w \quad (C \ll C^*). \tag{6.66a}$$

To find the concentration dependence, one may assume the following scaling form:

$$S(0, C) \sim C N_w f(C/C^*). \tag{6.66b}$$

For large concentrations, $C \gg C^*$, we expect the intensity to be independent of the average mass of the distribution and to depend only on concentration. Assuming that $f(x)$ in (6.66b) behaves as a power law, and using (6.63), we obtain

$$S(0, C \gg C^*) \sim C^{-5/3}. \tag{6.66c}$$

This scaling relation was found to be in good agreement with the experimental data [6.84], as shown in Fig. 6.13.

Finally, one may assume a similar scaling form for the osmotic pressure, π. There is, however, an important difference that comes from considering dilute solutions. The ideal-gas contribution for a very dilute solution is merely the concentration in polymers. This gives

$$\pi = \frac{C}{N_n} g(C/C^*), \tag{6.67}$$

where the prefactor is the concentration of polymers irrespective of their mass, and N_n is the average molecular weight. Thus, one does not expect any simple concentration power-law behavior for the osmotic pressure in the semi-dilute regime. This also implies that there is no simple relationship between the osmotic pressure and the zero-angle scattered intensity. This may be understood

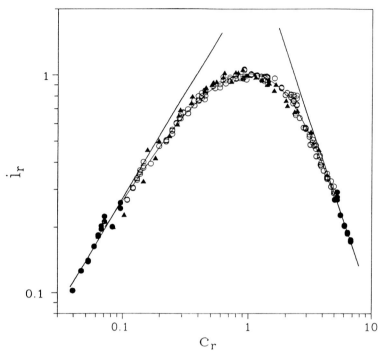

Fig. 6.13. The normalized scattered intensity for a semi-dilute polydisperse solution of randomly branched polymers as a function of normalized concentration for various q values. The maximum indicates the screening length (6.65) (after [6.84])

if one considers the system as a multicomponent one, where each of the masses acts as an independent component. For such systems, we know that $S(0)$ is not related simply to the compressibility [6.85].

6.4.5 Dynamics

a) Rheology. Because of polydispersity, the dynamics of the branched polymers become rather interesting. The reason is that we find that polydispersity implies a broad distribution of relaxation times. In order to discuss this distribution, we need to return to the behavior of the viscosity η and the modulus G in the vicinity of the gelation transition. It has been shown that one can define two exponents [6.86-90] that describe the divergence of the viscosity and the vanishing of the modulus:

$$\eta \sim \varepsilon^{-s} \quad (\varepsilon < 0) \tag{6.68}$$

and

$$G \sim \varepsilon^{\mu} \quad (\varepsilon > 0). \tag{6.69}$$

Both (6.68) and (6.69) are valid for vanishing frequencies. For nonvanishing frequencies, one may introduce a complex modulus, $\tilde{G}(\omega)$, or a complex viscosity, $\tilde{\eta}(\omega)$, such that

$$\tilde{G}(\omega) = G(\omega) + j\omega\eta(\omega) = j\omega\tilde{\eta}(\omega), \tag{6.70}$$

where $j = \sqrt{-1}$ and ω the frequency.

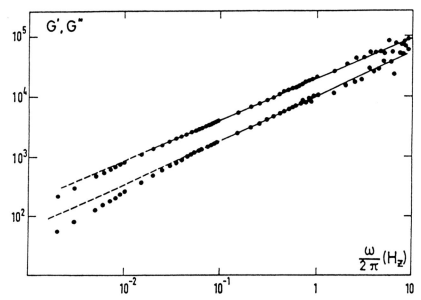

Fig. 6.14. The real and imaginary parts G' and G'' of the complex modulus behave as power laws for high frequencies (after [6.95])

It was shown in Refs. [6.91-94] that it is possible to express the complex response function, $\tilde{G}(\omega)$, in the scaling form

$$\tilde{G}(\omega) \sim \varepsilon^{\mu} f_{\pm}(j\omega\varepsilon^{-s-\mu}), \tag{6.71a}$$

where $f_{\pm}(x)$ are the functions above and below the threshold. For high frequencies, when one is probing finite regions, the response function should be independent of ε and be a function of frequency only. Assuming that $f(x)$ behaves as a power law, we find that

$$\tilde{G}(\omega) \sim \{j\omega\}^{\mu/(s+\mu)} \quad (\omega\varepsilon^{-s-\mu} \gg 1). \tag{6.71b}$$

This implies that both the real and the imaginary parts of the complex modulus behave as power-laws with the same exponent and that the loss angle is a constant. The scaling assumption (6.71a), as well as the power-law behavior (6.71b), were tested experimentally by Durand *et al.* [6.95], Rubinstein *et al.* [6.96], Martin *et al.* [6.97], Axelos *et al.* [6.98] and Devreux *et al.* [6.99] respectively on polyurethane, polyester, epoxy resins, pectin, and silica gels. They all find a value of 0.7 for the exponent $\mu/(s+\mu)$. Figure 6.14 shows the results of Durand *et al.* Thus, the scaling assumption for the complex viscosity and modulus seems to give a good description of the experimental results.

b) Distribution of relaxation times. Their interpretation for this exponent is, however, different as we discuss below. The complex viscosity is directly related to the distribution $H(\tau)$ of relaxation times [6.100] by

$$\eta(\omega) = \int \frac{H(\tau)}{1 - j\omega\tau} d\tau. \tag{6.72}$$

Using (6.71a) and (6.72), we get [6.101,102] the scaling form for the distribution of relaxation times:

$$H(\tau) \sim \tau^{-\mu/(s+\mu)} h_{\pm}(\tau/T_1) \tag{6.73}$$

and

$$T_2 \sim \varepsilon^{-s-\mu}. \tag{6.74a}$$

The quantities T_1 and T_2 are the first and second moments of the distribution $H(\tau)$, defined as

$$T_1 \sim \frac{\int \tau H(\tau) d\ln \tau}{\int H(\tau) d\ln \tau} \sim \varepsilon^{-s} \tag{6.74b}$$

and

$$T_2 \sim \frac{\int \tau^2 H(\tau) d\ln \tau}{\int \tau H(\tau) d\ln \tau}. \tag{6.74c}$$

Note that all normalized higher order moments diverge as T_2. The reason for this is that both the numerator and the denominator diverge, whereas for T_1, the denominator does not diverge. Thus the distribution may be characterized by two diverging times. The shortest one diverges as the viscosity, as may be seen directly from (6.72). The second one is the longest cut-off time.

Both characteristic times were measured by light-scattering and rheological experiments such as those discussed above. The results are not as universal as for the high-frequency complex modulus. In these experiments, what is measured is not a ratio between exponents but the exponents s and μ themselves. Two sets of results have been obtained, corresponding roughly to the Rouse and Zimm assumptions for the hydrodynamics. It may be shown that if one neglects interactions between monomers, the exponents are [6.87,88,103]

$$\tilde{s}_R = s/\nu = 2 - \tilde{\beta} = 2 - d + d_f \approx (6-d)/2, \tag{6.75a}$$

and

$$\tilde{\mu}_R = d. \tag{6.75b}$$

When hydrodynamic interactions are present [6.104], it was conjectured that [6.105-108]

$$\tilde{s}_Z \approx 1 - \tilde{\beta}/2 = 1 - (d - d_f)/2 \approx (6-d)/4 \tag{6.75c}$$

and

$$\tilde{\mu}_Z = d - 2 + \tilde{\zeta}/2 \approx (5d-6)/4, \tag{6.75d}$$

where $\tilde{\zeta}$ is the exponent for the shortest, or elastic path. Note that the last approximate terms in the above equalities are the Flory approximations.

Both sets of values were observed experimentally. The first was found by Martin et al. on silica gels [6.109], by Colby et al. on polyesters [6.110], and by Sorensen et al. on gelatin [6.111]. The second was observed by Adam et al. on polyurethane [6.112], and by Axelos et al. on pectin [6.98].

Finally, Devreux et al. found a crossover from scalar to vectorial percolation [6.113-117] as one goes further from the gelation threshold. It is not clear at present whether there are two universality classes for the dynamics or if highly nontrivial spurious effects affect some of the measurements. Arbabi and Sahimi [6.118] gave the first attempt to unify the various points of view. It is possible to consider the time-dependent properties, as we did above for the linear chains. For literature on the dynamical properties, see also [6.119-122].

Acknowledgements. The author is much indebted to M. Adam, E. Bouchaud, R. Colby, M. Delsanti, D. Durand, B. Farnoux, P.G. de Gennes, O. Guiselin, G. Jannink, L.T. Lee, L. Leibler, J. E. Martin, and M. Rubinstein for many discussions.

References

6.1   B.B. Mandelbrot: *The Fractal Geometry of Nature* (Freeman, San Francisco 1977)
6.2   P.G. de Gennes: *Scaling Concepts in Polymer Physics* (Cornell University Press, Ithaca 1979)
6.3   J. des Cloizeaux, G. Jannink: *Polymers in Solution* (Clarendon Press, Oxford 1987)
6.4   P.J. Flory: *Principles of Polymer Chemistry* (Cornell University Press, Ithaca 1953)
6.5   M. Gordon, S.B. Ross-Murphy: Pure Appl. Chem. **43**, 1 (1975)
6.6   D. Stauffer: Phys. Rep. **54**, 1 (1979); *Introduction to Percolation Theory* (Taylor and Francis, London 1985)
6.7   A. Bunde, S. Havlin, eds.: *Fractals and Disordered Systems* (Springer, Berlin 1991)
6.8   J. Teixeira, in: *On Growth and Form* , ed. by H.E. Stanley, N. Ostrowski (Martinus, Nijhoff 1985)
6.9   J.P. Cotton, D. Decker, H. Benoit, B. Farnoux, J. Higgins, G. Jannink, R. Ober, C. Picot, J. des Cloizeaux: Macromolecules **7**, 863 (1974)
6.10   S.F. Edwards: Proc. Phys. Soc. **65**, 613 (1965)
6.11   P.G. de Gennes: Phys. Lett. **38A**, 339 (1972)
6.12   M.J. Stephen: Phys Lett. **53A**, 363 (1975)
6.13   M. Daoud, G. Jannink: J. Physique Lett. **37**, 973 (1976)
6.14   M. Moore: J. Phys. A **10**, 305 (1977)
6.15   I.M. Lifshitz, A. Grosberg, A. Khakholov: Rev. Mod. Phys. **50**, 685 (1978)
6.16   M. Nierlich, J.P. Cotton, B. Farnoux: J. Chem. Phys. **69**, 1379 (1978)
6.17   C. Williams, F. Brochard, H.L. Frisch: Ann. Rev. Phys. Chem. **32**, 433 (1981)
6.18   B. Duplantier: Phys. Rev. Lett, **59**, 539 (1987)
6.19   B. Derrida: J. Phys. A **14**, L5 (1981)
6.20   H.J. Hilhorst: Phys. Rev. B **16**, 1253 (1977)
6.21   S.F. Edwards: Proc. Phys. Soc. **88**, 265 (1966)
6.22   M. Daoud, J.P. Cotton, B. Farnoux, G. Jannink, G. Sarma, H. Benoit, R. Duplessix, C. Picot, P.G. de Gennes: Macromolecules, **8**, 804 (1975)
6.23   J. des Cloizeaux: J. Physique, **36**, 281 (1975)
6.24   I. Noda, N. Kato, T. Kitano, M. Nagasawa: Macromolecules **14**, 668 (1981)
6.25   M. Doi, S.F. Edwards: *The Theory of Polymer Dynamics* (Oxford Science, London 1986)
6.26   J.F. Joanny, S.J. Candau, in: *Comprehensive Polymer Science*, ed by G. Allen (Pergamon, New York 1989)
6.27   T.P. Lodge, N.A. Rotstein, S. Prager: Adv. Chem. Phys. **79**, 1 (1990)
6.28   W.W. Graessley: Adv. Pol. Sci. **47**, 68 (1982)
6.29   G.H. Weiss, J.T. Bendler, M.F. Slesinger: Macromolecules **21**, 317 (1988); J.T. Bender, M. Dishon, H. Kesten, G.H. Weiss: J. Stat. Phys. **50**, 1069 (1988)
6.30   L. Leger, H. Hervet, F. Rondelez: Macromolecules **14**, 1732 (1982)
6.31   A. Baumgartner, K. Binder: J. Chem. Phys. **75**, 2994 (1981)
6.32   K. Kremer, G.S. Grest: J. Chem. Phys. **92**, 5057 (1990)
6.33   A. Kolinski, J. Skolnick, R. Yaris: J. Chem. Phys. **86**, 7174 (1987)
6.34   D. Richter, K. Binder, B. Ewen, B. Stühn: J. Phys. Chem. **88**, 6618 (1984)

194   Mohamed Daoud

6.35   J.S. Higgins, J.E. Roots: J. Chem. Soc. Faraday Trans. II **81**, 755 (1985)
6.36   R.H. Ottewill, C.H. Rochester, A.L. Smith, eds.: *Adsorption from Solutions* (Academic, New York 1983)
6.37   T.C. Lubensky, M.H. Rubin: Phys. Rev. B **11**, 543 (1975); B **12**, 3885 (1975)
6.38   K. Binder, in: *Phase Transitions and Critical Phenomena* , vol. 8, ed. by C. Domb, J. Lebowitz (Academic, New York 1983)
6.39   A.J. Bray, M.A. Moore: J. Phys. A **11**, 1927 (1977)
6.40   E. Eisenriegler, K. Binder, K. Kremer: J. Chem. Phys. **77**, 6296 (1982)
6.41   P.G. de Gennes: J. Physique **37**, 1445 (1976)
6.42   E. Bouchaud, M. Daoud: J. Physique **48**, 1991 (1987)
6.43   P.G. de Gennes: Macromolecules **13**, 1069 (1980)
6.44   P.G. de Gennes, P. Pincus: J. Physique Lett. **44**, 241 (1983)
6.45   P.G. de Gennes: Comptes Rendus Ac. Sci. (Paris) II, **294**, 1317 (1982)
6.46   E. Bouchaud, M. Daoud: J. Phys. A **20**, 1463 (1987)
6.47   M.F. Shlesinger: Ann. Rev. Phys. Chem. **39**, 269 (1988);
       M.F. Klafter, in: *On Growth and Form: Fractal and Non-Fractal Patterns in Physics*, ed. by H.E. Stanley and N. Ostrowsky (Nijhoff, Dordrecht 1985);
       M.F. Shlesinger, G.M. Zaslavshy, J. Klafter: Nature **363**, 31 (1993)
6.48   M. Kawaguchi, M. Mikura, A. Takahashi: Macromolecules **17**, 2063 (1984)
6.49   R. Varoqui, P. Dejardin: J. Chem. Phys. **66**, 4395 (1977);
       E. Pefferkorn, A. Carroy, R. Varoqui: J. Pol. Sci. Pol. Phys. Ed. **23**, 1997 (1985)
6.50   L. Auvray, J.P. Cotton: Macromolecules, **20**, 202 (1986)
6.51   X. Sun, B. Farnoux, E. Bouchaud, A. Lapp, M. Daoud, G. Jannink: Europhys. Lett. **6**, 207 (1988)
6.52   T. Cosgrove, T.G. Heath, K. Ryan, T.L. Crowley: Macromolecules **20**, 2879 (1987)
6.53   D. Hone, H. Ji, P. Pincus: Macromolecules **20**, 2543 (1987);
       F. Brochard: J. Physique **46**, 2117 (1985)
6.54   D. Stauffer: J. Chem. Soc. Faraday Trans. II **72**, 1354 (1976)
6.55   H. Nakanishi, H.E. Stanley: Phys. Rev. B **22**, 2466 (1980)
6.56   P.J. Reynolds, W. Klein, H.E. Stanley: J. Phys. C **10**, L167 (1977);
       P.J. Reynolds, H.E. Stanley, W. Klein: Phys. Rev. B **21**, 1223 (1980)
6.57   T.C. Lubensky, J. Isaacson: Phys. Rev. Lett. **41**, 829 (1978); Phys. Rev. A **20**, 2130 (1979)
6.58   B. Derrida, J. Vannimenus: J. Physique Lett. **41**, 473 (1980)
6.59   G. Parisi, N. Sourlas: Phys. Rev. Lett. **46**, 891 (1981)
6.60   B. Derrida, L. de Seze: J. de Phys. **43**, 475 (1982);
       F. Family: J. Phys. A **16**, L97 (1983);
       B. Derrida, D. Stauffer, H.J. Herrmann, J. Vannimenus: J. de Phys. Lett. **44**, 701 (1983);
       H.J. Herrmann, B. Derrida, J. Vannimenus: Phys. Rev. B **30**, 4080 (1984)
6.61   J. Isaacson, T.C. Lubensky: J. Phys. **42**, 175 (1981)
6.62   M.E. Cates: J. de Phys. Lett. **38**, 2957 (1985)
6.63   M. Daoud, J.E. Martin, in: *The Fractal Approach to Heterogeneous Chemistry* ed. by D. Avnir (Wiley, New York 1990)
6.64   P.G. de Gennes: J. Pol. Sci. Pol., Symp. **61**, 313 (1977)
6.65   B.H. Zimm, W.H. Stockmayer: J. Chem. Phys. **17**, 1301 (1949)
6.66   E.V. Patton, J.A. Wesson, M. Rubinstein, J.C. Wilson, L.E. Oppenheimer: Macromolecules **22**, 1946 (1989)
6.67   M. Daoud, J.F. Joanny: J. de Phys. **42**, 1359 (1981)
6.68   M. Daoud, F. Family, G. Jannink: J. de Phys. Lett. **45**, 119 (1984)
6.69   S.J. Candau, M. Ankrim, J.P. Munch, P. Rempp, G. Hild, R. Osaka, in: *Physical Optics of Dynamical Phenomena in Macromolecular Systems* (W. De Gruyter, Berlin 1985), p. 145
6.70   F. Schosseler, L. Leibler: J. Phys. Lett. **45**, 501 (1984)
6.71   F. Schosseler, L. Leibler: Macromolecules **18**, 398 (1985)
6.72   J.E. Martin, B.J. Ackerson, Phys. Rev. A **31**, 1180 (1985)
6.73   E. Bouchaud, M. Delsanti, M. Adam, M. Daoud, D. Durand: J. Physique Lett. **47**, 1273 (1986)
6.74   M. Daoud, L. Leibler: Macromolecules, **41**, 1497 (1988)
6.75   P.G. de Gennes: J. de Phys. Lett. **38**, 355 (1977)
6.76   M. Daoud: J. de Phys. Lett. **40,** 201 (1979)

6.77   C. Allain, L. Salome: Macromolecules, **20**, 2957 (1987)

6.78   L. Salome: Ph.D. Thesis, Universite Paris 11 (1987)

6.79   M. Daoud, A. Johner: J. Physique **50**, 2147 (1989)

6.80   S.J. Candau, J. Bastide, M. Delsanti: Adv. Pol. Sci. **44**, 27 (1982)

6.81   R.W. Richards, N.S. Davidson: Macromolecules, **19**, 1381 (1986)

6.82   J.P. Cohen Addad, C. Scmit: J. Pol. Sci., Pol. Lett. **25**, 487 (1987)

6.83   S. Mallam, F. Horkay, A.M. Hecht, E. Geissler: Macromolecules, **22**, 3356 (1989)

6.84   J.P. Munch, M. Delsanti, D. Durand: Europhys. Lett. **18**, 557 (1992)

6.85   J. des Cloizeaux, G. Jannink: Physica, **102** A, 120 (1980)

6.86   P.G. de Gennes: J. Physique Lett. **37**, 1 (1976)

6.87   P.G. de Gennes: J. Physique Lett. **40**, 197 (1979)

6.88   M.J. Stephen: Phys. Rev. B **17**, 4444 (1978)

6.89   A. Coniglio, H.E. Stanley: Phys. Rev. Lett. **52**, 1068 (1984)

6.90   J. Kertesz: J. Phys. A **16**, L471 (1983)

6.91   A.L. Efros, B.I. Shklovskii: Physica Status Solidi, B **76** 475 (1976)

6.92   D.C. Hong, H.E. Stanley, A. Coniglio, A. Bunde: Phys. Rev. B **33**, 4564 (1984)

6.93   J.M. Laugier, J.P. Clerc, G. Giraud, J.M. Luck: J. Phys. A **19**, 3153 (1986)

6.94   M. Daoud, D.C. Hong, F. Family: J. Phys, A **21**, L917 (1988)

6.95   D. Durand, M. Delsanti, M. Adam, J.M. Luck: Europhys. Lett. **3**, 297 (1987)

6.96   M. Rubinstein, R.H. Colby, J.R. Gillmor: Pol. Preprint, **30**, 1 (1989)

6.97   J.E. Martin, D. Adolf, J.P. Wilcoxon: Phys. Rev. Lett. **61**, 2620 (1988)

6.98   M. Axelos, M. Kolb: Phys. Rev. Lett. **64**, 1457 (1990)

6.99   F. Devreux, J.P. Boilot, F. Chaput, L. Malier, M.A.V. Axelos, to be published.

6.100 J.D. Ferry: *Viscoelastic Properties of Polymers* (Wiley, New York 1980)

6.101 M. Daoud: J. Phys. A **21**, L973 (1988)

6.102 J.E. Martin, J. Wilcoxon, D. Adolf: Phys. Rev. A **36**, 3415 (1987)

6.103 M. Daoud, A. Coniglio: J. Phys. A **14**, L301 (1981)

6.104 W.H. Stockmayer, in: *Molecular Fluids*, ed. by R. Balian and G. Weill (Gordon and Breach, New York 1977)

6.105 J. Kertesz: J. Phys. A **16**, L471 (1983)

6.106 A. Coniglio, H.E. Stanley: Phys. Rev. Lett. **52**, 1068 (1984)

6.107 A. Coniglio, F. Family: J. Physique Lett. **46**, 9 (1985)

6.108 S. Roux: C.R. Ac. Sci. (Paris) **301**, 367 (1985)

6.109 J.E. Martin, J. Wilcoxon, J. Odine: Phys Rev. A **43**, 858 (1991); see also J. E. Martin, in: *Time Dependent Effects in Disordered Systems*, ed. by R. Pynn and T. Riste (Plenum, New York 1987), p. 425

6.110 M. Rubinstein, R. Colby, J.R. Gillmor: *Space-Time Organization in Complex Fluids*, ed. by F. Tanaka, M. Doi, T. Ohta, (Springer Series in Chemical Physics 51, 1989)

6.111 W.F. Shi, W.B. Wang, C.M. Sorensen: unpublished

6.112 M. Adam, M. Delsanti: Contemporary Physics **3**, 203 (1989)

6.113 S. Alexander, in: *The Physics of Finely Divided Matter*, ed. by M. Daoud and N. Boccara (Springer Verlag, Berlin 1985)

6.114 D.J. Bergman, Y. Kantor: Phys. Rev. Lett. **53**, 511 (1984)

6.115 Y. Kantor, I. Webman: Phys. Rev. Lett. **52**, 1891 (1984)

6.116 S. Feng, P. Sen: Phys. Rev. Lett. **52**, 216 (1984)

6.117 S. Alexander: J. Physique **45**, 1939 (1984)

6.118 S. Arbabi, M. Sahimi: Phys. Rev. Lett. **65**, 725 (1990)

6.119 S. Alexander, R. Orbach: J. de Phys. Lett. **43**, 625 (1982)

6.120 R. Rammal, G. Toulouse: J. de Phys. Lett. **44**, 13 (1983)

6.121 E. Courtens, J. Pelous, J. Phalippou, R. Vacher, T. Woignier: Phys. Rev. Lett. **58,** 128 (1987)

6.122 S. Havlin, D. Ben Avraham: Adv. in Phys. **36**, 695 (1987); H.E. Roman, M. Schwartz, A. Bunde, S. Havlin: Europhys. Lett. **7**, 389 (1988)

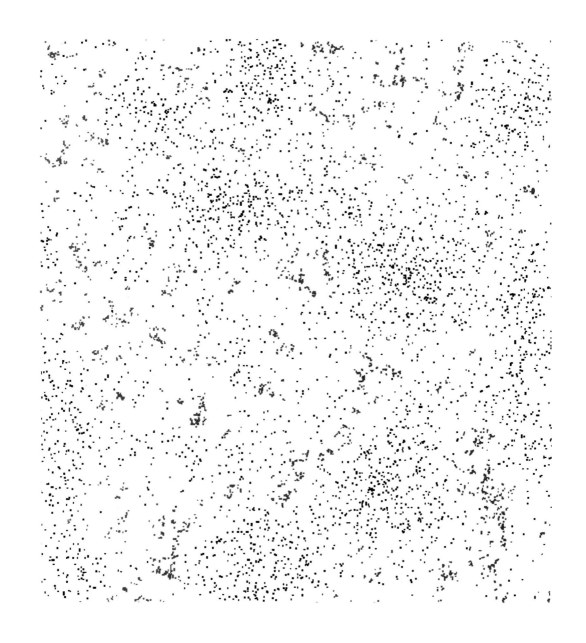

7 Kinetics and Spatial Organization of Competitive Reactions

Sidney Redner and Francois Leyvraz

7.1 Introduction

In this chapter, we review some kinetic and geometric properties of two-species annihilation in which an encounter between two distinct species A and B leads to the formation of an inert product I, $A + B \rightarrow I$ [7.1]. Examples of this reaction include electron-hole recombination in irradiated semiconductors [7.2], catalytic reactions on surfaces [7.3], exciton dynamics [7.4], and annihilation of primordial monopoles in the early universe [7.5]. At a more abstract level, two-species annihilation is a realization of an interacting Brownian particle system, a connection which has been helpful in establishing rigorous results [7.6]. For detailed discussion of Brownian motion see Chap. 5. For other types of reactions see Chap. 8.

Under homogeneous symmetric conditions, that is, where the two species are initially present at equal, spatially uniform concentrations, or are fed into the system homogeneously at equal rates, there is a spontaneous symmetry-breaking in which large-scale single-species heterogeneities form [7.7-18]. This spatial organization invalidates the mean-field approximation and its corresponding predictions. One of our goals is to outline mechanisms that lead to this spatial organization in two-species annihilation. In addition to the microscopic reaction itself, the rate of transport (usually diffusion) and the rate of input of new reactants play a fundamental role in governing the reaction kinetics. We shall discuss how the interplay between external input, the microscopic reaction, and transport influence the overall reaction kinetics and the growth of spatial heterogeneities.

◀ **Fig. 7.0.** Position of A and B particles in two species annihilation on a square 1000×1000 lattice after 1000 time steps. Here one of the species is static. To make each individual particle visible, 9 colored pixels are used to mark the particle position. Notice the clearly defined domains in which the mobile species are smoothly distributed while the immobile species have a singular spatial distribution. Since the immobile species are strongly clustered, the display algorithm distorts their apparent concentration. Courtesy of Stephen Harrington

To analyze two-species annihilation, a crucial observation is that the difference in the concentration of As and Bs is conserved, both globally and in each individual reaction. Thus the dynamical equations which describe the densities of As and Bs contain the same reaction term. Therefore the equation for the concentration difference is a linear one in which only transport and input terms appear. This observation, together with plausible physical hypotheses about the system, provides a solution for the individual species concentrations.

When there is no input or back reaction, the reactant densities decay monotonically. If the microscopic reaction rate is small compared to the rate of collisions between reactants, the process is *reaction-limited*. In this case, many collisions occur before a reaction event so that there is time for substantial mixing. Thus one expects the mean-field predictions to hold. In the opposite *diffusion-limited* process, the microscopic reaction is fast and the overall process is limited by the rate at which reactive pairs are brought together. In low spatial dimension, this implies lack of mixing, a feature which is responsible for large-scale coarsening of the reactant distributions. To quantify the distinction between these two limiting situations, consider the relation between the two fundamental time scales t_R, the typical time for a given particle to react with an arbitrary particle, and t_D, the time required for a particle to diffuse to a nearest neighbor. If the reactant concentration is $c(t)$, then within mean-field theory t_R is of the order of $1/(kc)$, where k is the rate constant, whereas for diffusing reactants t_D is of the order of $c^{-2/d}/D$, where d is the spatial dimension. Thus for $d \leq 2$, the system eventually reaches the limiting situation where $t_R \ll t_D$, that is, the diffusion-limited case. If $k \ll 1$, however, there may be a large crossover regime before diffusion-limited behavior is reached [7.11]. In fact, for two-species annihilation, the asymptotic diffusion-limited regime extends up to $d = 4$ since domain formation greatly increases t_D relative to a homogeneous system.

When the reactants are externally fed into the system at equal rates, the balance between the input and annihilation may lead to a steady state [7.15-18]. However, diffusive fluctuations can still promote a continuously coarsening pattern of A-rich and B-rich domains and steady behavior occurs only when the input is sufficiently coupled in both time and space to counteract diffusive fluctuations [7.16]. An important limiting case arises when the reactants are static after input. This can be viewed as a model catalytic reaction in which the input corresponds to the adsorption of reactants onto vacant sites of a catalytic surface (whose dimension need not be two). The adsorbates remain immobile, a characteristic feature of the adsorption of many gases on metal surfaces. A nearest-neighbor adsorbed AB pair forms a product at some finite rate which then immediately desorbs. These steps define the *monomer-monomer*-catalysis model [7.19-22], in which the formation of the AB product is assisted by the mere presence of the catalytic surface. Even when the adsorption rates of the two species are equal, this deposition of static particles leads to growing domains of A and B form owing to fluctuation in the reactant input [7.19-21]. This coarsening terminates in *saturation*, where a finite-size surface is fully occupied by only one of the two species, and further reactions cannot occur [7.22].

The tendency for domain formation in two-species annihilation confines the microscopic reaction to the interfacial regions between domains. This motivates the investigation of the reaction under heterogeneous conditions which promote the existence of a localized interface [7.23-35]. Two such examples include a system with initial semi-infinite domains of As and Bs which meet at a common boundary, or a finite system with input of As and Bs at opposite edges. Such geometries may also be more amenable to experimental realizations than a homogeneous system [7.27]. For these situations, the investigation of the reactive interface opens up a rich variety of phenomena from which useful insights about the microscopic reaction process can be obtained.

Our goal is to elucidate kinetic and spatial organization effects in these situations. We first discuss irreversible two-species annihilation for spatially homogeneous systems in Sect. 7.2. For equal initial densities of the two species, some basic geometric properties of the mosaic of continuously growing A and B domains are outlined. Although domains grow diffusively, a multiplicity of independent length scales is needed to characterize the spatial distribution of reactants properly [7.36,37]. In Sect. 7.3, we discuss the kinetics of the reaction under steady, spatially homogeneous input, where an exact solution for the continuum system can be derived [7.16-18]. For the special case of no transport (the monomer-monomer model) and infinitely large input rate, the lattice system can be mapped onto a kinetic Ising model [7.38], a birth-death process (in one dimension only) [7.39], and the "voter" model [7.40]. Through these mappings, the exact solution for the kinetics and the spatial correlations can be found. These results account for the coarsening of domains and the ultimate saturation of the surface by a single species. In Sect. 7.4, we discuss the reaction kinetics of the spatially inhomogeneous system [7.23-35]. Dimensional arguments of a mean-field spirit are used to characterize the geometry of the "reactive zone" at the interface. This approach is inadequate in low spatial dimensions, however, and we discuss analytical approaches and numerical investigations of interfacial profiles in one dimension. We give a brief summary in Sect. 7.5.

7.2 Irreversible Homogeneous Reactions

7.2.1 Decay of the Density

Consider two diffusing species A and B which are originally distributed at random with respective concentrations $c_A(0)$ and $c_B(0)$. When two particles of opposite species approach within the (fixed) reaction radius, they react irreversibly to form a third, inert species, which is then disregarded. For unequal initial concentrations, the minority concentration quickly decays to zero, whereas the majority concentration approaches a constant. For equal initial densities, the concentration of each species, $c(t)$, decays as a power law in time. Associated with this relatively slow kinetics is a large-scale spatial organization of reactants in low spatial dimensions. These features are the focus of this section.

To estimate the decay of the density within a mean-field approximation, note that in a time interval of order $1/c$, each particle will come into contact with another particle, on average. Consequently, in a time interval $\Delta t \propto 1/kc$, where k is the reaction rate, the concentration will decrease by an amount of order c, that is, $\Delta c \propto -c$. This gives the mean-field rate equation

$$\dot{c} \cong \frac{\Delta c}{\Delta t} \propto -kc^2, \tag{7.1}$$

with solution $c(t) = c(0)/(1 + kc(0)t) \sim (kt)^{-1}$. Thus the exponent of the decay is -1, and the time scale is set by k. This exponent value turns out to be correct only for spatial dimension $d \geq 4$, corresponding to the regime of validity of the mean-field prediction. However, this prediction for the exponent is clearly incorrect for spatial dimension $d < 2$ because the time interval for reactions to occur is now $\Delta t \propto \ell^2/D$, where $\ell \propto c^{-1/d}$ is the typical interparticle spacing and D is the diffusion coefficient. Notice that the reaction rate does not enter into this time scale because random walk trajectories are compact. Thus if two reactants collide once, they will collide an infinite number of times and the reaction rate rescales to a large value. The rate equation now becomes

$$\dot{c} \cong \frac{\Delta c}{\Delta t} \propto -Dc^{1+2/d}, \tag{7.2}$$

with solution $c(t) \sim (Dt)^{-d/2}$. This approach shows how microscopic details determine the effective "order" of the reaction.

Both of the above predictions for $c(t)$ are incorrect for $d \leq 4$ because domains containing only one of the two species form [7.10-14,37]. This invalidates the homogeneity assumption of the mean-field approximation. Explicit accounts of the local density fluctuations are needed to understand the long-time behavior of $c(t)$. Within a scaling formulation, this can be accomplished by considering a system with slightly different initial densities [7.11]. For this system, there exist regions of characteristic size $\xi \propto (\delta c)^{-2/d}$, where δc is the initial density difference, below which the identity of the local majority species is ambiguous because of local density fluctuations. Consequently, for times less than a characteristic time scale $t_\xi \sim \xi^2/D$, the density decays as a power law, while a faster-than-power-law decay is expected at longer times. Since this time scale is simply related to the conserved density difference, it is possible to infer the form of the decay through scaling ideas.

An explicit, but qualitative, version of the above approach is to note that the number difference between As and Bs in a finite volume Ω of linear dimension $\mathcal{L}$ remains nearly constant during the time for a particle to traverse the volume by diffusion, $t_\mathcal{L} \sim \mathcal{L}^2/D$. However, at $t = 0$, this difference is of the order of the square root of the initial particle number,

$$N_A - N_B \approx \pm\sqrt{c(0)}\mathcal{L}^{d/2}. \tag{7.3}$$

If we assume that As are initially the local majority in Ω, then for $d \leq 4$, essentially no Bs will remain at time $t_{\mathcal{L}}$. Thus $N_A(t_{\mathcal{L}})$ is approximately equal to $\sqrt{c(0)}\mathcal{L}^{d/2}$. Elimination of $\mathcal{L}$ in favor of t gives

$$c(t) \approx N(t)/\mathcal{L}^d \sim \sqrt{c(0)}\,(Dt)^{-d/4} \qquad (d \leq 4). \qquad (7.4)$$

Thus a homogeneous system is predicted to evolve into a mosaic of domains whose individual identities are determined by the species in the local majority in the initial state. At time t, these domains will be of typical length $\sqrt{Dt}$ within which a single species of density $\sqrt{c(0)}(Dt)^{-d/4}$ remains.

The above line of reasoning is invalid for $d > 4$, however. To appreciate what occurs in this case, suppose that a reaction-generated domain mosaic is used as an initial condition. For domains of linear dimension $\mathcal{L}$ and local concentration of order $\mathcal{L}^{-d/2}$, let us estimate the probability that an A particle is unsuccessful in crossing a typical B domain. For $d > 4$, the particle needs $\mathcal{L}^2$ time steps to cross and will visit $\mathcal{L}^2$ distinct B-occupied sites during the traversal. At each site, the A will react with a probability that is of the order of the B concentration, $\mathcal{L}^{-d/2}$. Therefore the probability that an A particle is unsuccessful in traversing a B domain is of the order of $\mathcal{L}^{(4-d)/2}$. Since this vanishes as $\mathcal{L} \to \infty$ if $d > 4$, the domain mosaic is unstable to diffusion for $d > 4$.

A more rigorous argument for the fluctuation-dominated behavior focuses on the evolution of the local concentration difference $c_-(\boldsymbol{x},t) \equiv c_A(\boldsymbol{x},t) - c_B(\boldsymbol{x},t)$. This difference evolves by pure diffusion [7.10], so that the time dependence of the Fourier transform is simply $c_-(\boldsymbol{k},t) = c_-(\boldsymbol{k},t=0)e^{-Dk^2t}$. At long times, the existence of domains implies that there is minimal coexistence of As and Bs. Thus the product $c_A(\boldsymbol{x},t)c_B(\boldsymbol{x},t) = 0$, which implies that

$$\int d\boldsymbol{x}\, c_A(\boldsymbol{x},t)^2 \sim \frac{1}{2}\int d\boldsymbol{k}\, |c_-(\boldsymbol{k},0)|^2 \exp(-Dk^2t)$$
$$\propto (Dt)^{-d/2}\int d\boldsymbol{q}\, |c_-(\boldsymbol{q}/(Dt)^{1/2},t)|^2 \exp(-q^2). \qquad (7.5)$$

For a random initial condition, $|c_-(\boldsymbol{k},t=0)|^2 = N$ for all $\boldsymbol{k}$, since the mean-square involves the sum of N random unit vectors. Thus the integral over $\boldsymbol{q}$ in (7.5) is independent of t, so that $\langle c_A(\boldsymbol{x},t)^2\rangle \sim \frac{N}{V}(Dt)^{-d/2}$. Finally the assumption of no cross correlations implies that $\langle c_A(\boldsymbol{x},t)^2\rangle \cong \langle c_A(\boldsymbol{x},t)\rangle^2$ and (7.4) is reproduced. Notice that the random initial condition is a crucial aspect for obtaining the anomalous slow decay. In particular, for correlated initial conditions with no long-wavelength fluctuations in $c_-(\boldsymbol{x},t)$, the integral over $\boldsymbol{q}$ will vanish as $t \to \infty$, thus invalidating the above reasoning [7.12,41].

It is also instructive to give a dimensional argument for the $\sqrt{c(0)}$ prefactor in the expression (7.4) for the density. There are only two dimensionless parameters that can be constructed using the available model parameters D, k, and $c(0)$, namely

$$\tau = Dc(0)^{2/d}t, \qquad x = \frac{Dc(0)^{(2-d)/d}}{k}. \qquad (7.6)$$

Thus on purely dimensional grounds, $c(t)$ must be equal to $c(0)\Phi(x,\tau)$ for arbitrary x and τ. If $c(t) \propto t^{-d/4}$ as $t \to \infty$, then $\Phi(x,\tau)$ must have the form $C(x)\tau^{-d/4}$ as $\tau \to \infty$. This reproduces (7.4), except for a possible prefactor which depends on x. If one further assumes that the asymptotic behavior of the concentration is independent of k, one obtains exactly (7.4). The numerical verification of the $\sqrt{c(0)}$ prefactor is not fully settled, however. While it is reasonably confirmed in one dimension [7.11], Cornell *et al.* appear to find a different initial density dependence in their two-dimensional simulations [7.42]. Further work is needed to clarify the origin of this discrepancy.

By further assuming that the limit $k \to 0$ corresponds to the mean-field behavior, one obtains $\Phi(x,\tau) \approx x/\tau$, as $x \to \infty$. This indicates that the crossover from mean-field to diffusion-limited behavior occurs when $\tau \approx x^{4/(4-d)}$. Thus if $k \ll 1$, there is a large temporal domain of mean-field behavior [7.11]. This is physically reasonable, since the low reaction probability renders small-scale heterogeneities transparent. For long times, however, sufficiently large heterogeneities occur which are opaque to the local minority species. Thus the fluctuation-dominated behavior of (7.4) should result in sufficiently low spatial dimension.

The fluctuation-controlled kinetics is not qualitatively modified by varying the diffusion constants of the two species, even in the extreme case where one species is immobile. The basic analysis about fluctuations can also be generalized to fractals [7.43-45,37], where one obtains that $c(t_{\mathcal{L}})$ is approximately given by $\sqrt{c(0)}\mathcal{L}^{-d_f/2}$, and $t_{\mathcal{L}}$ is equal to $\mathcal{L}^{d_w}$, leading to the time dependence, $c(t) \sim t^{-d_s/4}$. Here, d_f is the fractal dimension, d_w is the dimension of a random walk on the fractal, and d_s the spectral dimension. While an alternative line of reasoning has recently been given [7.44], numerical results do appear to suggest that the decay kinetics is determined by simply making the replacement $d \to d_s$ [7.45]. For a short introduction to fractals see Chap. 1. For dynamical processes on fractals see Chap. 3 in [7.46].

Thus the kinetics of the $A + B \to I$ reaction is determined by the competition between A and B domains. This process depends fundamentally on geometric features of these domains, such as the interparticle distances between closest-neighbor like and unlike species and the domain size distribution. Some unexpected properties emerge from these quantities as we discuss in the rest of this section.

7.2.2 Interparticle Distances

A surprising feature of the spatial distribution of reactants is that the typical distances between AA and AB closest-neighbor pairs, see Fig. 7.1, ℓ_{AA} and ℓ_{AB}, respectively, grow with different powers of t for $d < 3$ [7.37]. This indicates that there is a nontrivial modulation in the reactant density over the extent of a domain.

To determine the time evolution of ℓ_{AB} in one dimension, consider the behavior of c_{AB}, the concentration of closest-neighbor AB pairs. In a time

Fig. 7.1. Definition of fundamental interparticle distances in one dimension: the typical distance between closest-neighbor same-species particles, ℓ_{AA}, the distance between closest-neighbor unlike species, ℓ_{AB} (the gap between domains), and the typical domain length, L

increment $\Delta t \sim \ell_{AB}^2/D$, there is sufficient time for essentially all such AB pairs to react, since one-dimensional random walks are compact. Consequently, the number of reactions per unit length is of the order of c_{AB}. Thus the rate of change of the concentration is

$$\frac{\Delta c}{\Delta t} \approx -k \frac{c_{AB}}{\ell_{AB}^2/D}. \tag{7.7}$$

This relation holds only for spatial dimension $d < 2$, since the compactness of random walks is an essential hypothesis. Since we independently know the left-hand side of (7.7) from $c(t)$ itself, the behavior of c_{AB} determines the time dependence of ℓ_{AB}. In one dimension, c_{AB} scales as $(Dt)^{-1/2}$, since the typical domain size is $(Dt)^{1/2}$ and there is one AB pair per domain. Therefore [7.37]

$$\ell_{AB} \sim c(0)^{-1/4} (Dt)^{3/8}. \tag{7.8}$$

Thus at least three lengths are needed to characterize the spatial distribution of reactants: the average domain size, which scales as $(Dt)^{1/2}$; the typical interparticle spacing, which scales as $c(t)^{-1} \propto t^{1/4}$; the interdomain gap ℓ_{AB}. (see Figs. 7.1 and 7.2) The stronger time dependence of ℓ_{AB} compared to that of the typical interparticle spacing is a manifestation of an effective "repulsion" between opposite species particles. Nearby opposite-species pairs annihilate preferentially, leaving behind a population where opposite species are separated by a distance which is larger than the typical interparticle separation.

The time dependence of the *average* distance between same-species nearest-neighbor pairs, $\langle \ell_{AA} \rangle$, is characterized by an exponent which is close to, but measurably larger than, the value of $1/4$ that one might naively expect. This larger exponent value arises because the local density vanishes near the domain edge and causes the average distance between neighboring particles to be larger than the typical distance. Below, we show that this modulation in the domain profile leads to a multiplicative logarithmic factor in the scaling behavior of $\langle \ell_{AA} \rangle$ compared to the typical value.

Remarkably, the average interparticle distances depend on the mobility ratio of both species. The most interesting situation is that of one immobile species (B), where the effective exponent for $\langle \ell_{BB} \rangle$ slowly approaches its asymptotic value, whereas the effective exponent for $\langle \ell_{AA} \rangle$ is roughly the same as in the

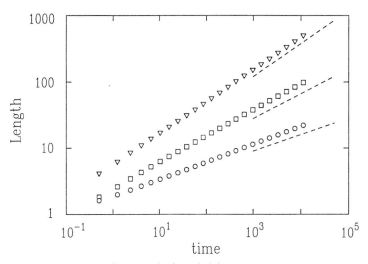

Fig. 7.2. Time dependence of $\langle \ell_{AA} \rangle$ (○), $\langle \ell_{AB} \rangle$ (□), and the average domain length, $\langle L \rangle$ (▽) for a one-dimensional system when both species have the same mobility. The *dashed* lines of respective slopes of 1/4, 3/8, and 1/2 serve as guides to the eye

equal mobility case. Interestingly the effective exponent for $\langle \ell_{BB} \rangle$ remains *below* 1/4 over a large temporal range, even though the asymptotic value of this exponent can be bounded from below by 1/7. This anomaly stems from a short-distance power-law tail in the distribution of separations between neighboring Bs, a feature which is a vestige of the initial distribution of BB separations.

To generalize to spatial dimension $1 \leq d \leq 2$, note that the time dependence of ℓ_{AB} should still follow by applying (7.7), since it holds whenever random walks are compact. To estimate c_{AB}, we assume a smooth domain perimeter of length $t^{(d-1)/2}$ and that particles in the perimeter zone are separated by a distance of the order of ℓ_{AB}, irrespective of identity. This approach leads to the following generalization of (7.8):

$$\ell_{AB} \propto t^{\frac{(d+2)}{4(d+1)}}, \qquad c_{AB}(t) \propto t^{-\frac{d(d+3)}{4(d+1)}}, \qquad (7.9)$$

which yields $\ell_{AB} \sim t^{1/3}$ and $c_{AB}(t) \sim t^{-5/6}$ in $d = 2$.

In three and higher dimensions, the transience of random walks implies that two neighboring opposite species particles which are confined to a region of linear dimension ℓ_{AB} will react within a time of the order of ℓ_{AB}^d (rather than ℓ_{AB}^2). Consequently, (7.7) must be modified to

$$\frac{\Delta c}{\Delta t} \approx -k \frac{c_{AB}}{\ell_{AB}^d}. \qquad (7.10)$$

This relation, together with the assumption of a smooth interfacial region between domains, gives, in d dimensions:

$$c_{AB} \approx t^{-\frac{d^2+5d-4}{4(2d-1)}}, \qquad \ell_{AB} \approx t^{\frac{d+2}{4(2d-1)}}, \qquad (7.11)$$

which coincides with (7.9) at $d = 2$. For $d = 3$ (7.11) yields

$$c_{AB} \approx t^{-1}, \qquad \ell_{AB} \approx t^{1/4}. \tag{7.12}$$

These exponents represent limiting values: indeed, there is no mechanism whereby $\langle \ell_{AB} \rangle$ could become less than $\langle \ell_{AA} \rangle$, and the non-trivial scaling of interparticle distances disappears in three dimensions and above. This has been confirmed in large-scale three-dimensional numerical simulations [7.47]. Thus for $d \geq 3$, there is no appreciable depletion in the average density in the interfacial region between domains. Thus the predictions of (7.11) are expected to be valid only below an upper critical dimension which is equal to 3. The existence of this critical dimension for the behavior of the interparticle distances is remarkable, as the upper critical dimension equals 4 for the behavior of $c(t)$ and equals 2 for the case of two-species annihilation with input or back reactions.

7.2.3 The Domain-Size Distribution in One Dimension

A more complete understanding of the spatial organization of reactants can be gained by considering the distribution of domain sizes. This distribution can be deduced indirectly by considering the dynamics of the domain walls (W), roughly defined as the midpoint between two unlike neighboring particles. Since individual particles diffuse, it is natural to postulate that the walls also move diffusively. This approximation is questionable, however, since annihilation events can lead to large distance jumps in the wall position. Nevertheless, we adopt the hypothesis of diffusing domain walls, since it leads to several correct predictions about the domain size distribution. Since domain walls annihilate on contact, their dynamics should coincide with that of the density in the exactly soluble single-species annihilation model, $W + W \rightarrow I$ [7.48-50]. Thus the distribution of domain sizes in two-species annihilation corresponds to the interparticle distance distribution in single-species annihilation. This latter quantity is known to decay exponentially for large separations and vary linearly in separation in the small-distance limit [7.48].

Consequently the number of domains of length L at time t, $N(L, t)$, may be written in the scaling form

$$N(L, t) \sim \frac{1}{t} \Phi(L/\sqrt{t}), \tag{7.13}$$

where the scaling function $\Phi(x)$ has the following asymptotic limits:

$$\begin{aligned} \Phi(x) &\sim x & x &\rightarrow 0 \\ &\sim e^{-Cx}, & x &\rightarrow \infty. \end{aligned} \tag{7.14}$$

The prefactor $1/t$ in (7.13) ensures that the system length, $\sum L N(L, t)$, is time independent. This scaling form has several non-trivial consequences which have been verified numerically. For example, the small-x behavior of $\Phi(x)$, leads to a $t^{-3/2}$ decay of the number of fixed-size domains.

When the Bs are immobile, the domain size distribution develops an $x^{-1/2}$ power law tail in the small-size limit. This exponent value can be accounted for by the following random walk argument. Consider the initial density difference $c_-(x,0) = c_A(x,0) - c_B(x,0)$ which traces a random walk along x. As a function of time, this function relaxes by diffusion in the regions where the density difference is positive, while only the reactions influence the evolution of regions of negative density difference. Thus any region where $c_-(x,0)$ was initially positive will ultimately leave a gap of the order of its length between two B particles. Thus the gaps between consecutive Bs are distributed in the same way as first return times to the origin of an ordinary one-dimensional random walk [7.51], that is, as $x^{-3/2}$. From this, it follows that the domain lengths, which are the sum of the gaps of the distances between the constituent B particles, are distributed as $x^{-1/2}$.

Thus for immobile Bs, the scaling form for the distribution of B domain sizes is

$$N(L,t) \sim \frac{1}{t}\Psi(L/\sqrt{t}), \qquad (7.15)$$

with $\Psi(x) \sim x^{-1/2}$ as $x \to 0$ to account for the power-law tail for small domain sizes. Consequently the number of domains of fixed size L vanishes as $t^{-3/4}$ as $t \to \infty$, as has been confirmed numerically [7.37].

7.2.4 The Domain Profile

A revealing physical picture of the spatial distribution of the reactants can be obtained from the average density profile of the domains. Within a scaling formulation, it is natural to define the density profile as the probability of finding a particle at a fixed scaled distance from the domain midpoint, using either the typical domain size (the canonical profile) or the size of the domain to which the particle actually belongs (the microcanonical profile) as the rescaling factor. For the canonical profile, the ordinate is therefore proportional to the density times the number of domains. By dividing by these factors, one obtains the canonical density profile, $P^{(C)}(x)$, the probability of finding a particle at a distance $x\sqrt{Dt}$ from the center of the domain at time t. In contrast, the microcanonical profile $P^{(M)}(x)$ is obtained by superposing the density profiles of all domains, when each domain is rescaled to a fixed length, and then dividing by the product of the average concentration, the number of domains, and the average domain size. For numerical simulations see Fig. 7.3.

The roughly sinusoidal microcanonical profile is reminiscent of the probability distribution for pure diffusion in a fixed-size absorbing domain. In fact, the $t^{1/2}$ growth of the average domain size in two-species annihilation is at the limit of applicability of the adiabatic approximation. Thus the adiabatic approximation should provide a valid prediction for the domain profile [7.52]. For the absorbing domain $[-L(t)/2, L(t)/2]$ with $L(t)$ growing deterministically as $t^{1/2}$, the adiabatic approximation predicts the density profile $\cos(\pi x/L(t))$ (as in the case of a static absorbing domain). A more complete model would ac-

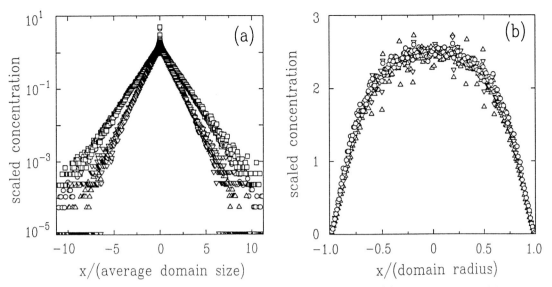

Fig. 7.3. (a) The canonical density profile of As at $t = 194$ (o) and $t = 1477$ (□), and Bs at $t = 194$ (△) and $t = 1477$ (▽) for the case of immobile Bs. (b) The corresponding microcanonical density profile for the Bs at $t = 194$ (△), $t = 1477$ (▽), and $t = 11222$ (o)

count for the stochastic motion of the domain walls by considering the extreme particle in each of the two enclosing domains to be wall particles W, with the As and Ws reacting via $W + W \rightarrow I$ and $W + A \rightarrow W$ to describe two-species annihilation. This mapping to an exactly soluble three-particle system of a single A between two Ws [7.53] neglects the possibility that a W particle can disappear in an encounter with another W before reacting with the A. Proceeding nevertheless, we postulate that the Ws diffuse with a diffusion constant D_W, so that the determination of the density profile of the As reduces to obtaining the probability distribution of a random walker between the two absorbing wall particles. The exact solution to this three-particle problem again predicts the profile $\cos(\pi x/L(t))$, independent of the diffusion coefficients of the two species.

The linear decay of the tail of the microcanonical profile and the roughly constant concentration in the domain core provides a revealing alternative approach for determining the $t^{3/8}$ growth of the gap length ℓ_{AB} [7.54]. Within a quasi-static approximation, the density near the domain edge varies as $u/t^{3/4}$. Here u is the distance to the domain edge (defined to be at $u = 0$), and the time dependence of the denominator ensures that this density profile matches smoothly onto the core density of order $t^{-1/4}$, when $u \sim t^{1/2}$. The location of the "first" particle in this distribution, u_1, is given by the condition that there is one particle in the range $[0, u_1]$, that is, $\int_0^{u_1} du' \, u'/t^{3/4} = 1$. This immediately gives $u_1 \sim t^{3/8}$, which also gives the time dependence of ℓ_{AB}.

Although the canonical and microcanonical density profiles are obtained by very different prescriptions, they are simply related. Namely, the contribution to the canonical density profile at scaled position x is equal to the microcanonical density profile at scaled position x/α, times the probability of finding a domain of scaled length α, summed over the possible values of α. Therefore

$$P^{(C)}(x) = \int_x^\infty d\alpha \, \Phi(\alpha) \, P^{(M)}(x/\alpha), \qquad (7.16)$$

where $\Phi(\alpha)$ is the scaling function for the probability of finding a domain of size $\alpha t^{1/2}$. From the large-distance asymptotic form of Φ, we immediately conclude that the large-distance tail of the canonical distribution should decay exponentially in x as has been observed numerically over many decades [7.34]. An additional noteworthy feature of the canonical profile for the immobile species is the sharp peak near the origin due to the $x^{-1/2}$ singularity in the distribution of small-sized B domains. In contrast, the canonical profile for the As has a smooth peak at the origin.

It is worth mentioning that the spatial distribution of the reaction product for initially separated reactants (Sect. 7.4) appears to have the same exponential decay as the canonical profile. This feature suggests that there may be a simple explanation for the nature of the reaction zone in terms of a mapping to an equivalent single-species annihilation process.

7.2.5 The Interparticle-Distance Distribution

The inhomogeneity of the domain profile also governs the distribution of interparticle distances. Particles are separated by the typical interparticle separation within the core of the domain, but systematically become more sparse as the domain edge is approached. This large-scale modulation in the interparticle spacings leads to an interparticle-distance distribution which is controlled both by the gap length ℓ_{AB} and the typical spacing between same species particles ℓ_{AA}.

The crucial element in the determination of the spatially averaged distance distribution is the incorporation of the density modulation within a domain. Thus to calculate the average probability of finding a local AA separation of length ℓ_{AA}, we assume a local Poisson distribution of distances, $\rho e^{-\rho \ell_{AA}}$, in which ρ is of the order of $t^{-1/4}$ in the domain core and vanishes *linearly* in the distance to the domain edge. Upon averaging the local Poisson distribution over this slow density variation, the spatially-averaged distribution of nearest-neighbor distances is

$$P_{AA}(x,t) \simeq x^{-2} \exp(-x t^{-1/8}), \qquad (7.17)$$

where $x = \ell_{AA} t^{-1/4}$ is the scaled distance. In unscaled units, the decay length in the exponential increases as $t^{3/8}$ rather than the $t^{1/4}$ growth that would be expected naively.

From this interparticle-distance distribution, the corresponding moments are

$$M_n(t) \equiv \langle \ell_{AA}^n(t) \rangle^{1/n} = \left[\int_0^\infty x^n \, P_{AA}(x,t) \, dx \right]^{1/n},$$

$$\sim \begin{cases} t^{1/4}, & n < 1; \\ t^{1/4} \ln t, & n = 1; \\ t^{(3n-1)/8n}, & n > 1. \end{cases} \qquad (7.18)$$

While numerical simulations are consistent with most of these predictions, the case $n = 1$ is somewhat problematical and the source of the discrepancy has not yet been resolved.

Rather unusual behavior occurs when one species (B) is immobile. Regions of initially closely spaced Bs contribute to the $x^{-3/2}$ tail in the probability of finding a BB closest-neighbor distance equal to x for $x < t^{1/2}$. For such a distribution, the time dependence for the reduced moments of interparticle distances between the immobile species is

$$M_n(t) \sim \begin{cases} \text{const.,} & n < 1/2; \\ \ln t, & n = 1/2; \\ t^{(2n-1)/4n}, & n > 1/2. \end{cases} \qquad (7.19)$$

Owing to the relatively large number of small BB distances, the reduced moments of order less than $1/2$ have a finite limiting value as $t \to \infty$, while higher-order moments grow indefinitely and are dominated by very large BB distances of the order of $t^{1/2}$. These results indicate that immobile Bs have a fractal spatial distribution (see Fig. 7.0).

To generalize to higher dimensions for equally mobile species, we treat the density profile as the probability distribution of a particle diffusing within an absorbing sphere whose radius expands as $\sqrt{t}$. The adiabatic approximation still applies in this case, so that the density should decay linearly to zero in the radial coordinate near the extremity of the domain. By following exactly the same calculation as in one dimension, the reduced moments $M_n(t)$ are

$$M_n(t) \sim \begin{cases} t^{1/4}, & n < 2; \\ t^{1/4}(\ln t)^{1/2}, & n = 2; \\ t^{(2n-1)/6n}, & n > 2. \end{cases} \qquad (7.20)$$

These predictions have yet to be adequately tested, however. This stems, in part, from the algorithmic difficulty in defining a domain in higher dimensions. It is natural to define a microcanonical density profile by superposing the profiles from one-dimensional slices through the system. This gives, however, a density which varies non-linearly in the distance to the edge of the domain. The reason for this untenable feature is as yet unresolved. It should also be noted that the limiting case of immobile Bs has not been treated adequately in greater than one dimension. Indeed the Bs develop a self-similar distribution (Fig. 7.0) which has not yet been properly analyzed.

7.3 Reactions with Particle Input

An important realization of two-species annihilation is the situation where reactants are continuously fed into the system. This input may arise from a constant external source [7.15-18] or from the back reaction $I \to A + B$ [7.11,55-61]. In the former case, the long-time behavior depends sensitively on the interplay between the input, the rate of transport, and the reaction itself. Although a steady state is to be anticipated when there is input, transient behavior persists if the input of the two species are not sufficiently correlated with each other. To determine the long-time behavior of the driven reaction, we make use of the fact that the equation for the concentration difference is reaction independent. This simplification permits a simple analysis of whether a steady state is actually reached. Interestingly, the approach to the long-time asymptotic behavior for both a steadily driven reaction and a reversible system is not exponential, as in mean-field theory, but occurs at a power-law rate [7.11,55-61]. This behavior originates from the diffusive relaxation of a conserved reaction variable ($c_A - c_B$ for the steadily driven system and $c_A + c_B - 2c_I$ for the reversible system) and is therefore independent on microscopic details.

When the species are immobile, two-species annihilation with particle input is closely related to the monomer-monomer model of surface catalysis [7.19-22]. This limiting situation merits attention because of the connection with catalysis and related interacting-particle models [7.40], and also because of the variety of interesting methods for analyzing this system. The input of particles now corresponds to their adsorption onto a reactive substrate whose presence promotes the transformation of neighboring A and B reactants into an AB product, which immediately desorbs from the surface. Since the input in the catalytic system is typically uncorrelated, there will be no steady state, but rather the surface eventually fills with only a single species and the reaction stops. We will discuss the rate at which this saturation phenomenon occurs.

7.3.1 Steady Input and Diffusing Reactants

For a constant feed of mobile reactants, the reaction-diffusion equations for the concentrations are

$$\frac{\partial c_{A,B}(\boldsymbol{x}, t)}{\partial t} = D\nabla^2 c_{A,B}(\boldsymbol{x}, t) - \mathcal{R} + \eta_{A,B}(\boldsymbol{x}, t), \tag{7.21}$$

where $\mathcal{R}$ represents the reaction term whose precise form is unknown, and $\eta_{A,B}(\boldsymbol{x}, t)$ are the input rates of A and B, respectively, at position $\boldsymbol{x}$ and time t. Since the reaction term is the same for both species, the concentration difference $c_-(\boldsymbol{x}, t) \equiv c_A(\boldsymbol{x}, t) - c_B(\boldsymbol{x}, t)$ satisfies

$$\frac{\partial c_-(\boldsymbol{x}, t)}{\partial t} = D\nabla^2 c_-(\boldsymbol{x}, t) + \eta_-(\boldsymbol{x}, t), \tag{7.22}$$

where $\eta_-(\boldsymbol{x}, t) \equiv \eta_A(\boldsymbol{x}, t) - \eta_B(\boldsymbol{x}, t)$, and the (unknown) form of the reaction term does not enter. The solution for the instantaneous concentration difference can now be found by standard Fourier transform methods. Following ben-Avraham and Doering [7.16], it is more revealing to consider the temporally-averaged two-particle correlation function

$$\langle c_-(\boldsymbol{k}, t)c_-(\boldsymbol{k}', t) \rangle = \int_0^t dt_1 \int_0^t dt_2\, e^{-Dk^2(t-t_1)-Dk'^2(t-t_2)} \langle \eta_-(\boldsymbol{k}, t_1)\eta_-(\boldsymbol{k}', t_2) \rangle, \tag{7.23}$$

whose behavior depends on the precise form of the particle input.

Consider first spatially and temporally uncorrelated input with a Poisson distribution for the number of particles. Thus the noise correlation is

$$\langle \eta_A(\boldsymbol{x},t)\eta_A(\boldsymbol{x}',t')\rangle = \mathcal{N}^2 + \mathcal{N}\delta(\boldsymbol{x} - \boldsymbol{x}')\delta(t - t'). \tag{7.24}$$

Here $\mathcal{N} = \langle \eta_A(\boldsymbol{x},t)\rangle = \langle \eta_B(\boldsymbol{x},t)\rangle$ is the average number of particles injected per unit volume per unit time, and the Poissonian nature of the input is reflected by the fact that $\langle \eta_A(\boldsymbol{x},t)^2\rangle = \mathcal{N}^2 + \mathcal{N}$. Substitution of the Fourier transform $\langle \eta_-(\boldsymbol{k},t)\eta_-(\boldsymbol{k}',t')\rangle = 2\mathcal{N}\delta(\boldsymbol{k} + \boldsymbol{k}')\delta(t - t')$ into (7.23) and inverting, gives, for the concentration difference squared at a given point in an infinite volume,

$$\langle c_-^2(\boldsymbol{x},t)\rangle = \frac{\mathcal{N}}{D}\int \frac{d\boldsymbol{k}}{(2\pi)^d}\frac{1 - e^{-2Dk^2 t}}{k^2} \sim \begin{cases} \text{const.,} & d > 2; \\ \ln t, & d = 2; \\ t^{(2-d)/2}, & d < 2. \end{cases} \tag{7.25}$$

In a finite volume, however, the spectrum of wavenumbers is discrete. Thus at long times only a single (smallest) wavenumber dominates in the sum corresponding to the finite-volume analogue of (7.25). The long-time behavior of $\langle c_-^2(\boldsymbol{x},t)\rangle$ can therefore be estimated as the $k = 0$ component of (7.25), yielding

$$\langle c_-^2(\boldsymbol{x},t)\rangle \sim t. \tag{7.26}$$

These results have a simple interpretation. In a finite volume, diffusive relaxation to spatial uniformity has time to occur and the concentration difference at a given position is governed only by imbalance in the input of As and Bs. In an infinite volume, however, diffusional relaxation modes of all time scales exist, and they aid in reducing the fluctuations compared to those characteristic of a finite volume system. Thus the growth of $\langle c_-^2(\boldsymbol{x},t)\rangle$ is slower than linear in t for $d \leq 2$, while diffusive relaxation is sufficiently fast that a steady state is achieved for $d > 2$.

However, a steady state is more likely to be reached for an input that explicitly conserves the particle number difference. For example, when As and Bs are introduced simultaneously (e.g., from the back reaction $I \rightarrow A + B$) and at a constant rate, the noise correlation obeys $\langle \eta_-(\boldsymbol{x},t)\eta_-(\boldsymbol{x}',t')\rangle = 2\langle \eta_A(\boldsymbol{x},t)\eta_A(\boldsymbol{x}',t')\rangle - 2\langle \eta_A(\boldsymbol{x},t)\eta_B(\boldsymbol{x}',t')\rangle$, with the second term equal to $2\mathcal{N}^2 + 2\mathcal{N}\delta(t - t')p(\boldsymbol{x} - \boldsymbol{x}')$, rather than vanishing as in the case of temporally uncorrelated input. Here $p(\boldsymbol{r})$ is the probability that an AB pair is created with separation $\boldsymbol{r}$. From the Fourier transform for the noise correlation function $\langle \eta_-(\boldsymbol{k},t)\eta_-(\boldsymbol{k}',t')\rangle = 2\mathcal{N}\delta(\boldsymbol{k} + \boldsymbol{k}')[1 - p(\boldsymbol{k})]\delta(t - t')$, where $p(\boldsymbol{k})$ is the Fourier transform of $p(\boldsymbol{r})$, the correlation function for $c_-(\boldsymbol{x},t)$ is

$$\langle c_-^2(\boldsymbol{x},t)\rangle = \frac{\mathcal{N}}{D}\int \frac{d\boldsymbol{k}}{(2\pi)^d}[1 - p(\boldsymbol{k})]\frac{1 - e^{-2Dk^2 t}}{k^2}. \tag{7.27}$$

The long-time behavior of the correlation function in an infinite volume is now controlled by the small-k behavior of $p(\boldsymbol{k})$. The integral converges only if $|1 - p(\boldsymbol{k})| < |\boldsymbol{k}|^\alpha$, with $\alpha > 2 - d$. This, in turn, implies that $p(\boldsymbol{x}) < |\boldsymbol{x}|^{-\beta}$, with $\beta > 3 - d$. Thus AB pairs must be fed in with sufficient spatial correlation to achieve steady behavior in an infinite system. For a finite-volume system, the factor $1 - p(\boldsymbol{k})$ vanishes at $k = 0$, which implies that steady behavior is always achieved, independent of the extent of the spatial correlation of AB pairs.

7.3.2 The Approach to Asymptotic Behavior

The approach to the long-time behavior for both two-species annihilation with steady input and in the presence of back reactions follows the power law $|c_{A,B}(t) - c_\infty| \sim t^{-d/2}$ for any spatial dimension d [7.11,55-61]. This result is essentially independent of microscopic details. In contrast, the asymptotic density is approached exponentially in the mean-field limit, suggesting that the upper critical dimension of such systems is infinite. The power-law approach to the asymptotic state arises from the microscopic conservation law. The conserved quantity equals $c_A - c_B$ in the case of $A + B \to I$ with steady input, and equals $c_A + c_B - 2c_I$ for the case of $A + B \rightleftharpoons I$. Since this conserved quantity relaxes only by diffusion (although driven by the external noise in the case of $A - B$ for $A + B \to 0$ with input) an excess of the conserved quantity in a given region decays as $t^{-d/2}$.

More quantitatively, consider $P(\boldsymbol{x}, t) \equiv \langle c_-(\boldsymbol{x}, t)c_-(0,0) \rangle$. From the initial reaction-diffusion equations for the system, the equation of motion of this correlation function is

$$\frac{\partial P(\boldsymbol{x}, t)}{\partial t} = D\nabla^2 P(\boldsymbol{x}, t) \qquad \text{with} \quad P(\boldsymbol{x}, 0) = \delta(\boldsymbol{x}). \qquad (7.28)$$

Such an equation holds for any linear quantity which is unaffected, on average, by the reaction (statistical conservation law). It therefore follows that $P(\boldsymbol{x}, t)$ decays as $t^{-d/2}$ in any dimension d. A similar equation holds for the case of $A + B \rightleftharpoons I$, so that identical conclusions may be drawn. A remarkable aspect of the latter result is that the system with back reactions satisfies detailed balance and tends to an uncorrelated equilibrium state, corresponding to a complete lack of interaction. Thus the existence of diffusive dynamics and a conservation law generally implies a slow approach to an equilibrium state. One can therefore conjecture a similar approach to equilibrium for the kinetic Ising model with Kawasaki dynamics, a feature which has in fact been proved by Spohn [7.62].

7.3.3 Immobile Reactants; Equivalence to Catalysis, Kinetic Ising Models, and Branching Random Walks

When the reactants are immobile, driven two-species annihilation can be regarded as an idealized catalysis model in which the input corresponds to the irreversible adsorption of gaseous reactants onto a catalytic surface, and a reaction occurs only when a nearest neighbor AB pair is created by the input [7.19-22]. These steps can be represented as

$$A + S \xrightarrow{k_A} A_s$$

$$B + S \xrightarrow{k_B} B_s \qquad (7.29)$$

$$A_s + B_s \xrightarrow{k_r} (AB) \uparrow + 2S.$$

If the adsorption rates for each species, k_A, k_B, are much greater than the surface reaction rate k_r, the process is *reaction-limited*, while the process is *adsorption-limited* otherwise. In the adsorption-limited case, a newly-formed nearest-neighbor $A_s B_s$ pair immediately reacts and desorbs. In the complementary reaction-limited process, an initially empty substrate quickly fills up, as any nearest-neighbor reactive pairs are quickly replaced by the (relatively) rapid input.

In the context of catalysis, the input is temporally uncorrelated, corresponding to the catalytic surface being placed in a large reservoir of reactants. The resulting non-conservation in the concentration difference implies that a finite-size system is unsteady, even for equal rates of A and B input. This leads to coarsening domains of A and B, a process which ends when the surface is completely covered by only A or B and no further reaction is possible [7.22].

We will focus on the reaction-limited process because it can be solved exactly and because of its connection to disparate statistical models. (Qualitatively similar behavior occurs in the adsorption-limited case.) In an elementary reaction event, a nearest-neighbor AB pair reacts and desorbs, and the two vacated sites are immediately replenished by AA, BB, BA, or AB. When the input rates of the two species are equal, each of these possibilities occurs with probability 1/4. This system can be mapped onto a kinetic Ising model by identifying an A as a spin up and a B as a spin down [7.38]. In the spin representation, the first two of the above pair updates correspond to single-spin flip events, while the third corresponds to spin exchange. Thus the evolution of the catalytic system is equivalent to a kinetic Ising model with mixed Glauber [7.63] and Kawasaki [7.64] dynamics at $T = 0$ and $T = \infty$, respectively. The monomer-monomer catalysis model is also related to the Voter model [7.40], an interacting-particle system in which each site of a lattice is in one of two states. The time evolution is defined by picking a site at random and defining its state to coincide with that of a randomly chosen nearest neighbor. This can be viewed as a population of infinitely gullible "voters" in which a randomly selected voter adopts the opinion of one of its nearest neighbors in an elemental time increment. In spin language, this updating corresponds to a zero temperature Glauber dynamics, since spin exchange $AB \to BA$ cannot occur.

a) **Mean-field theory.** In the mean-field limit, the concentration of As evolves according to

$$\dot{x} = 2(p-q)x(1-x). \qquad (7.30)$$

Here $p = k_A/(k_A + k_B)$ and $q = 1 - p$ are the relative adsorption probabilities of A and B. For $p \neq q$, the system approaches saturation exponentially quickly, with a time constant $\tau = 1/(p-q)$. When $p = q$, $\dot{x} = 0$; thus at the level of average densities the system is static.

More detailed information within the mean-field approximation is provided by the master equation for the probability distribution of the particle numbers, $P(n_A, n_B)$. To write this equation, consider the reaction on an N-site complete graph in which each pair of sites is connected. When an AB pair reacts, it is replaced by AA, BB, AB, or BA, with probabilities p^2, q^2, pq, and pq, respectively. Thus $P(n_A, n_B)$ evolves according to a stochastic process with the corresponding hopping probabilities

$$W(n_A, n_B \to n_A \pm 1, n_B \mp 1) = 2 \left\{ \frac{p^2}{q^2} \left(\frac{n_A}{N} \right) \left(1 - \frac{n_A}{N} \right), \right. \tag{7.31}$$

with $W(n_A, n_B \to n_A, n_B) = 1 - W(n_A, n_B \to n_A + 1, n_B - 1) - W(n_A, n_B \to n_A - 1, n_B + 1)$. From the master equation for $P(n_A, n_B)$, the corresponding Fokker-Planck equation for $P(x)$ in the continuum limit, when $p = q$, is

$$\frac{\partial P(x,t)}{\partial t} = \frac{1}{2N} \frac{\partial^2}{\partial x^2} \left(x(1-x) P(x,t) \right). \tag{7.32}$$

The state-dependent diffusion coefficient $D(x) = x(1-x)$ reflects the probability of a reaction event being proportional to the concentration of AB pairs, $x(1-x)$. Thus the evolution of the surface concentration is analogous to diffusion in a medium that is increasingly "sticky" near the extremities of a finite absorbing interval.

The solution to the Fokker-Planck equation can be written as an eigenfunction expansion over the Gegenbauer polynomials of order $3/2$ [7.22]. Interestingly, the lowest mode of this expansion is a constant, compared to the sinusoidal profile associated with diffusion on a finite absorbing interval with a spatially constant diffusion coefficient. The uniform probability distribution reflects a balance between the decrease in the diffusion coefficient near saturation and the depletion of probability at $x = \pm 1$. From the eigenfunction expansion, the survival probability $S(t) \equiv \int_0^{+1} P(x,t)\,dx$ decays exponentially at long times with a characteristic decay time equal to N; thus the typical time until a system saturates is proportional to N. The mean time until saturation for a given initial condition can be computed from the adjoint recursion formula for the first passage time to reach saturation when starting from an arbitrary initial-state site [7.65]. In the continuum limit, the solution to this adjoint equation is

$$t(x) = -2N \left[(1-x) \ln(1-x) + x \ln x \right], \tag{7.33}$$

where $t(x)$ denotes the first-passage time to reach saturation, $x = 0, 1$, when the initial concentration of As equals x. Thus the mean saturation time for system with equal initial concentration of As and Bs ($x = 1/2$) equals $2N \ln 2$, while the mean saturation time is proportional to $\ln N$ when starting very close to saturation, $x \cong 1 - 1/N$.

b) Mapping to a kinetic Ising model. Despite the inherent approximations of the mean-field approach, it provides an attractive intuitive picture of the kinetics for both the reaction-limited and adsorption-limited processes which turns out to be quantitatively valid when the substrate dimensionality d is greater than 2. However, the reaction-limited process can be solved exactly in all dimensions through the mapping onto the kinetic Ising model [7.38]. We present here the approach given by Krapivsky. The crucial aspect that leads to the solution is that the master equation for the probability density of a spin configuration is *linear*, and the corresponding hopping rates can be written directly in terms of the spins. Due to the constraint that the square of a spin value is unity, one can decouple and thus solve the hierarchy of equations for the time dependence of spin correlations.

According to the reaction-limited monomer-monomer dynamics, the probability of a given spin configuration, $P(\{s\}, t)$, evolves according to the master equation.

$$\frac{d}{dt} P(\{s\}, t) = \sum_{\boldsymbol{r}} [U_{\boldsymbol{r}}(\{s\}_{\boldsymbol{r}}) P(\{s\}_{\boldsymbol{r}}, t) - U_{\boldsymbol{r}}(\{s\}) P(\{s\}, t)]$$
$$+ \sum_{\boldsymbol{r}, \boldsymbol{r}'} [V_{\boldsymbol{r}, \boldsymbol{r}'}(\{s\}_{\boldsymbol{r}, \boldsymbol{r}'}) P(\{s\}_{\boldsymbol{r}, \boldsymbol{r}'}, t) - V_{\boldsymbol{r}, \boldsymbol{r}'}(\{s\}) P(\{s\}, t)]. \tag{7.34}$$

Here $\{s\}_{\boldsymbol{r}}$ denotes the configuration derived from $\{s\}$ with the spin at $\boldsymbol{r}$ reversed, while $\{s\}_{\boldsymbol{r}, \boldsymbol{r}'}$ has the spins at $\boldsymbol{r}$ and $\boldsymbol{r}'$ reversed. Further, $U_{\boldsymbol{r}}$ is the rate at which single spin flips occur at $\boldsymbol{r}$ while $V_{\boldsymbol{r}, \boldsymbol{r}'}$ is the rate at which the spin at $\boldsymbol{r}$ changes due to spin exchange between $\boldsymbol{r}$ and $\boldsymbol{r}'$, with $\boldsymbol{r}'$ a nearest-neighbor of $\boldsymbol{r}$. For the monomer-monomer catalysis model, these rates are given by

$$U_{\boldsymbol{r}}(\{s\}) = A[2z - s_{\boldsymbol{r}} \sum_{\boldsymbol{r}'} s_{\boldsymbol{r}'}] \quad V_{\boldsymbol{r}}(\{s\}) = A[1 - s_{\boldsymbol{r}} \sum_{\boldsymbol{r}'} s_{\boldsymbol{r}'}], \tag{7.35}$$

with z the lattice coordination number, and A an arbitrary overall rate which may be set equal to unity. Thus the first term in (7.34) gives the change in $P(\{s\}, t)$ due to single spin flip events ($AB \to AA$ or $AB \to BB$), while the second term gives the contribution due to spin exchange ($AB \to BA$).

From the master equation, it can be shown that the mean magnetization at a given site $\boldsymbol{r}$ obeys the discrete Laplace equation

$$\frac{d}{dt} \langle s(\boldsymbol{r}) \rangle = \Delta \langle s(\boldsymbol{r}) \rangle, \tag{7.36}$$

where the discrete Laplace operator is defined as $\Delta \langle s(\boldsymbol{r}) \rangle = -z \langle s(\boldsymbol{r}) \rangle + \sum_{\boldsymbol{r}'} \langle s(\boldsymbol{r}') \rangle$, where $\boldsymbol{r}'$ is a nearest-neighbor of $\boldsymbol{r}$. Using standard methods, the solution for the magnetization can be expressed in terms of modified Bessel functions [7.38]. The basic feature of this solution is that the magnetization of a translationally invariant initial state remains at its initial value.

More useful information is obtained by considering the two-spin correlation function $C(x, y, t) \equiv \langle s_{ij} s_{i+x, j+y} \rangle$. In the large-distance continuum limit the correlation function obeys the diffusion equation

$$\frac{\partial C(\boldsymbol{r}, t)}{\partial t} = \frac{1}{2} \nabla^2 C(\boldsymbol{r}, t), \tag{7.37}$$

subject to the initial condition of $C(\boldsymbol{r}, t = 0) = 0$ for all non-zero $\boldsymbol{r}$, and the boundary condition $C(\boldsymbol{r} = 0, t) = C(|\boldsymbol{r}| = L, t) = 1$, where L is the system length. Due to the fixed boundary condition at $\boldsymbol{r} = 0$, there will be a diffusive spread of non-zero correlation function to increasing $|\boldsymbol{r}|$. This is the underlying mechanism that leads to a coarsening pattern of A- and B-domains.

In greater than two dimensions (the mean-field limit), the correlation function approximately obeys the Laplace equation, $C(\boldsymbol{r}, t) \propto 1/r^{(d-2)}$, for $r < \sqrt{t}$ and $C(\boldsymbol{r}, t) \to 0$ rapidly for $r > \sqrt{t}$. Consequently, the spatial integral of the correlation function, the "correlation volume", varies as $\int^{\sqrt{t}} r^{(d-1)} \, dr \times 1/r^{(d-2)} \propto t$. When this quantity becomes of the order of the system volume, saturation

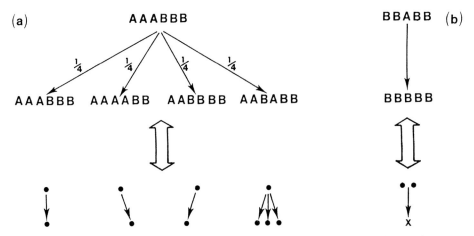

Fig. 7.4. Illustration of the mapping between the dynamics of the reaction-limited monomer-monomer catalysis model and branching annihilating random walks. (a) In a single update of the reactive AB pair, the possibilities $AB \rightarrow AB$, AA, BB, and BA each occur with probability 1/4. These events are equivalent to the domain wall (*open circle*) remaining fixed, hopping to the right, hopping to the left, or branching, respectively. (b) When two domain walls meet, they annihilate

occurs. Thus we conclude that the saturation time is proportional to the system volume Ω in the mean-field limit. In two dimensions and below, there is no steady solution for the correlation function for $r < \sqrt{t}$ and one must explicitly consider the full time-dependent solution. In two dimensions, this leads to the logarithmic correction, $\langle t \rangle \propto \Omega \ln \Omega$. In fact, detailed information exists about domain coarsening and spatial correlations in the closely related Voter model [7.40]. In one dimension, one finds that the spread of unit correlation function is governed by an error-function profile which advances as $\sqrt{t}$. Saturation occurs when this unit correlation extends to the system length L, which leads to a saturation time which is proportional to L^2.

c) Mapping onto branching random walks. Another creative exact solution in one dimension was developed by Takayasu and Inui in which the dynamics of domain walls (the point between two oppositely oriented sites) is mapped onto a soluble birth-death process [7.39]. In this mapping, (see Fig. 7.4) a domain wall is a "particle" which can hop to a neighboring site, "triplicate", or annihilate with a neighboring wall upon contact in a single reaction event. These interactions can also be viewed as a mass-conserving aggregation process modulo 2. Thus the mass in a finite interval is simply given by the number of domain walls in the interval (mod 2). In a single reaction event, the mass inside a fixed interval changes only by diffusion of mass into or out of the interval. Since the particles move by one lattice spacing, the mass of an interval of length r depends only on the mass within intervals of length r and $r \pm 1$ at the previous time step.

By enumerating all the ways in which the total mass within an r interval, $M_r(t) \equiv [m_j(t) + \ldots + m_{j+r-1}(t)](\text{mod}2)$, changes in a single reaction event, it is straightforward to show that the probability of finding an odd number of particles in an interval of length r at time t obeys the recursion relation

$$P_r(t+1) = (1 - \frac{1}{N})P_r(t) + \frac{1}{2N}P_{r+1}(t) + \frac{1}{2N}P_{r-1}(t) \qquad (7.38)$$

for $r \geq 2$, where N is the number of sites in the system. (A slightly modified version of this equation holds for the special case $r = 1$.) The kinetics of the catalytic system is naturally expressed in terms of the two-particle correlation function, $C_r(t) \equiv \langle \sigma_j(t)\sigma_{j+r}(t) \rangle$, where $\sigma_j(t) = 1$ if site j is occupied by an A at time t and $\sigma_j(t) = -1$ if j is occupied by a B. Then $\sigma_j(t)\sigma_{j+r}(t) = +1$ if there is an odd number of particles in the interval $[j, j + r]$ and $\sigma_j(t)\sigma_{j+r}(t) = -1$ if there is an even number of particles in $[j, j + r]$. As a consequence,

$$\langle \sigma_j(t)\sigma_{j+r}(t) \rangle = \text{prob}\{M_r(t) \text{ even}\} - \text{prob}\{M_r(t) \text{ odd}\} = 1 - 2P_r(t). \quad (7.39)$$

This linear relation between the occupation number (mod 2) inside an r-interval and the correlation function implies that these two quantities obey the same same dynamical recursion relation. In the continuum limit, this leads to the correlation function satisfying the diffusion equation, thus reproducing the results found by the Ising mapping.

7.4 Heterogeneous Reaction Conditions

Under homogeneous initial conditions, the reactants in two-species annihilation evolve into a coarsening mosaic of A- and B-domains in which the reaction rate at domain interfaces controls the kinetics. Because of the central role of these interfaces, considerable effort has recently been devoted to studying a system which is prepared with a single reactive interface, $i.$ $e.$, initially separated components. This classic problem was apparently first studied by Zeldovich in his investigation of the combustion of unmixed gases in a virtually unknown paper [7.23]. A more recent pioneering investigation by Gálfi and Rácz [7.25] has stimulated considerable investigation [7.23-35]. This reactive configuration is also important because initially separated components are relatively simple to prepare experimentally [7.27]. The interesting geometric and dynamical features of this system provide a useful laboratory for the microscopic investigation of two-species annihilation.

There are two complementary situations under which the reaction interface may be investigated. Transient response occurs when the two components are initially separated, and then some form of mixing occurs. Due to the effective reaction-induced repulsion of opposite species, there is a weak tendency for the two components to recede from each other when diffusion is the only mode of transport. Consequently, the reaction zone, the region within which the inert product is formed, gradually increases in width, while the overall rate of reaction decreases. On the other hand, the two components may be continuously fed into the system from localized sources. For many geometries, a steady state is reached where reactions at the interface compensate for the input of new reactants. We will discuss the dynamical properties of the interface as a function of the geometry of the reactive system.

7.4.1 Transient Response

Consider a d-dimensional system which is divided in two by a $(d-1)$-dimensional hyperplane at $x = 0$. Initially As are homogeneously distributed in the region $x > 0$ and Bs are placed in the region $x < 0$. In a mean-field description, the kinetics is described by the reaction-diffusion equations

$$\frac{\partial}{\partial t}c_i(\boldsymbol{x}, t) = D_i \nabla^2 c_i(\boldsymbol{x}, t) - kc_A(\boldsymbol{x}, t)c_B(\boldsymbol{x}, t), \qquad (7.40)$$

where i refers to either the A or B species and k is the reaction constant. The appropriate initial conditions are $c_A(\boldsymbol{x}, t=0) = c_{A0}H(x)$, and $c_B(\boldsymbol{x}, t=0) = c_{B0}H(-x)$, where $H(x)$ is the Heaviside step function.

A straightforward dimensional analysis predicts the width of the reaction zone w and the overall reaction rate [7.25]. For simplicity, consider a one-dimensional geometry with no dependence on transverse coordinates, with $c_{A0} = c_{B0}$, $D_A = D_B$, and $k \gg 1$. Because of this latter condition, the domain of one species acts as a nearly fixed absorbing boundary condition for the opposite species. Accordingly, the density profile of the As, for example, can be well approximated by that associated with independently diffusing particles in the presence of a fixed absorbing boundary at $x = 0$, namely, $c_0\,\mathrm{erf}(x/\sqrt{4Dt})$. The essential feature of this profile is a depletion layer of width $\sqrt{Dt}$, where the density decays linearly to zero as the absorber is approached, which matches smoothly onto the constant concentration c_0 far from the interface. The corresponding asymptotic forms of this density profile are

$$c(x, t) = \begin{cases} c_0\, x/\sqrt{4Dt} & \text{for } 0 < x \ll \sqrt{4Dt} \\ c_0 & \text{for } \sqrt{4Dt} \ll x. \end{cases} \qquad (7.41)$$

The properties of the reaction interface are determined by balancing the concentration-gradient driven flux into the reaction zone, $j = -Dc' = \sqrt{D/t}\,c_0$, with the rate at which particles are annihilated. Within the mean-field approximation, the total number of reactions per unit time equals the integral of $kc_A(x, t)c_B(x, t)$ over the extent of the reaction zone. We estimate this integral as the square of a typical concentration within the zone times the zone width. The typical concentration, in turn, is of the order of $c_A(x, t)$ evaluated at $x = w$. Therefore, from (7.41), the total reaction rate is $kc(w, t)^2 w \sim kc_0^2 w^3/(Dt)$. Equating this to the flux, gives the reaction zone width:

$$w \sim (D^3 t/k^2 c_0^2)^{1/6}. \qquad (7.42)$$

Correspondingly, the local production rate within the reaction zone ($\propto c(w, t)^2$) vanishes as $t^{-2/3}$. The geometrical properties of the reactive interface are shown in Fig. 7.5.

Interestingly, w can be recast as a geometric average of basic lengths

$$w \sim (Dt)^{1/6}\, c_0^{-1/3}\, \left(\frac{D}{k}\right)^{1/3} \equiv (\ell_D \ell_0 \ell_R)^{1/3}. \qquad (7.43)$$

Here $\ell_D \equiv \sqrt{Dt}$ is the diffusion length, $1/c_0 \equiv \ell_0$ is the initial interparticle

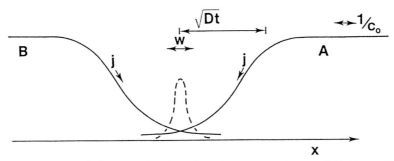

Fig. 7.5. The geometry of the reactive interface in two-species annihilation with initially separated components. For diffusive transport, there is a depletion zone of width $\sqrt{Dt}$, where the concentration decays as the opposite species domain is approached. The reaction zone is typically deep within the depletion zone

separation (in one dimension), and $\ell_R \equiv D/k$ is related to the "extrapolation length" associated with a semi-infinite population of reactants which obeys a radiation boundary condition at an interface [7.66]. Very roughly, this length is related to the penetration depth of As into a partially absorbing medium whose trapping rate is related to k. This conspiracy of three independent lengths may offer a clue toward understanding the essential nature of the reaction interface.

The spatial distribution of the local production rate within mean-field theory can be obtained by employing a quasi-static approximation to (7.40). Assuming that the concentration profile is nearly linear in the reaction zone, as in (7.41), one may then write an approximation equation for the deviation of the profile from linearity by a perturbative expansion of (7.40). This resulting equation is linear and the solution can be expressed in terms of Airy functions. In the spatial region $t^{1/6} \ll x \ll t^{1/2}$, the asymptotic form of the production rate is [7.30]

$$\mathcal{R}(x,t) \sim t^{-2/3} z^{3/4} \exp(-az^{3/2}), \tag{7.44}$$

where $z = x/t^{1/6}$ and a is a constant.

Numerical simulations confirm the mean-field predictions in two dimensions [7.26,29], and quantitatively similar behavior is anticipated to hold in greater than two dimensions. However, the situation in one dimension is problematical. A plausible first guess is that the width of the reaction zone should be of the order of the typical distance between the nearest AB pair in the system, a distance which is expected to scale as $t^{1/4}$ [7.54]. However, simulations in one dimension indicate that $w \sim t^\alpha$ with $\alpha \cong 0.3$, while the local reaction rate varies as $t^{-\beta}$ with $\beta \cong 0.8$ [7.28,30,34]. The source of this discrepancy is not yet fully resolved. Cornell and Droz [7.34] have recently suggested that it is a finite-time effect and that the correct transient behavior can be inferred by analysis of the steady-state response (see the next section). Araujo *et al.* [7.35] argue that the spatial distribution of the product exhibits multiscaling in which the reduced moments $[\int dx\, x^n \mathcal{R}(x,t)]^{1/n} \sim t^{\alpha(n)}$ with $\alpha(n)$ ranging from $1/4$ for $n \to 0$ to $3/8$ for $n \to \infty$. Here $\mathcal{R}(x,t)$ denotes the spatial distribution of the production rate at time t. Thus they conclude that there is no unique measure of the reaction zone width, very much reminiscent of the multiscaling exhibited by the interparticle-distance distribution in the homogeneous system.

When one of the species is static, the preceding mean-field approach suggest that the reaction-zone width is constant, while the local production rate vanishes as $t^{-1/2}$. In one dimension, it is possible to formulate an appealing model for the stochastic recession of the boundary between the static and mobile species as reactions occur. In this approach, the static particles recede at an average rate of $1/\sqrt{t}$ owing to the flux of the mobile species but with a superimposed random component that accounts for the discreteness of the system. By taking the simplest possible form for this fluctuating contribution and solving the resulting random-walk process, one finds a reaction zone which translates at a rate of $\sqrt{t}$ and whose width grows as $t^{1/4}$. This length is a lower bound to the width of the reaction zone in the case where both species diffuse. Correspondingly, the rate production of the inert species in one dimension varies as $t^{-3/4}$.

An intriguing feature of the transient system is that it can exhibit short-time behavior which is markedly different from that of the asymptotic limit for appropriate values of basic system parameters. For example, if the reaction rate is vanishingly small, then the two species will initially penetrate over a region whose extent grows as $\sqrt{Dt}$. Within this region, reactions will occur homogeneously, leading to an initial reaction rate which also grows as $\sqrt{Dt}$ [7.28]. However, for a low-dimensional system this penetration is eventually damped by the reaction because the effective reaction rate is infinite due to the recurrent nature of one-dimensional random walks. Another interesting phenomenon is that non-monotonic interface motion can be realized for a system with a small reaction rate when $D_A > D_B$ but $c_{A0} < c_{B0}$, as was recently shown by Taitelbaum et al. [7.33]. At early times, the diffusive penetration is dominated by the faster motion of the As and the reaction interface moves toward the Bs. However, the interface motion at long times is ultimately controlled by the majority species, so that the interface moves toward the As. This non-monotonicity has been observed on a time scale of the order of hours in a simple bimolecular reaction scheme.

7.4.2 Steady-State Behavior

Complementary results for the reactive interface can be obtained for a steady system [7.32] where particles are confined to a finite d-dimensional bar with equal fluxes of As and Bs injected from opposite directions on $(d-1)$-dimensional hyperplanes. In the steady state, we can redefine the concentrations by $c_A \Rightarrow D_B c_A$ and $c_B \Rightarrow D_A c_B$ to map the general problem to the case of equal diffusion coefficients for the two species without loss of generality. In the mean-field approximation, this system is described by the steady-state version of (7.41) ,

$$D\nabla^2 c_A = D\nabla^2 c_B = k c_A c_B, \qquad (7.45)$$

with boundary conditions

$$Dc'_A\big|_{x=L} = -Dc'_B\big|_{x=-L} = -j \quad \text{and} \quad Dc'_A\big|_{x=-L} = Dc'_B\big|_{x=L} = 0,$$

corresponding to a constant flux of a species at one end of the system and reflection at the opposite end.

Properties of the reactive interface can be obtained by the same dimensional considerations as in the transient case. Consider first the large-flux limit, which leads to a linear concentration profile near the domain boundaries, with the magnitude of the slope proportional to j/D. Since the reaction zone is the region for which the concentrations of both species are non-negligible, then the typical concentration in the reaction zone is of the order of $c_{zone} \sim jw/D$. Consequently, the number of annihilation events per unit time is of the order of $kc_{zone}^2 w$, which is obtained by integrating the reaction term over the reaction zone. Equating this number to the particle flux j then gives

$$ w \sim \left(\frac{D^2}{jk} \right)^{1/3}, \tag{7.46} $$

while the typical concentration in the reaction zone is

$$ c_{zone} \sim jw/D \sim \left(\frac{j^2}{Dk} \right)^{1/3}. \tag{7.47} $$

These results apply as long as the width w is much less than L, which corresponds, from (7.46), to $j > j_0 \sim D^2/kL^3$.

It is reassuring that the previously derived transient behavior can be recovered by applying a quasi-static approximation to the steady-state description [7.32]. In this approach, the time derivative in the diffusion equation is neglected but the time dependence is implicitly incorporated through a moving boundary condition. For the transient system, the growing depletion layer of the order of $L \sim \sqrt{Dt}$ provides the requisite moving boundary and corresponds to a time-dependent flux that is proportional to $\sqrt{Dc_0^2/t}$. Using this flux in the steady-state expression for w, (7.46), the transient result $w \sim (D^3t/k^2c_0^2)^{1/6}$ is recovered. *A posteriori*, the high-flux limit is appropriate, since the flux $j \sim t^{-1/2}$ is much greater than $j_0 \sim L^{-3} \sim t^{-3/2}$.

The steady behavior in both the low- and high-flux limits can be obtained by a direct analysis of the reaction-diffusion equations for $c_\pm = c_A \pm c_B$. From $c_-(x)'' = 0$, with the boundary conditions $Dc_-(x = \pm L)' = j$, one has $c_-(x) = jx/D$. Then $c_+(x)$ obeys $Dc_+(x)'' = 2kc_Ac_B$, which, using $4c_Ac_B = c_+^2 - c_-^2$, can be rewritten as

$$ c_+(x)'' = \frac{k}{2D} \left[c_+^2 - \left(\frac{jx}{D} \right)^2 \right], \tag{7.48} $$

with the boundary conditions $Dc_+(x = \pm L)' = \pm j$. Notice that this can be brought into the dimensionless form $c_+''(x) = c_+^2 - x^2$ when the length is written in units of $w_0 = (D^2/jk)^{1/3}$ and the concentration is expressed in units of $c_0 = (j^2/kD)^{1/3}$, thereby reproducing the conclusions from the large-flux dimensional analysis.

For large flux, $j > j_0$, it is easy to verify that c_+ has the power-law form

$$ c_+(z)/c_0 = (\frac{4}{5})^{1/3} + \frac{z^2}{(10)^{2/3}} - \frac{z^4}{40} + \dots, \tag{7.49} $$

where $z \equiv x/w_0 \ll 1$. This series for c_+ matches c_- at $w \propto w_0$, thus providing an independent determination of the reaction zone width.

Outside the reaction zone, the concentration of the minority species can be obtained directly by substituting the large-x approximation $c_A \cong c_B + (jx/D)$ in the steady-state equation for c_B to give $Dc_B(x)'' = kc_B(c_B + jx/D)$. This reduces to the Airy equation $c_B(z)'' = z\,c_B(z)$ for $z \gg 1$ with solution

$$c_B(x)/c_0 \sim z^{-1/4} \exp\left(-2z^{3/2}/3\right). \tag{7.50}$$

Thus in the high flux limit, the concentration of Bs is approximately $-jx/D$ for $x < -w$, exponentially small for $x > w$, and equal to $(c_+ - c_-)/2 \propto j^{2/3}$ in the reaction zone. The concentration of As is the mirror image of $c_B(x)$ about $x = 0$.

In the low-flux limit, the density varies weakly in space, and it can be easily verified that

$$c_+(x) \cong \sqrt{\frac{2j}{kL}} + \left(\frac{j}{2DL}\right)x^2 \tag{7.51}$$

is an accurate approximation for the entire domain $[-L, L]$. The second term is determined by the boundary conditions on c_+ at $x = \pm L$, and the leading term is then determined from (7.48). Notice that the magnitude of the concentration variation ($\sim j$) is small compared to the value of the concentration itself ($\sqrt{j}$). An interesting feature which emerges is that the local reaction rate $\mathcal{R}(x) \equiv kc_A(x)c_B(x)$ is unimodal for large flux and bimodal otherwise. That is, for $j > j_0$, $\mathcal{R}(x) \propto (jw)^2 - \text{const.} \times (jx)^2$, so that the reaction rate is large only within a region of width w_0 about $x = 0$, while for $j < j_0$, $\mathcal{R}(x) \propto j + \text{const.} \times j^{3/2}x^2$, so that the reaction rate is large only near the edges of the system.

The validity of the mean-field approximation and the nature of the steady state in a one-dimensional system has been investigated only very recently, and the results may also shed light on the puzzling behavior of the corresponding transient system [7.34]. On dimensional grounds, the system is characterized by three basic parameters, namely j, D, and k. In spatial dimension $d \leq 2$, the recurrence of random walks implies that particles meet infinitely often, and hence react, irrespective of k. Thus the reaction rate can be disregarded. It then follows that the only length in the system is $(D/j)^{1/(d+1)}$. In one dimension, this gives an interface width which scales as $j^{-1/2}$. Furthermore, the mean-field result of $w \propto j^{-1/3}$ is recovered for $d = 2$, indicating that $d = 2$ is indeed the upper critical dimension for this problem.

This line of reasoning relies crucially on the time independence of the reaction, so that the width of the reaction zone is fixed by the unique dimensional length scale. In contrast, several independent lengths exist for the transient system, and the interface width cannot be obtained unambiguously through dimensional analysis alone. However, one can infer the transient behavior by again applying the quasi-static approximation to the steady-state results. As discussed above, the flux in the transient system decreases as $t^{-1/2}$ which, when substituted into $w \propto j^{-1/2}$, gives $w \propto t^{1/4}$ [7.34].

Another remarkable property of the one-dimensional system is the large-distance behavior of the reaction profile. Whereas mean-field theory predicts an $\exp(-x^{3/2})$ decay of the density of the reaction product at large distances, it is found numerically that the decay is a pure exponential [7.30]. The exponential decay is particularly puzzling in light of the fact that in the large-distance asymptotic region there is a well-defined majority species with little fluctuation. Under these circumstances, the mean-field prediction is expected to hold in this region (S. Cornell, private communication).

The above approaches can be extended in several interesting ways [7.32]. One physically relevant situation is when reactants drift toward each other, as would occur if two oppositely charged species were in a uniform electric field. For relatively weak drift, the density profile of each species is nearly constant and of the order of j/v outside a boundary layer of thickness D/v, and varies linearly with position inside the boundary layer where diffusion predominates. This reasoning is valid as long as the thickness of the boundary layer D/v is larger than the reaction-zone width $(D^2/jk)^{1/3}$, or, equivalently, $v < v_0 \sim (Djk)^{1/3}$.

Another example is an annular geometry in which a flux j of Bs is injected at an inner radius r_1, and a flux $-j$ of As is injected at outer radius r_2. When both radii are large with their difference remaining finite, the solution to the radial Laplace equation for the concentration difference predicts that the reaction zone center is at $r_0 \cong (r_1 + r_2)/2$, with the density profiles and the reaction zone width behaving in the same manner as in one dimension. However, if $r_2/r_1 \to \infty$, then the reaction zone center remains very close to the inner radius with $r_0 \cong r_2/\sqrt{e}$ for $d = 2$ and $r_0 \cong 2r_2/3$ for $d = 3$ [7.32]. When one species is injected at a point in a background medium of the opposite species rather different behavior occurs. When both species are mobile, the reaction zone center is at $r_0 \sim \sqrt{t \ln t}$ and the width of the zone is proportional to $t^{1/6}$ times a logarithmic correction, within mean-field theory [7.67]. When the background consists of immobile particles, then the number of injected particles which remain unreacted increases slower than linear in time for $d = 3$ [7.68]. These radial geometries offer a new range of phenomena that merit further exploration.

7.5 Concluding Remarks

In diffusion-limited two-species annihilation, an initially homogeneous distribution of As and Bs at equal initial densities evolves into a mosaic of growing single-species domains. When both species are equally mobile, the concentration within the domain core is roughly uniform and proportional to $t^{-d/4}$, while the concentration vanishes linearly in the distance to the domain edge for spatial dimension $d \leq 2$. This linear decay leads to a new length scale, ℓ_{AB}, intermediate to the typical interparticle spacing and the typical domain size, as well as multiscaling in the reduced moments $\langle \ell_{AA}^n(t) \rangle^{1/n}$. When one species is immobile, the domains of immobile species evolve only by infiltration of mobile species from the exterior, and vestiges of the initial distribution persist in the domain interior. This leads to a power law distribution for the separations of the immobile particles in which $\langle \ell_{BB}^n(t) \rangle^{1/n}$ appears to increase as a power law in time for $n \geq 1/2$, but approaches a finite limiting value for $n < 1/2$.

In higher dimensions, there is less understanding of domain structure and the spatial organization of reactants. A good algorithm is needed to unambiguously define domains in greater than one dimension. This would aid in addressing questions such as the distribution of domain sizes, the nature of the density profile within a domain, and the spatial organization of the domains. Numerical evidence indicates that anomalous scaling of interparticle distances exists in $d = 2$ but not in $d = 3$. Thus $d = 3$ appears to be a critical dimension above which there is no substantial depletion layer at the periphery of a domain.

There are several generalizations of two-species annihilation that raise a variety of interesting questions. For example, consider immobile reactants in which the reaction is mediated only by long-range exchange [7.69,70]. For an exchange rate which decreases exponentially in the particle separation, the density decays as $(\ln t/\tau)^{-d/2}$ and large-scale reactant segregation ensues. Another potentially important situation is two-species annihilation in the presence of (unscreened) Coulomb forces. A scaling argument suggests that the decay is mean-field-like in all dimensions [7.71], but this has yet to be tested numerically. Anomalous initial conditions, such as a lamellar configuration, are also found to influence the kinetics [7.72,73]. Many of the interesting properties of two-species annihilation also arise in related stoichiometries such as higher-order reactions [7.74] and n-species reactions [7.75]. In particular, in two-species competition [7.76-78] species A and B evolve both by logistic growth (with the associated rate equation $\dot{c} = k_1 c - k_2 c^2$) and mutual annihilation. This can be regarded as a prototypical model for interspecies competition in a biological context. A homogeneous initial state evolves into single-species domains which coarsen logarithmically in one dimension and at a power-law rate in two dimensions. This latter growth strongly resembles that observed in spinodal decomposition [7.79].

With constant input of reactants, there is a relatively good understanding of the conditions needed to achieve a steady state. Typically, a steady state is reached when there is complete temporal correlation and relatively strong spatial correlation of particle-antiparticle pairs. The situation where the reactants are immobile is particularly noteworthy because of the connection to the monomer-monomer catalysis model. This model is exactly soluble in all dimensions by a mapping onto a kinetic Ising model. Close connections also exist with the voter model and with branching random walks. All of these connections provide useful insights for understanding the basic mechanism of domain coarsening and saturation in the monomer-monomer model.

At the microscopic level, two-species annihilation is controlled by the rate at which reactions occur at the interface between A and B domains. This observation, together with recent experimental work, provides the motivation for considering geometries which possess a single reaction interface. From analysis of the reaction-diffusion equations or by dimensional analysis, geometrical properties and the reactivity of the reaction zone at an AB interface can be determined. The one-dimensional case is especially intriguing as there are indications that many length scales are needed to characterize the reaction zone. This focus on the reaction interface in microscopic modeling may prove fruitful for other reactions. A classic example is that of the growth and wave-like spread of a localized population whose dynamics is governed by logistic growth [7.74,80,81]. While mean-field theory predicts the existence of a soliton-like propagating wave front, an exact solution in one dimension shows that the width grows as $\sqrt{t}$ [7.82], while numerical simulations [7.83] indicate that the width grows in the same manner as that KPZ-like growth [7.84].

Acknowledgements. We thank D. ben-Avraham, E. Ben-Naim, D. Considine, K. Kang, P. Meakin, G. Murthy, H. Takayasu, and J. Zhuo for pleasant collaborations on various aspects of reaction kinetics. We have also greatly benefitted from helpful discussions with M. Araujo, E. Clément S. Cornell, M. Droz, S. Havlin, H. Larralde, and G. Zumofen. We also thank E. Ben-Naim, S. Havlin, and H. Larralde for a critical reading of the manuscript and helpful suggestions. Finally, we gratefully acknowledge grants DAAL03-86-K-0025 and DAAL03-89-K-0025 from the ARO, INT-8815438 from the NSF, DGAPA project IN100491, and project #903922 from CONACYT for partial financial support of this research.

References

7.1   For recent reviews on two species annihilation see *e. g.*, Ya.B. Zeldovich, A.S. Mikhailov: Sov. Phys. – Usp. **30**, 23 (1988); V. Kuzovkov, E. Kotomin: Rep. Prog. Phys. **51**, 1479 (1988); R. Kopelman: Science **241**, 1620 (1988); A.S. Mikhailov: Phys. Repts. **184**, 307 (1989); A.A. Ovchinnikov, S.F. Timashev, A.A. Belyy: *Kinetics of Diffusion Controlled Chemical Processes* (Nova Science Publishers 1990)
7.2   Z. Vardeny, P.O'Connor, S. Ray, J. Tauc: Phys. Rev. Lett. **44**, 1267 (1980)
7.3   I.M. Campbell: *Catalysis at Surfaces* (Chapman and Hall, New York 1988); G.C. Bond: *Heterogeneous Catalysis: Principles and Applications* (Clarendon Press, Oxford 1987)
7.4   W.M. Yen, P.M. Selzer, eds.: *Laser Spectroscopy of Solids, 2nd ed.* (Springer-Verlag, Berlin 1986)
7.5   J. Preskill: Phys. Rev. Lett. **43**, 1365 (1979)
7.6   M. Bramson, J.L. Lebowitz: Phys. Rev. Lett. **61**, 2397 (1988); M. Bramson, J.L. Lebowitz: J. Stat. Phys. **62**, 297 (1991); M. Bramson, J.L. Lebowitz: J. Stat. Phys. **65**, 941 (1991)
7.7   Ya.B. Zeldovich, A.A. Ovchinnikov: Chem. Phys. **28**, 215 (1978)
7.8   S.F. Burlatskii: Sov. Theor. Exp. Chem. **14**, 483 (1978)
7.9   S.F. Burlatskii, A.A. Ovchinnikov: Russ. J. Phys. Chem. **52**, 1635 (1978)
7.10   D. Toussaint, F. Wilczek: J. Chem. Phys. **78**, 2642 (1983)
7.11   K. Kang, S. Redner: Phys. Rev. Lett. **52**, 955 (1984); Phys. Rev. A **32**, 435 (1985)
7.12   K. Lee, E.J. Weinberg: Nucl. Phys. B **246**, 354 (1984)
7.13   P. Meakin, H.E. Stanley: J. Phys. A **17**, L173 (1984)
7.14   G. Zumofen, A. Blumen, J. Klafter: J. Chem. Phys. **82**, 3198 (1985)
7.15   L.W. Anacker, R. Kopelman: Phys. Rev. Lett. **58**, 289 (1987); L.W. Anacker, R. Kopelman: J. Chem. Phys. **91**, 5555 (1987)

7.16   D. ben-Avraham, C.R. Doering: Phys. Rev. A **37**, 5007 (1988)

7.17   K. Lindenberg, B.J. West, R. Kopelman: Phys. Rev. Lett. **60**, 1777 (1988)

7.18   E. Clément, L.M. Sander, R. Kopelman: Phys. Rev. A **39**, 6455 (1989); Phys. Rev. A **39**, 6466 (1989)

7.19   R.M. Ziff, K. Fichthorn: Phys. Rev. B **34**, 2038 (1986)

7.20   P. Meakin, D. Scalapino: J. Chem. Phys. **87**, 731 (1987)

7.21   K. Fichthorn, E. Gulari, R.M. Ziff: Phys. Rev. Lett. **63**, 1527 (1989)

7.22   D. ben-Avraham, D. Considine, P. Meakin, S. Redner, H. Takayasu: J. Phys. A **23**, 4297 (1990); D. ben-Avraham, S. Redner, D. Considine, P. Meakin: J. Phys. A **23**, L613 (1990)

7.23   Ya.B. Zeldovich: Zh. Tekh. Fiz. **19**, 1199 (1949); A.S. Mikhailov: in [7.1]

7.24   I.M. Sokolov: Sov. Phys. JETP Letters **44**, 67 (1986)

7.25   L. Gálfi, Z. Rácz: Phys. Rev. A **38**, 3151 (1988)

7.26   Z. Jiang, C. Ebner: Phys. Rev. A **42**, 7483 (1990)

7.27   Y.-E.L. Koo, L. Li, R. Kopleman: Mol. Cryst. Liq. Cryst. **183**, 187 (1990); Y.-E.L. Koo, R. Kopleman: J. Stat. Phys. **65**, 893 (1991)

7.28   H. Taitelbaum, S. Havlin, J.E. Kiefer, B. Trus, G.H. Weiss: J. Stat. Phys. **65**, 873 (1991)

7.29   S. Cornell, M. Droz, B. Chopard: Phys. Rev. A **44**, 4826 (1991)

7.30   M. Araujo, S. Havlin, H. Larralde, H.E. Stanley: Phys. Rev. Lett. **68**, 1791 (1992); H. Larralde, M. Araujo, S. Havlin, H.E. Stanley: Phys. Rev. A **46**, 855 (1992)

7.31   H. Larralde, M. Araujo, S. Havlin, H.E. Stanley: Phys. Rev. A **46**, 6121 (1992)

7.32   E. Ben-Naim, S. Redner: J. Phys. A **25**, L575 (1992)

7.33   H. Taitelbaum, Y.-E.L. Koo, S. Havlin, R. Kopelman, G. H. Weiss: Phys. Rev. A **46**, 2151 (1992)

7.34   S. Cornell, M. Droz: Phys. Rev. Lett. **70**, 3824 (1993)

7.35   M. Araujo, H. Larralde, S. Havlin, H.E. Stanley: Phys. Rev. Lett. **71**, 3592 (1993); S. Havlin, M. Araujo, H. Larralde, H.E. Stanley: Physica A **191**, 143 (1992)

7.36   The importance of of interparticle distributions in two-species annihilation was apparently first raised in P. Argyrakis, R. Kopelman: Phys. Rev. A **41**, 2121 (1990)

7.37   F. Leyvraz, S. Redner: Phys. Rev. Lett. **66**, 2168 (1991); Phys. Rev. A **46**, 3132 (1992)

7.38   E. Clément, P. Leroux-Hugon, L.M. Sander: Phys. Rev. Lett. **67**, 1661 (1991) P.L. Krapivsky: Phys. Rev. A **45**, 1067 (1992); J. Phys. A **25**, 5831 (1992)

7.39   H. Takayasu, N. Inui: J. Phys. A **25**, L585 (1992)

7.40   T.J. Cox, D. Griffeath: Ann. Probab. **14**, 347 (1986); for general results about the voter model, see *e. g.*, R. Durrett: *Lecture Notes on Particle Systems and Percolation* (Wadsworth & Brooks/Cole, Pacific Grove, CA 1988); M. Scheucher, H. Spohn: J. Stat. Phys. **53**, 279 (1988)

7.41   K. Lindenberg, B.J. West, R. Kopelman: Phys. Rev. A **42**, 890 (1990)

7.42   S. Cornell, M. Droz, B. Chopard: Physica A **188**, 322 (1992)

7.43   See *e. g.*, S. Havlin, D. ben-Avraham: Adv. Phys. **36**, 695 (1987)

7.44   W.-S. Sheu, K. Lindenberg, R. Kopelman: Phys. Rev. A **42**, 2279 (1990); K. Lindenberg, W.-S. Sheu, R. Kopelman: Phys. Rev. A **43**, 7070 (1991); K. Lindenberg, W.-S. Sheu, R. Kopelman: J. Stat. Phys. **65**, 1285 (1991);

7.45   G. Zumofen, J. Klafter, A. Blumen: Phys. Rev. A **43**, 7068 (1991); G. Zumofen, J. Klafter, A. Blumen: J. Stat. Phys. **65**, 1015 (1991)

7.46   A. Bunde, S. Havlin, eds: *Fractals and Disordered Systems* (Springer, Heidelberg 1991)

7.47   F. Leyvraz: J. Phys. A **25**, 3205 (1992)

7.48   M. Bramson, D. Griffeath: Z. Wahrsch. verw. Gebiete **53**, 183 (1980)

7.49   D.C. Torney, H.M. McConnell: Proc. Roy. Soc. London Ser. A **387**, 147 (1983); A.A. Lushnikov: Sov. Phys. JETP **64**, 811 (1986); J.L. Spouge: Phys. Rev. Lett. **60**, 873 (1988)

7.50   C.R. Doering, D. ben-Avraham: Phys. Rev. A **38**, 3035 (1988); D. ben-Avraham, M.A. Burschka, C.R. Doering: J. Stat. Phys. **60**, 695 (1990)

7.51   See *e. g.*, W. Feller: *An Introduction to Probability Theory and its Applications* (Wiley, New York 1968), Vol. 1

7.52   See *e. g.*, L.D. Landau, E.M. Lifshitz: *Quantum Mechanics* (Pergamon Press, New York 1977).

7.53   F. Leyvraz: unpublished notes; D. ben-Avraham: J. Chem. Phys. **88**, 941 (1988); M.E. Fisher, M.P. Gelfand: J. Stat. Phys. **53**, 175 (1988)

7.54 This approach for finding the "first" particle in a distribution was first considered for the case of a fixed trap by G.H. Weiss, S. Havlin, R. Kopelman: Phys. Rev. A **39**, 466 (1989); see also S. Redner, D. ben-Avraham: J. Phys. A **23**, L1169 (1990); S. Havlin, H. Larralde, R. Kopelman, G.H. Weiss: Physica A **169**, 337 (1990); H. Taitelbaum, R. Kopelman, G. H. Weiss, S. Havlin: Phys. Rev. A **41**, 3116 (1990); H. Taitelbaum: Phys. Rev. A **43**, 6592 (1991)

7.55 Ya.B. Zel'dovich, A.A. Ovchinnikov: Sov. Phys. JETP Letters **26**, 440 (1977)

7.56 Ya.B. Zel'dovich, A.A. Ovchinnikov: Sov. Phys. JETP **47**, 829 (1978)

7.57 S.F. Burlatskii, A.A. Ovchinnikov: Sov. Phys. JETP **65**, 908 (1987)

7.58 Y.-C. Zhang: Phys. Rev. Lett. **59**, 1726 (1987)

7.59 S.F. Burlatskii, A.A. Ovchinnikov, G.S. Oshanin: Sov. Phys. JETP **68**, 1153 (1989)

7.60 N. Agmon, A. Szabo: J. Chem. Phys. **92**, 570 (1990)

7.61 D. Huppert, S.Y. Goldberg, A. Masad, N. Agmon: Phys. Rev. Lett. **68**, 3932 (1992)

7.62 H. Spohn, in: *Statistical Physics and Dynamical Systems: Rigorous Results*, eds. L. Fritz, A. Jaffe, D. Szász, (Birkhäuser 1985)

7.63 R.J. Glauber: J. Math. Phys. **4**, 294 (1963)

7.64 K. Kawasaki, in: *Phase Transitions and Critical Phenomena,* eds. C. Domb and M.S. Green (Academic, New York 1976)

7.65 G.H. Weiss: Adv. Chem. Phys. **13**, 1 (1966); J. Stat. Phys. **24**, 587 (1981)

7.66 E. Ben-Naim, S. Redner, G.H. Weiss: J. Stat. Phys. **71**, 75 (1993)

7.67 E. Ben-Naim, S. Redner: unpublished

7.68 H. Larralde, Y. Lereah, P. Trunfio, J. Dror, S. Havlin, R. Rosenbaum, H.E. Stanley: Phys. Rev. Lett. **70**, 1461 (1993).

7.69 J.R. Eggert: Phys. Rev. B **29**, 6664 (1984);
      T.M. Searle, J.E.L. Bishop: Philos. Mag. B **53**, L9 (1986)

7.70 H. Schnörer, V. Kuzovkov, A. Blumen: Phys. Rev. Lett. **63**, 805 (1989)

7.71 T. Ohtsuki: Phys. Lett. A **106**, 224 (1984)

7.72 F.J. Muzzio, J.M. Ottino Phys. Rev. Lett. **63**, 47 (1989); Phys. Rev. A **42**, 5873 (1990)

7.73 I.M. Sokolov, A. Blumen: Phys. Rev. A **43**, 2714 (1991)

7.74 K. Kang, P. Meakin, J.H. Oh, S. Redner: J. Phys. A **17**, L665 (1984)

7.75 D. ben-Avraham, S. Redner: Phys. Rev. A **34**, 501 (1986)

7.76 J.D. Murray, *Mathematical Biology*, (Springer-Verlag, Berlin 1989)

7.77 S.F. Burlatskii, K. A. Pronin: J. Phys. A **22**, 531 (1989)

7.78 J. Zhuo, G. Murthy, S. Redner: J. Phys. A **25**, 5889 (1992)

7.79 See: *e. g.*, J.D. Gunton, M. San Miguel, P.S. Sahni: in *Phase Transitions and Critical Phenomena*, Vol. 8, C. Domb, J.L. Lebowitz, Editors (Academic Press, London 1984)

7.80 A. Kolmogorov, I. Petrovsky, N. Piscounov: Moscow Univ. Bull. Math. **1**, 1 (1937)

7.81 R.A. Fisher: Ann. Eugenics **7**, 353 (1937)

7.82 C.R. Doering, M.A. Burschka, W. Horsthemke: J. Stat. Phys. **65**, 953 (1991)

7.83 E. Ben-Naim, S. Redner: unpublished

7.84 M. Kardar, G. Parisi, Y.-C. Zhang: Phys. Rev. Lett. **56**, 889 (1986)

8 Fractal Analysis in Heterogeneous Chemistry

David Avnir, Ricardo Gutfraind, and Dina Farin

8.1 Introduction

Chemical reactions are traditionally treated in terms of three major parameters: the nature of the chemical bond which is formed between the reacting molecules; the stereochemical requirements or restrictions which govern this molecular association; and the energetic profile of the reaction. *Heterogeneous* chemistry introduces a fourth parameter: the structure and geometry of the environment in which the reaction takes place. The geometry parameter is as important as the three other parameters to the extent that it alone can dictate whether a reaction will take place at all.

Despite its importance, the quantitative treatment of this parameter and of its relation to the other reaction parameters has been sluggish and cumbersome. The main reason for this situation has been the very nature of the structures involved: the majority of materials and their surfaces is characterized by extremely complex geometry (Fig. 8.0). Metallic catalysts, for instance, are disordered compact aggregates, the building blocks of which are imperfect crystallites with broken faces, steps, and kinks. Many of the supports of heterogeneous reactions (silica, alumina, carbon-blacks) are high-surface-area porous materials with convoluted surfaces. Zeolites, those highly ordered materials, are an exception which illuminates the geometry problem prevailing in most industrial and laboratory-scale chemically reactive materials.

The geometry problem has two important consequences: first, on the basic-science level, it becomes exceedingly difficult to study geometry–efficiency relations, both phenomenologically and in a predictive fashion; and second, on the practical side, quality control has routinely focused on parameters such as particle size, surface area and pore size but neglected the geometric parameter which is not less important, namely, the surface morphologies of the reactive material.

◀ **Fig. 8.0.** Electrodeposition of copper. (From [1.14], courtesy of V. Fleury)

A traditional approach to the problem of complex morphologies has been to select some "ideal" reference shape (a straight line, a flat plane, a sponge with equally spaced, single-sized pores, etc.) and try to quantify by how much the material under study deviates from that ideal reference. Quite often, the result is a very cumbersome collection of many parameters needed for the task [8.1]. In addition to this inherent difficulty, one must also question whether the analysis of a natural phenomenon in terms of a deviation from an arbitrarily chosen ideal reference is the best approach one can take. Although many felt that the answer to this is negative, better general solutions were not at hand until the recent advent of fractal geometry and the theory of chaos. These seem to provide a principle new framework of discussion for some old problems in chemistry.

A general theme of this book is that fractal geometry treats structures which have a certain repetitive morphological feature over a certain size range. Figure 8.1 shows one such structure: it is a theoretical object known as the Devil's Staircase, and one can see that it is characterized by a typical feature, a step, which appears on many scales.

One can already point at this stage to the apriori attractiveness that fractal geometry offers: its theoretical objects are strikingly similar to objects found in reality. For instance, the Devil's Staircase structure resembles very much the edges of imperfect crystals [8.2] and the edges of laminar structures like clays and graphites. Another example is the Sierpinski Carpet (see e.g., [8.2,3] and Chap. 1) in which the repetitive feature is a triangle. A randomized form of it can be found on defective surfaces of diamonds (features known as trigons) [8.4].

An important empirical observation was that the similarity of real objects to the theoretical fractal objects goes beyond superficial resemblance: both obey the simple scaling relation

Fig. 8.1. The Devil's Staircase. The step edges form a fractal set of points, with a fractal dimension, d_f, of 0.63. This object is a convenient model of a catalytic fructured metal crystallite. (Shown is a random version - c.f. Fig. 8.5)

$$number\ of\ features \sim resolution\ of\ measurement^D. \qquad (8.1)$$

Fractal geometry emerges, therefore, as a natural descriptive tool of complex structures, which takes the very complexity as a starting point and not as a deviation. It should be noted that the descriptive applicability of fractal geometry crosses over virtually all domains of the natural sciences, from molecular assemblies [8.5] through geological features [8.6] and up to galaxies [8.7]. Why is it so common in nature is a key enigma yet to be solved.

Returning to (8.1), we see that another important generalization has emerged [8.5]: it became clear that one cannot talk about the geometry of a structure, say a catalytic surface, but that, instead, one has to talk about a plethora of effective surfaces for specific processes. For instance, in catalysis an immediate well-known example is the fact that quite often not all of the surface of the catalyst participates in the reaction but only a subset of reactive sites do so. One can envisage the structure in Fig. 8.1 as a metallic imperfect crystallite (such as obtained by dispersing a metal catalyst on an inert support) in which not all of the stepped surface is active but only the step vertices. So for that specific catalytic process the geometry of relevance is the pattern of the collection of the vertices. Similarly, one can ask what is the effective geometry of that object with regard to weak adsorption of say, nitrogen molecules - then the full boundary line in Fig. 8.1 is probably the relevant surface. And another example follows. Consider a high-surface area porous catalyst or porous adsorbent. Here one may distinguish between two types of surfaces which may or may not coincide with each other. One is the surface which is available for incoming molecules: only the molecularly accessible portions of the surface are relevant here [8.8]. The other is the surface which can scatter radiation such as X-rays (at low angles) [8.9]. Since X-rays penetrate the material everywhere, they probe surface zones which are totally inaccessible to molecules such as closed pores or pores with bottle-neck entrances. The generalized form of (8.1) is therefore

$$amount\ of\ material\ property \sim resolution\ of\ measurement^D. \qquad (8.2)$$

Examples of "amount of material property" could be the accessible surface area for adsorption [8.10], the *reactive* area or the reaction rate itself [8.11], the intensity of scattered X-rays [8.9], etc. Examples of various yardsticks which can be used in order to vary the resolution of measurement are magnification changes in microscopy [8.12], the classical compass step-size as used in geographical distance measurement [8.13], the size of the object itself (to which we return below) [8.21], diffusional distances to the surface [8.14], the scattering angle of X-rays [8.9], and varying the size of the molecules which interact with the surface [8.15]. It is therefore helpful to mark the D with a subscript which points to the property and method of measurement applied. Thus, d_{ad} would be the fractal dimension of the surface accessible for adsorption [8.10], d_{re} would be the fractal dimension of the area participating in a reaction [8.11], etc. As mentioned above, $d_{ad} = d_{re}$ is a possibility, but not a necessity, and useful information can be elucidated from the comparison of these two values (see Sect. 8.2).

In the following sections we have collected together a number of specific examples in order to illustrate how fractal-geometry considerations can be used to approach typical problems in chemistry, which involve complex geometry. We begin (Sect. 8.2) with a description of the reaction dimension, namely the fractal dimension of the subset of sites of an object which participates in a reaction. In Sects. 8.3 and 8.4 we describe the applications of that concept to pharmacology and catalysis. We continue with catalysis in Sect. 8.5 and fine-tune the analysis with the aid of the multifractal approach. And finally, in Sect. 8.6, we describe the surface accessibility problem, and exemplify it for the case of chromatographic materials.

8.2 The Reaction Dimension

The parameters which dictate the chemical reactivity of a fractal object are both numerous and difficult to separate; yet it is possible to describe the sum of these effects in an empirical scaling exponent, the reaction dimension, d_{re}. In many cases d_{re} can be interpreted as the effective fractal dimension of the object towards a reaction. A priori, one would expect that, in general, d_{re} would be smaller than the surface fractal dimension because of selective participation of surface sites in the reaction. However, heterogeneous reactions are phenomenologically much richer, creating also, as we shall see, cases with the opposite relation.

If the surface is a fractal, then the surface area, A (m^2/g), is related to the particle radius, r, as

$$A \sim r^{d_f - 3}, \tag{8.3}$$

where d_f is the fractal dimension determined by that method; (it not always coincides with d_{ad} [8.21]). Many materials obey this relation [8.10,15-21]. Replacing the total surface area, A, by the effective one for reaction, S, one similarly gets

$$S \sim r^{d_{re} - 3}, \tag{8.4}$$

in which d_{re} is the reaction dimension of the effective surface, S. Since the initial rate of a reaction with a surface, v (moles/time/g), is given by the very old Wenzel law (1777) [8.22], $v \sim S$, one can determine d_{re} from

$$v \sim r^{d_{re} - 3}. \tag{8.5}$$

In Table 8.1 (see also Figs. 8.2 - 4) we have collected the analyses of many surface reactions according to (8.5). It can be concluded that the whole spectrum of relations between the surface dimension and the reaction dimension is possible: $d_{re} \approx d_f$, $d_{re} < d_f$, $d_{re} > d_f$.

QUARTZ DISSOLUTION

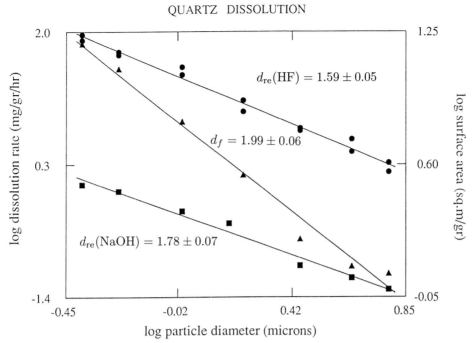

Fig. 8.2. Surface area (right axis, ▲, data from [8.32]) and dissolution rates (left axis, ●, HF, data from [8.32]; ■, NaOH, data from [8.92]) of quartz as a function of particle size

The relation between these two parameters is governed by one or more of the following effects:

(a) Screening [8.23 - 25]: Sterically hindered and inner parts of the surface are diffusionally inaccessible to the reactive molecules. This would lead to an effective smoothing of the surface, i.e., $d_{re} < d_f$.

(b) Chemical selectivity: Since reactive sites are a subset of all surface sites, one expects that $d_{re} < d_f$ when this effect operates.

(c) Roughening and smoothing: The morphology of a surface may change (not always - see below) in the course of the reaction [8.26]. If this change is very rapid, or if it occurs at the very first instances of the reaction, then the experimentally determined "initial rate" will actually refer to this modified morphology. d_{re} may then be larger (roughening) or smaller (smoothing) than d_f.

(d) Trapping [8.23,27]: Reactive molecules may be trapped in a small portion of cracks and narrow pores, finding it difficult to escape these cages either because of diffusional limitations or because of stronger adsorbate - adsorbent interactions.

It should be noted that many theoretical simulations [8.14,23 - 29] have indeed demonstrated that all of these effects, namely screening, trapping, selective reactivity, roughening and smoothing, when applied to fractal or smooth objects, result in (effective) structures which are still describable in terms of scaling fractal exponent. In view of this multiplicity of factors, it seems that it should be quite difficult to elucidate the source for the deviation of d_{re} from d_f. However, we briefly highlight a few examples from Table 8.1 to show that quite often the origin of the specific relation between these two dimensions can be explained by using other information about the reacting system.

Love and Whittaker [8.30] studied the acidic dissolution of limestones and found that "both total and reacting surface areas studied increased much less rapidly with particle size reduction than did the geometric surface". In our interpretation, this may indicate fractality of both types of surface. Indeed analysis of their data reveals that the limestones have fractal surfaces and fractal reaction areas ($1 - 4$ in Table 8.1). Furthermore, the hypothesis of the authors that "the total surface appears to consist of an interior surface that has little or no effect on the chemical reactivity of the limestone", i.e., a screening effect, is nicely indicated by the $d_{re} < d_f$ values. That the low d_{re} values originate mainly from screening and not so much from chemical selectivity is apparent from the fact that, in the absence of porosity, $d_f \approx d_{re}$ in this type of reaction (6 in Table 8.1). In another set of experiments, Love and Whittaker studied the effect of decrease in particle size (owing to the reactive dissolution) on the reaction rate. Interestingly we find (5 in Table 8.1) that d_{re} remains constant down to 80% consumption of the particles! This seems to indicate structural homogeneity of the investigated particles.

A case of $d_{re} > d_f$, which originates from a reaction which occurs selectively in micropores, is apparently the observation of Holdren and Speyer [8.31]. These authors studied the dissolution of feldspars (natural aluminosilicates) at various pH's. They observed the empirical relation

$$v \sim A^{\mu}, \tag{8.6}$$

in which μ, the reaction order in surface area, is < 1. Analysis of their data, ($10 - 13$ in Table 8.1) shows that both the surface and the reactive surface are fractals. From (8.3-5) it is readily seen that

$$v \sim A^{(d_{re}-3)/(d_f-3)}, \tag{8.7}$$

i.e.,

$$\mu = (d_{re} - 3)/(d_f - 3). \tag{8.8}$$

Equation (8.7) is a generalization of the original Wenzel law, in which the distinction between S and A was not made. $\mu < 1$ values indicate a weak dependency of S on particle size, r. The dependency of the total surface area, A, on r is composed of two contributions: the outer particle surface area, which depends strongly on r, and the hidden inner area, which is less sensitive to r. (At the extreme, A of highly porous materials is independent of r.) Consequently if a reaction takes place preferentially in internal parts, it will show up in $\mu < 1$.

It is thus possible to increase the efficiency of such reactions by increasing the relative contribution of these internal parts, i.e., by increasing particle size. And so $d_{re} > d_f$ situations are expected to occur in cases where there is a mechanism for preferential selection of cracks, defects, and narrow pores. Such surface zones are characterized by a higher density of adsorption or reactive sites, and consequently the residence time of the molecules near the surface increases, increasing the probability of reaction. This picture is in keeping with the defects mechanism suggested repeatedly by Holdren and Speyer in their paper [8.31]. An important conclusion made is therefore that reaction efficiency can be increased by reducing the particle size for $d_{re} < d_f$, $\mu > 1$ cases and increasing particle size for $d_{re} > d_f$, $\mu < 1$ reactions.

The dissolution of quartz provides an example for the role of chemical limitations: When nonporous quartz is dissolved in concentrated HF, one gets $D \sim d_{re} \sim 2.1$ (14, 15 in Table 8.1) [8.17]. However, in dilute acid (16 in Table 8.1) or base (17 in Table 8.1), the dissolution reactions become much more difficult [8.32]. Only selected surface sites (probably weak crystalline imperfections) are capable of reacting, and therefore one gets $d_{re} < 2$ (Fig. 8.2). We notice that $d_{re}(\text{HF})$ is somewhat smaller than $d_{re}(\text{NaOH})$ (1.59 and 1.78, respectively) and that the prefactors (the intercepts) are quite different: 25.3 g·h·mg^{-1} for dissolution in HF and 0.44 $\text{g}\,h\,\text{mg}^{-1}$ for NaOH. These observations seem to indicate that, despite the difference in reaction rates, a similar collection of active sites reacts both under the acidic conditions and under the basic conditions.

8.3 Surface Morphology Effects on Drug Dissolution

The approach outlined in Sect. 8.2 can be extended [8.33] to an important problem in pharmacology, namely the rate of dissolution of drug particles. The dissolution properties of a drug are crucial in determining its bioavailability. In addition to the solubility constant of the drug crystallites, the dissolution process, both in vivo and in vitro, is determined by a variety of other properties of the drug particles, such as surface morphology, degree of porosity, surface area, particle size, size distribution, particle shape, and diffusion constant of the dissolved material.

One of the oldest equations derived for the drug dissolution rate is the 1897 Noyes and Whitney equation [8.34], which was later reformulated as follows [8.35,36]:

$$-\frac{dw}{dt} = kA_t(w_e - w_0 + w), \tag{8.9}$$

where dw/dt is the dissolution rate (g/min), A_t is the total surface area (m^2), and w_e, w_0, and w are the weight of the drug necessary to saturate the solution, the initial weight of the drug, and the weight of the undissolved drug after time t, respectively. The constant k is a function of the drug diffusivity and

Table 8.1. Reaction dimension of catalytic and non-catalytic processes

No.	Reaction	Material or Catalyst	Particle Size Range[a] (no. of fractions)	$d_{re}(d_f)$	[Refs.]
a.	**Non-Catalytic Reactions**				
1	acidic dissolution in NH₄Cl	Upper Columbus dolomite[1]	163-2605 (5)	2.15 ± 0.10 (2.9)	[8.30]
2	acidic dissolution in oxalate buffer	Upper Columbus dolomite	163-2605 (5)	2.34 ± 0.04) (2.9)	[8.30]
3	acidic dissolution in NH₄Cl	Niagara dolomite	163-2605 (5)	2.07 ± 0.06 (2.6)	[8.30]
4	acidic dissolution in oxalate buffer	Niagara dolomite	163-2605 (5)	2.19 ± 0.05 (2.6)	[8.30]
5a	acidic dissolution in NH₄Cl	Niagara dolomite	81-325 (3)		
5b	after 10% decomposition	Niagara dolomite	81-325 (3)	2.40 ± 0.08 (2.6)	[8.30]
5c	after 20% decomposition	Niagara dolomite	81-325 (3)	2.32 ± 0.08 (2.6)	[8.30]
5d	after 30% decomposition	Niagara dolomite	81-325 (3)	2.37 ± 0.06 (2.6)	[8.30]
5e	after 50% decomposition	Niagara dolomite	81-325 (3)	2.40 ± 0.06 (2.6)	[8.30]
5f	after 80% decomposition	Niagara dolomite	81-325 (3)	2.37 ± 0.08 (2.6)	[8.30]
6	acidic dissolution in seawater	nonporous rombic calcite (CaCO₃)	5,81[b]	1.96 (2.0)	[8.93]
7	acidic dissolution in seawater	Halimeda skeletal carbonate (green algea)	81-513 (5)	2.05 ± 0.08 (3.0)	[8.93]
8	acidic dissolution in seawater	Fungia (coral) skeletal carbonate (CaCO₃)	51-513 (5)	1.98 ± 0.07 (2.7)	[8.93]
9	acidic dissolution in seawater	Clypeaster skeletal carbonate (CaCO₃/MgCO₃)	81-513 (5)	2.15 ± 0.07 (2.7)	[8.93]
10	dissolution in HCl (pH 3)	Hybla alkali feldspar (potassium aluminosilicate)	56-400 (4)	3.0 ± 0.1 (2.4)	[8.31]
11	dissolution in HCl (pH 5)	Hybla alkali feldspar (potassium aluminosilicate)	56-400 (4)	2.95 ± 0.16 (2.4)	[8.31]
12	dissolution in boric acid buffer (pH 9)	Hybla alkali feldspar (potassium aluminosilicate)	56-400 (4)	3.06 ± 0.06 (2.4)	[8.31]
13	dissolution in K-biphtalate KOH buffer (pH 6)	sodium aluminosilicate	56-400 (4)	2.81 ± 0.01 (2.4 − 2.6)	[8.31]
14	dissolution in HF (3.66M)	quartz	45-1000 (10)	2.14 ± 0.06 (2.1)	[8.94]
15	dissolution in HF (3.66M)	Ottawa sand (quartz)	89-711 (7)	2.15 ± 0.06 (2.1)	[8.94]
16	dissolution in HF (0.1M)	Madagascar quartz	0.4-6 (14)	1.59 ± 0.05 (2.0)	[8.32]
17	dissolution in dilute NaOH	Madagascar quartz	0.4-6 (4)	1.78 ± 0.07 (2.0)	[8.92]
18	dissolution in distilled water	Sulfisomezole	194-650 (6)	1.89 ± 0.04	[8.95]
19	dissolution in 0.1% and 1%sodium lauryl sulfate	Sulfisomezole	194-650 (6)	2.0	[8.96]
20	dissolution in distilled water	Sulfamethizole	81-650 (7)	2.07 ± 0.04	[8.95]
21	dissolution in 0.1% and 1% sodium lauryl sulfate	Sulfamethizole	97-650 (7,9)	2.0	[8.96]

[1]Dolomit = CaCO₃/MgCO₃

Table 8.1. Reaction dimension of catalytic and non-catalytic processes, (contd.)

No.	Reaction	Material or Catalyst	Particle Size Range$^{(a)}$ (no. of fractions)	$d_{re}(d_f)$	[Refs.]
b.	**Catalytic Reactions**				
22	coal liquefaction	Shell 324 (NiMO/Al$_2$O$_3$)	840-3400 (4)	2.0 – 2.2 (± 0.02)	[8.97]
23	propylene polymerization	NiCl$_3$/0.33AlCl$_3$	0.05-0.3 (9)	2.18 ± 0.05	[8.98]
24a	dehydrogenation of butene to butadiene	BiMo	160-3100 (6)	3.02 – 3.11	[8.99]
24b	dehydrogenation of butene to butadiene	Fe$_2$O$_3$	0.15-0.61 (3,6)	1.1 ± 0.2, 1.29 ± 0.04	[8.38]
25	oxidation of tetralin	Co^{+2}/SiO$_2$	50-500 (6)	2.41 ± 0.12	[8.100]
26a	re-esterification of ethyl-acetate with 1-propanol	sulfonated styrene ion-exchanger SS-2	50-405 (3)	2.97 ± 0.02	[8.101]
26b	re-esterification of ethyl-acetate with 1-propanol	sulfonated styrene ion-exchanger SS-25	50-405 (3)	2.83 ± 0.03	[8.101]
26c	re-esterification of ethyl-acetate with 1-propanol	sulfonated styrene ion-exchanger SS-50	50-405 (3)	2.68 ± 0.09	[8.101]
26d	re-esterification of ethyl-acetate with 1-propanol	sulfonated styrene ion-exchanger MS-10	50-405 (3)	2.79 ± 0.01	[8.101]
26e	re-esterification of ethyl-acetate with 1-propanol	sulfonated styrene ion-exchanger MS-15	50-405 (3)	2.93 ± 0.01	[8.101]
26f	re-esterification of ethyl-acetate with 1-propanol	sulfonated styrene ion-exchanger MS-60	50-405 (3)	3.02 ± 0.02	[8.101]
27a	oxidation of acetaldehyde to peracetic acid	Co^{+3}Dowex X2	50-200 (4)	2.65 ± 0.02	[8.102]
27b	oxidation of acetaldehyde to peracetic acid	Co^{+3}Dowex X8	140-830 (5)	2.61 ± 0.03	[8.102]
28	oxidation of 3,5-di-tert butyl catechol	PVP-Cu(II)	37-550 (4)	2.87 ± 0.04	[8.103]
c.	**Reactions on Dispersed Metal Catalysts**				
29	water cleavage	Pt/PVA	0.011-0.053 (4)	1.93 ± 0.08	[8.104]
30	ethane hydrogenolysis	Pt/SiO$_2$	(c)	0.75 ± 0.21	[8.48]
31	ethane hydrogenolysis	Pt/Al$_2$O$_3$	10-118 (8)	2.8 ± 0.2	[8.105]
32	ethane hydrogenolysis	Pt/Al$_2$O$_3$	32-147 (5)	2.9 ± 0.2	[8.105]
33	ethane hydrogenolysis	Pt/Al$_2$O$_3$	23-150 (9)	3.1 ± 0.3	[8.105]
34	cyclopropane hydrogenolysis	Pt/Al$_2$O$_3$	13-175 (7)	2.29 ± 0.07	[8.106]
35	cyclopentane hydrogenolysis	Pt/Al$_2$O$_3$	19-175 (6)	2.1 ± 0.1	[8.107]
36	cyclopentane hydrogenolysis	Pt/Al$_2$O$_3$	66-150 (5)	2.02 ± 0.05	[8.107]
37	cyclopentane hydrogenolysis	Rh/Al$_2$O$_3$	(c) (6)	3.1 ± 0.2	[8.108]
38	propene hydrogenation	Ni/Al$_2$O$_3$	8-224 (4)	1.65 ± 0.05	[8.109]
39	benzene hydrogenation	Pd/charcoal	23-213 (7)	1.59 ± 0.04	[8.110]
40	benzene hydrogenation	Pt/SiO$_2$	66-379 (5)	1.1 ± 0.1	[8.111]
41	ethylene oxidation to ethylene oxide	Ag/Cab-O-Sil	61-290 (5)	1.2 ± 0.1	[8.112]
42	ethylene oxidation to ethylene oxide	Ag/Cab-O-Sil	200-450 (7)	0.7 ± 0.2	[8.113]
43	ethylene oxidation to ethylene oxide	Ag/silica Z	66-392 (5)	1.6 ± 0.1	[8.112]
44	ethylene oxidation to CO$_2$	Ag/Cab-O-Sil	61-290 (5)	0.4 ± 0.2	[8.112]
45	ethylene oxidation to CO$_2$	Ag/Cab-O-Sil	200-450 (7)	0.2 ± 0.2	[8.113]
46	ethylene oxidation to CO$_2$	Ag/silica Z	66-392 (5)	0.7 ± 0.2	[8.112]
47	ammonia synthesis	Fe/MgO	10-110 (17)	5.8 ± 0.5	[8.114]
48	CO methanation	Ni/Al$_2$O$_3$	21-144 (57)	2.8 ± 0.2	[8.115]
49	epimerization of cis-1,2-dimethylcyclohexane	Pt/Al$_2$O$_3$	11-118 (8)	1.98 ± 0.04	[8.116]

$^{(a)}$ Particle sizes are given in μm for reactions 1–29 and in Å (metal particle size) for reactions 31–36, 38–49.

$^{(b)}$ Reference tests for examples 7–9.

$^{(c)}$ Original particle size data in dispersion units.

a reciprocal function of the sample solution volume and of the diffusion layer thickness [8.36]. (The symbol k is used here and below as a "constant" in general, regardless of its different units in the various equations.)

If the drug particles have a smooth surface and are spheroidal, with size r and with a narrow dispersion, then one can insert into (8.9)

$$A_t \sim r^2 \sim w^{2/3}, \tag{8.10}$$

obtaining

$$-\frac{dw}{dt} = kw^{2/3}(w_e - w_0 + w). \tag{8.11}$$

Equation (8.11) is a widely used form of the Noyes - Whitney equation. However, (8.11) does not apply to surfaces which are not smooth. By employing the more general relation

$$S_t = kw^{d_{re}}, \tag{8.4a}$$

in which S_t (m^2) is the total area which is reactive in the dissolution process (the difference between (8.4) and (8.4a) is only in units - m^2/gr and m^2, respectively, as the different trades customarily use it), one obtains the fractal form of the Noyes - Whitney equation:

$$-\frac{dw}{dt} = kw^{d_{re}/3}(w_e - w_0 + w). \tag{8.12}$$

For low-porosity drug particles which are not sparingly soluble, one may assume $d_{re} = d_f$.

Following Hixon and Crowell [8.35], two special cases of (8.12) are treated next.

Case 1. This case was developed for dissolution under far-from-saturation conditions. Such conditions are fulfilled at short times or when the drug is sparingly soluble. One may then assume that in (8.12) $w = w_0$ (i.e., $w_e - w_0 + w = w_e$). Equations (8.11) and (8.12) simplify then to (8.13) and (8.14), respectively:

$$-\frac{dw}{dt} = kw_e w^{2/3} \tag{8.13}$$

and

$$-\frac{dw}{dt} = kw_e w^{d_{re}/3}. \tag{8.14}$$

Notice that the far-from-saturation conditions imply that $w_e \gg w$ and therefore that also $w_e \gg w_0$. The meaning of (8.13) is simple. Since the area of a spheroidal particle is proportional to $w^{2/3}$, then (8.13) states that the rate is proportional to the area which undergoes the dissolution reaction. This is Wenzel's law of the 18th-century [8.22,33]. A similar interpretation is also applicable for the fractal form (8.14), since the area of a spheroidal particle with fractal surface irregularity is $\sim w^{D/3}$.

Notice that for $d_{re} = 3$, (8.14) becomes a first-order decay equation

$$-\frac{dw}{dt} = kw_e w. \tag{8.15}$$

A classical interpretation of (8.15) would be that dissolution occurs at equal rates from all of the bulk of the particle. This may occur if the particle (or an outer shell of the particle) is so highly porous that most of the material is exposed at the pore interface. Alternatively $d_{re} = 3$ may represent a special case of growth rate of dissolution sites with particle size; here the resemblance of (8.15) to a first-order decay is artificial.

Case 2. In this case, by contrast with case 1, the initial weight is the one necessary to obtain saturation after full dissolution (i.e., $w_e = w_0$). This means that the dissolution is carried out under conditions which are allowed to develop up to saturation. In this case, (8.11) and (8.12) reduce, respectively, to (8.16) and (8.17)

$$-\frac{dw}{dt} = kw^{5/3}, \tag{8.16}$$

and

$$-\frac{dw}{dt} = kw^{(d_{re}+3)/3}. \tag{8.17}$$

Since in this case $w < w_e(= w_0)$, the dissolution rate is slower than for case 1. The physical picture reflected by the faster behavior of (8.14) is that it represents a situation which is still far away from saturation. Case 2, on the other hand, still holds the properties of the general case (8.12), which treats dissolution in solutions which already contain some dissolved drug, namely, a system which is already on the way to saturation hence, its slower behavior. (For comparison purposes, $w_0 = w_e$ was taken for the two cases. Under these conditions, the difference between the two equations is only the condition $w = w_0$ for case 1.)

The integrated form of the Noyes - Whitney equation is known as the Hixon - Crowell cube root law [8.35]. For case 1 integration of (8.14) yields the fractal form of that law (Fig. 8.3a):

$$\frac{3}{d_{re} - 3}(w^{(3-d_{re})/3} - w_0^{(3-d_{re})/3}) = w_e kt. \tag{8.18}$$

Notice that, since w is positive, the following physical condition must be obeyed throughout the dissolution:

$$\left|\frac{d_{re} - 3}{3}kw_e t\right| \leq \left|w_0^{(3-d_{re})/3}\right| \tag{8.19}$$

For case 2 integration of (8.17) yields (Fig. 8.3b)

$$\frac{3}{d_{re}}(w^{-d_{re}/3} - w_0^{-d_{re}/3}) = kt, \tag{8.20}$$

with the condition

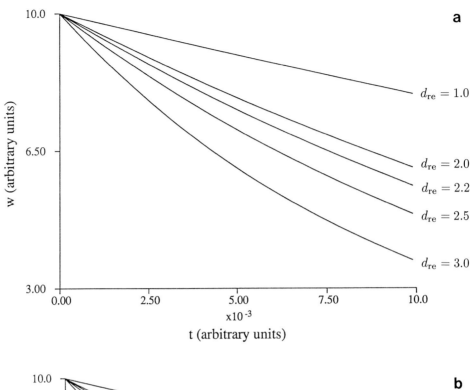

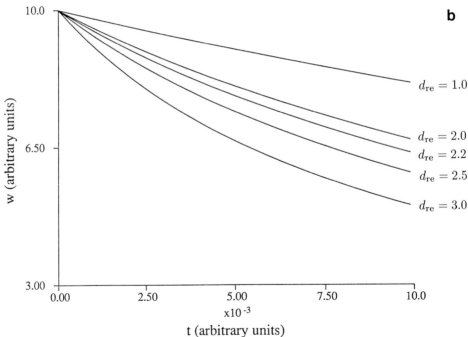

Fig. 8.3. The amount of undissolved drug w as a function of time t according to (a) Case 1 (Eq. (8.18) with $w_0 = 10$, $w_c = 100$, $k = 1$) or (b) Case 2 (Eq. (8.20) with $w_0 = w_c = 10$, $k = 1$). All data are in arbitrary units. For $d_{re} = 3.0$ in Case 1, the integrated form of (8.15) is plotted

$$\left|\frac{d_{\text{re}}}{3}kt\right| \leq \left|w_0^{-d_{\text{re}}/3}\right|. \tag{8.21}$$

It is seen (Fig. 8.3) that for a given d_{re}, the dissolution rate is faster for case 1 than for case 2. The explanation for this was proposed above. To the best of our knowledge, the treatment described in this section is the first attempt to quantify drug-particle surface morphology effects on the rate of active dissolution.

8.4 Size Effects in Catalysis

A third important class of reactions which can be analyzed by the approach of Sect. 8.2 is the catalytic reactions. As an example, we consider the dehydrogenation of butenes to butadienes: the catalytic dehydrogenation of n-butenes to butadiene is a widely used reaction, which has been carried out industrially since 1943 on various types of catalysts. We found that different catalysts are also characterized by different d_{re} values. Thus a commonly used catalyst for butene dehydrogenation is (supported and unsupported) Fe_2O_3. The particle size effect on the reaction rate of this catalyst was studied by Yang et al. [8.38] in order to reveal what are the likely causes of the size dependence of activity and selectivity. They could not, however, provide a definite answer and suggested a number of possible explanations (see below). The analysis of the scaling relations between the reaction rate and particle size (8.5), based on the data in Table 8.1 in [8.38], gave $d_{\text{re}} = 1.08 \pm 0.20$ and $d_{\text{re}} = 1.29 \pm 0.04$ for $T = 300$ and $375°C$, respectively (24b in Table 8.1, Fig. 8.4). These very low d_{re} values are indicative of a subset of active sites and support two of the

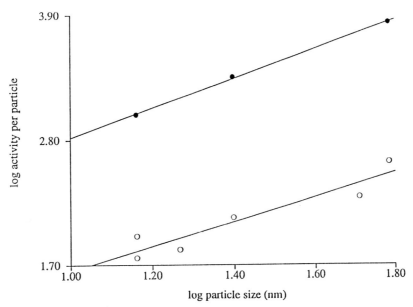

Fig. 8.4. The oxidation rate of butene to butadiene on Fe_2O_3 as a function of catalyst particle size and of reaction temperature (○ : 300°C; ● : 375°C)

authors' suggestions as to the origin of the particle size dependence: that the production of butadiene may require specific sites such as anion or cation vacancies, or that the reaction may be crystal face specific. In both cases only a subset of the surface sites participates in the reaction, a situation which leads to $d_{re} < 2$. Several other examples of catalytic reactions are collected in Table 8.1.

A related family of reactions which we found to be describable by (8.5) is catalysis on dispersed metal catalysts (on inert supports) and the associated well-known "structure sensitivity" issue [8.39]. Here the size of relevance is that of the metal crystallite [8.41] and not, as in the previous example, the whole catalyst pellet [8.40]. The characterization of these catalytic reactions by (8.5) has been found to be applicable in virtually all types of catalytic reactions. These include [8.16,21,41 - 46] hydrogenolyses with Pt, Ir and Rh supported on Al_2O_3, hydrogenations on, e.g., Ni/Al_2O_3, Pd/C, Pt/SiO_2, ethylene oxidations under various conditions, electrooxidations and electroreductions, an isomerization, photocatalytic reactions [8.37], CO methanations, and ammonia synthesis. A wide range of d_{re} values was found: from close to zero, indicating weak dependence of the activity on particle size, up to $d_{re} \approx 6$. It has been suggested that the observed d_{re} values reflect specific population distributions of surface active sites, and fits between observed d_{re} values and specific distributions of sites on model crystallites were shown [8.41,43].

In its simplest form, the catalytic activity of the surface atoms of a crystallite is determined by the type of coordination at the surface and all exposed surface atoms participate in the reaction: the reaction is insensitive to surface structure. A structure-insensitive [8.39] case is therefore recognized by $d_{re} \approx 2$. An example is the benzene hydrogenation study of Mutin et al. [8.47], carried out on Pt/Al_2O_3. These authors studied several catalysts prepared by different methods and analysis of all their experimental results reveals that $d_{re} = 2.03 \pm 0.03$, in agreement with their hypothesis that all atoms of the 100 and 110 surface planes of Pt participate in the reaction. This structure-insensitivity result for benzene hydrogenation seems to be quite general: we obtained $d_{re} \approx 2.0$ values from the analysis of various independent studies and on various additional catalysts [8.41,43].

In the absence of additional experimental data, it is difficult to attach a unique model fit to an observed $d_{re} \neq 2$ value: several surface site selections can result in similar d_{re}'s. In the process of choosing a model, we suggest following two principles: first, that the model is chemically and physically feasible and in agreement with the known features of the reaction under study; and second, that the distribution law is kept as simple as possible. The application of these principles is illustrated in the following case.

Guczi and Gudkov [8.48] studied dispersity effects on the hydrogenolysis of ethane on Pt/SiO_2. Analysis of their data by (8.5) yields $d_{re} = 0.75 \pm 0.21$ (30 in Table 8.1). Standard-geometry considerations predict $d_{re} \approx 0$ and $d_{re} \approx 1$ for a reaction which occurs on corner atoms or on edge atoms, respectively, of ideal crystallites. Guczi and Gudkov indeed "put forward the idea that the increase in the proportion of corner and edge atoms" may be responsible for the observed structure sensitivity, by virtue of enhancing the rate of C - C rupture on such sites. Since Cantor gap-set distributions (and Devil Staircases) have been found, at least macroscopically, in several cases [8.49,50], we suggest that a Cantor distribution of active strips with $d_{re} = 0.75$ is a feasible interpretation here too. For a detailed discussion, see [8.21,41,43]. Table 8.1 contains further examples.

8.5 Multifractal Analysis of Catalytic Reactions

In the previous section we analyzed size effects on catalytic reactions, using a single D value. A finer picture will now be presented in this section: we show now how *multifractal* analysis methods can be used to characterize heterogeneous catalytic reactions and, in particular, to elucidate the details of the influence of the surface geometry on specific catalytic processes [8.51].

Objects of similar D may exhibit different site-specific accessibilities, i.e., different degrees of steric screening for an incoming molecule (Sect. 8.6). Different connectivities may lead to different site-specific accessibilities or different reaction probabilities with an incoming molecule. Here we show how to characterize the ensuing reaction probability distribution (RPD) using multifractal functions. These functions proved to be useful in describing processes in environments of complex scale-invariant geometry [8.52 - 56].

The studied objects are Devil's Staircases (DS) and Cantor sets (CS) in random (Fig. 8.1) and deterministic (Figs. 8.5,6) versions. The former is a multiple-step structure in which the tips of the stairs constitute a fractal CS subset of the surface.

The CS and its corresponding DS are characterized by the same fractal dimension, so they allow the testing of morphological effects in the case of catalysts characterized by similar fractal dimension.

When the probability of reaction of a fractal set with diffusing molecules is measured, reaction probability-distribution (RPD) curves of the type shown in Fig. 8.7 are usually obtained.

It is seen that the observed probability is distributed in a nonhomogeneous form on the fractal substrate. The multifractal formalism provides the function which characterizes the dependence of these probability distributions on the yardstick used to measure them and enables one to transform the RPD profiles into a workable form.

We discuss first the origin of this type of nonhomogeneous distribution in the case of deterministic CS and DS catalytic structures, when they react under steady-state conditions with molecules that diffuse from the bulk and react upon collision with the surface active sites (Eley - Rideal mechanism). The peaks in the distributions and their relative size arise from two effects: (a) the reflection of the diffusing molecules on the nonactive gaps results in a higher reaction probability of the active sites near wide gaps, and (b) the inner sites in the DS are screened by the outer ones. Owing to effect (b), in

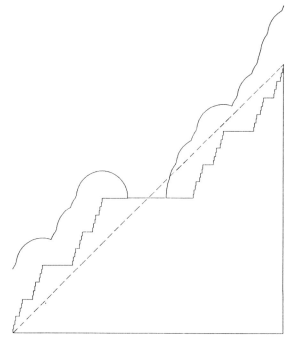

Fig. 8.5. A deterministic 6-iteration Devil's Staircase (DS) structure of 64 steps. The fractal dimension of the tips of the stairs is that of the representative Cantor sets of Fig. 8.6. The *dashed line* is the average slope. Also shown is the neighboring volume to the fractal set of active sites of the deterministic 6-iteration DS

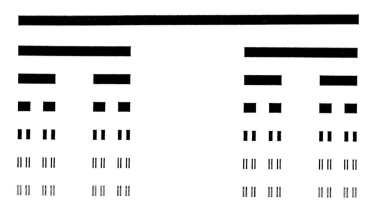

Fig. 8.6. A deterministic Cantor set (CS): representation of the iterative generation process. The fractal dimension (as measured by a box-counting technique) is 0.62

the deterministic DS the singularities of the RPD appear as sharper peaks than in the CS. Moreover, the symmetry shown by the RPD of the CS is lost in the case of the DS, which is due to the nonsymmetric screening. These two phenomena (higher singularity and nonsymmetric RPD of the DS) can be explained as follows. One draws a line connecting the ends of the DS structure (Fig. 8.5). Since the source is a line parallel to this line, the arrangement which is symmetric for the CS is no longer symmetric here: active sites on the left part of the DS are closer to the source than those on the right part of it and, in

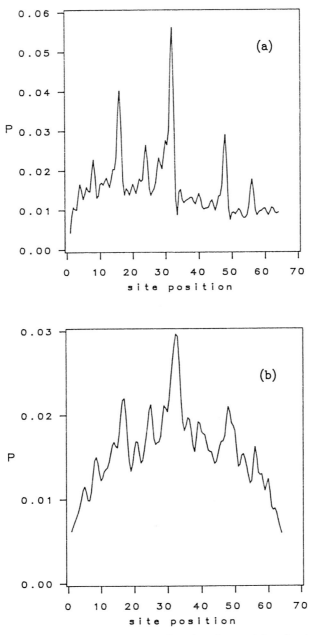

Fig. 8.7. Reaction probability as a function of active site position for (**a**) the deterministic DS and (**b**) the deterministic CS. There are 64 active sites which are numbered from left to right along the horizontal axis. The vertical axis shows the probability of reaction of such sites with the diffusing molecules

particular, of the two sites near the largest terrace, the right one is more buried than the left one. This difference leads to the sharper central peak in the RPD profile of the DS. The same phenomenon occurs on all scales, namely, sharper peaks are generated on the sides of the central peak.

The multifractal formalism enables the transformation of these RPD profiles into a useful condensed form through the $f(\alpha)$ curves, which we briefly review next [8.52,57]. In order to perform the multifractal analysis, the fractal substrate is divided into boxes of linear size r and the probability of reaction, the Eley - Rideal reaction in our case, associated with each box is measured. In the limit of small r values a scaling exponent α_i is defined:

$$p_i(r) \sim r^{\alpha_i}, \tag{8.22}$$

where p_i is the reaction probability at box i. The exponent α_i is dispersed owing to the different values the probability can take, depending on the location of i. If one then looks at the collection of α_i values that fall within a small region $[\alpha, \alpha + d\alpha]$, a second scaling assumption can be made, namely that the number $n(\alpha)d\alpha$ of regions which obey (8.22) with an α value between $[\alpha, \alpha + d\alpha]$ scales with r as

$$n(\alpha)d(\alpha) \sim r^{-f(\alpha)}. \tag{8.23}$$

To calculate $f(\alpha)$, we notr that it is related to measurable properties of the probability distribution, namely, to the moments M_q and to their scaling exponents $\tau(q)$,

$$M_q = \sum_i p_i{}^q = \sum_p n(p)p^q \sim r^{\tau(q)}, \tag{8.24}$$

where q is the moment order, $\tau(q)$ is a scaling exponent, and $n(p)$ is the number of boxes with reaction probability p. Assuming that $n(p)p^q$ has a sharp maximum at a value $p(q)$, then one can approximate M_q by the element of the sum that takes the maximum value. Then substituting (8.22) and (8.23) in (8.24) yields

$$\tau(q) = q\alpha(q) - f(\alpha). \tag{8.25}$$

The value that dominates the sum in (8.24) is associated with the smallest $\tau(q)$; thus $\alpha(q)$ is given by the extremal condition

$$\frac{d}{d\alpha}[q\alpha(q) - f(\alpha)] = 0. \tag{8.26}$$

It also follows from (8.25) and (8.26) that

$$\frac{d}{dq}[\tau(q)] = \alpha(q). \tag{8.27}$$

For details see [8.58].

In the $f(\alpha)$ spectra we first evaluate the range of α values, which indicates the range of reaction probablities (8.22). As expected from the RPD profiles one sees that the range is larger in the DS compared to that of the CS (Figs. 8.8 and 8.9). The difference between the deterministic and random DS (Figs. 8.8b and 8.9b) can be explained in terms of the random scrambling of the sites in the latter compared to the former, causing a decrease in the average distance of the sites from the average slope of the structure.

The structures differ at the low-probability end of the spectrum (CS: ≈ 0.7, DS: ≈ 0.9). This observation reflects the screening operating in the DS, which adds low-probability values and extends the α range.

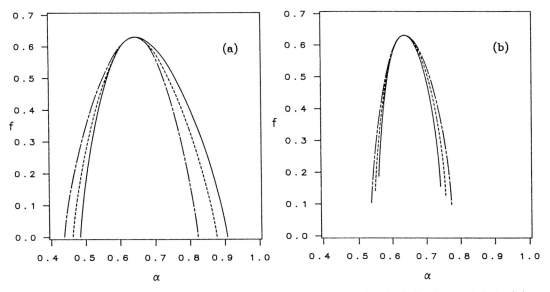

Fig. 8.8. $f(\alpha)$ curves and their dependence on the recursion level of the deterministic (**a**) DS and (**b**) CS. The *solid line* is for the structures of 4 iterations, the *dashed line* is for 5 iterations and the *solid-dashed* line is for 6 iterations

In the multifractal formalism, scaling assumptions are made for the moments and the singularities of the distributions (8.22 - 24), namely, that, at the limit of small r values, the $\tau(q)$ and $f(\alpha)$ curves converge to a unique r-independent curve. In the deterministic objects these assumptions were checked by measuring the RPD for objects of three different recursion levels (four, five and six), whereas for the random structures this was done by changing the minimum gap size with which the structure was constructed [8.58]. Convergence has not been observed at the recursion level studied for the deterministic structures (Fig. 8.8). However, in the low-α- value domain of the random DS a full convergence to a single curve has been observed (Fig. 8.9a).

We thus see how the combined analysis of catalytic reaction using both fractal (Sect. 8.4) and multifractal tools, provides insight of structure-sensitivity relations, which has been a classical, major issue in catalysis.

8.6 The Accessibility of Fractal Surfaces to Derivatization Reactions

Our final example deals with a basic property of fractal surfaces that has already been mentioned above: their accessibility to incoming molecules depends on their size [8.10,15,21,59 - 78]. We analyze here the implications of this property on the derivatization of surfaces. Derivatization of surfaces is a key process in the preparation of chromatographic materials, giving the ability to fine-tune the type of surface - adsorbate interactions by a suitable choice of the derivatizing agent [8.79 - 83]. This is particularly true for silylation reactions, which

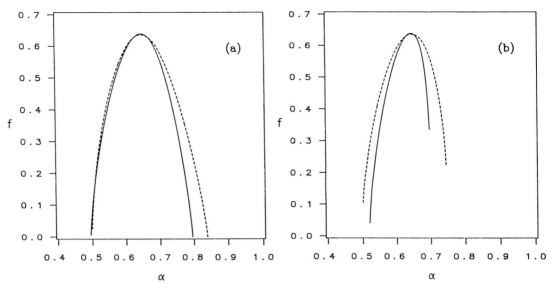

Fig. 8.9. The dependence of the f(α) curve on the set size of the random **(a)** DS and **(b)** CS. The *solid line* is for the 32-site structure (the analog to the 5-iteration deterministic object) and the *dashed line* is for the 64-site structure (the analog to the 6-iteration object). A complete convergence to a unique curve independent of size is observed in the low-α zone of the DS

are the focus of this section. It is well established that the performance of the chromatographic material is determined not only by the nature of the phase-reversing ligand, but also by other material properties, the most important of which are the packing density of the ligands on the surface, the surface concentration or coverage, the area per ligand, and the proportion of surface sites that undergo reaction. For silica in its various forms all of the above-mentioned parameters are determined by the availability of silanols to silylation reactions and by their distribution (the surface heterogeneity) [8.83,84]. There are three main factors that determine the accessibility of silanols to reaction. The first one is relatively simple to evaluate - the "umbrella" effect of the ligand. Most ligands have cross-sectional areas that shield more than the silanol(s) with which they react. The second factor is the surface irregularity, tortuosity, and connectivity and the degree to which these pore and surface features shield the silanols from an incoming reagent molecule. The third parameter is surface heterogeneity, from the point of view of both the possible clustering of reactive sites and the differences in reactivities of silanols due to inductive effects of surrounding moieties.

A routine practice in many chromatographic studies is the use of the nitrogen-BET value as a basic parameter of the material. However, the silylating reagent (SR) is much bulkier than nitrogen, and consequently the smaller molecule has a much better accessibility to the silanols than has SR. In other words, the nitrogen-BET value is of no relevance in calculating various grafting parameters when the accessibility of the surface is not equal for N_2 and SR. The relevant surface area in all cases is

$$A = N_0 m \sigma, \qquad (8.28)$$

where A is the apparent surface area (m^2/gr silica), m is the experimental monolayer value (mol/gr silica) for a ligand with cross-sectional area σ (m^2), and N_0 is Avogadro's number. (For the nontrivial task of evaluating σ, see [8.85].) For an irregular surface, the accessibility decreases with increasing σ, and therefore

$$A_{N_2} > A_{SR}. \tag{8.29}$$

An immediate consequence is that the standard procedure of calculating the area, σ_a, occupied by, e.g., an alkyl chain from

$$\sigma_a = A_{N_2}/(N_0 m_{SR}), \tag{8.30}$$

is correct only for flat surfaces and gives overestimates for an irregular surface. Quantitative estimation of the effects of surface irregularity and chemical selectivity on surface derivatization parameters is possible by applying

$$m_{SR} \sim \sigma^{-d_{re}/2}, \tag{8.31}$$

where m_{SR} is the number of moles of SR needed to densely coat the surface.

From (8.31) we obtain a measure for silanol accessibility to silylation (or any other) reactions [8.21,61,86]:

$$\text{accessibility} = m_{SR}/m_{OH} = (\sigma_{SR}/\sigma_{OH})^{-d_{re}/2}, \tag{8.32}$$

where the index OH refers to surface silanol and then, if one can assume that A_{N_2} probes "all" silanols,

$$\text{accessibility} = m_{SR}/m_{N_2} = (\sigma_{SR}/\sigma_{N_2})^{-d_{re}/2}. \tag{8.33}$$

The dependence of the accessibility on the ratio σ_{SR}/σ_{OH} for various d_{re} values is shown in Fig. 8.10.

In view of this picture, we analyze now the work of Larsen and Schou who studied the silylation of various silicas with trimethylchlorosilane (TMS) and with dimethyloctadecylchlorosilane (ODS) [8.87]. From their data we first calculate d_{re} for porous silica with an average pore size of 60Å (Si-60): $\sigma_{N_2} = 16.2$Å^2; $\sigma_{ODS} = 46$Å^2 (or 48Å^2 from the fractal analysis we described in [8.20], regarding the σ values of anchored n-alkanols); $A_{N_2} = 505$ m^2/gr; A_{ODS}(average)$=304$ m^2/gr. From (8.33), $d_{re}=2.98$. The agreement between this value and the value $d_{re} = 2.97$ obtained for a series of spherical alcohols adsorbed on Si-60 is indeed remarkable: furthermore, the agreement is not only in the exponent but also in the prefactor. For alcohols we obtained [8.20]

$$\log m = 2.50 - \frac{2.97}{2}\log\sigma, \tag{8.34}$$

whereas the prefactor for Larsen's N$_2$ value is 2.52 and the value for ODS is 2.51 (8.31). Equation (8.34) served as a calibration curve for determination of σ for n-alkanols and n-alkanoic acids. The good fit of Larsen's data to (8.34) (see Fig. 8.11) suggests that this equation can be used for silylation as well, at least for a preliminary estimate of σ_{SR}.

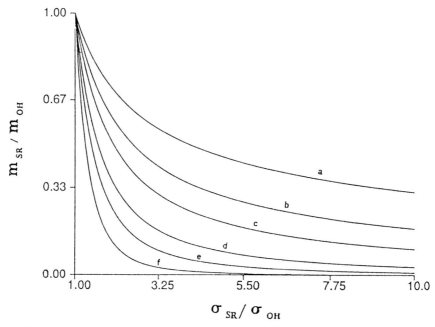

Fig. 8.10. Surface accessibility of silanols, m_{SR}/m_{OH}, as a function of the relative size of silylating reagent, σ_{SR}/σ_{OH}, for various reaction dimensions, $d_{re}=1.0$ (a), 1.5 (b), 2.0 (c), 3.0 (d), 4.0 (e), 6.0 (f)

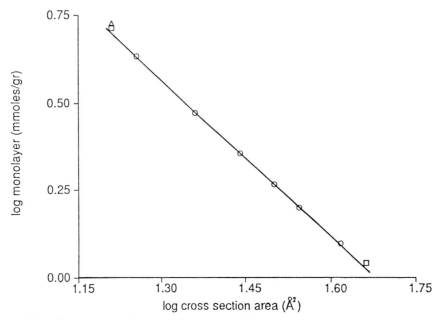

Fig. 8.11. Monolayer values for the adsorption of "spherical alcohol" (∘; from left to right, methanol, ethanol, 2-propanol, tert-butanol and triethylcarbinol) and for the adsorption of N_2 (△; by the BET method) vs. cross-sectional area, and the fit to the Larsen and Schou [8.87] data (□ ; left, N_2; right, ODS) to this line. Si-60, $d_{ad} = 2.97 \pm 0.02$

In other words, it seems that the two types of process (hydrogen bonding and silylation) have similar effective accessibilities for the silica surface, with $d_{ad} = d_{re}$. The result presented here adds to accumulated data that indicate that d_{ad} of Si-60 is in the range 2.8 - 3.1 [8.10]. (In a recent study it was also found that for the interaction of alcohols with vigorously dried silica, $d_{re} \approx 6$ [8.88]. This could be reflecting an increase in the surface chemical reactivity with decrease in the alcohol chain length [8.84,89,90].) Hydrothermal treatment of silica is known to smooth the texture and widen the pores. Indeed an analysis of Larsen's data on Si-60 after this treatment reveals that $d_{re} = 2.24$ (silica 3 in Table 8.1 of [8.87]; $A_{N_2}=216$ m^2/gr, A_{TMS}(average)= 193 m^2/gr, $\sigma_{TMS} = 38$Å^2).

Finally we comment on the area occupied by a grafted ligand. We discuss this issue by commenting on a study of Lochmuller and Wilder [8.91]. These authors derivatized Si-60 ($A_{N_2} = 550$ m^2/gr) with the alkyl chains indicated in Table 8.2, to maximum coverage. The first three columns in this table are reproduced from Table 8.2 in [8.91].

Table 8.2. Area per alkyl chain for a series of reversed-phase silicas

Stationary phase (no. of carbons)	m_{SR} (μmol/g)	Alkyl chain per 100 Å^2	$\sigma_a^{(a)}$ (Å^2/chain)	$\sigma_{SR}^{(b)}$ (Å^2/chain)	$\sigma^{(c)}$ (Å^2/chain)
Pentyl (C$_5$)	1.30	1.42	70.4	43.5	40.4
Hexyl (C$_6$)	1.23	1.34	74.6	45.2	42.0
Heptyl (C$_7$)	1.18	1.29	77.5	46.4	43.2
Octyl (C$_8$)	1.07	1.17	85.5	49.6	46.1
Decyl (C$_{10}$)	1.04	1.13	88.5	50.6	47.0
Dodecyl (C$_{12}$)	0.98	1.07	93.5	52.6	48.9
Octadecyl (C$_{18}$)	0.83	0.91	109.9	58.9	54.7
Docosyl (C$_{22}$)	0.75	0.82	122.0	63.0	58.6

[a] From Eq. (8.30) (the reciprocal of the value in the third column)
[b] From Eq. (8.33)
[c] From Eq. (8.34)

We focus our discussion on the third and fourth columns, which show how much area, σ_a, is available for each alkyl chain, if the silica surface is that of a "flat table", as shown in Fig. 8.2 of [8.91]. As discussed above [8.61] this leads to an overestimation of σ_a. The reference area for calculating σ_a must be the effective area for the chains investigated and not for N$_2$. Since m_{SR} is the maximum experimental capacity for the alkyl chain, the best one can say is

$$\sigma_a = \sigma_{SR} = A_{SR}/m_{SR}N_0 \qquad (8.35)$$

Thus the problem boils down to an independent estimation of σ_{SR}. Methods for evaluating cross-sectional areas of molecules have been discussed in detail in the literature and have recently been reviewed with the emphasis on associated difficulties [8.10,85]. Here we show how knowing d_{re} can be useful for this purpose by use of two (related) methods. In the first method, we calculate each of the σ_{SR} values by using (8.33) with $d_{\mathrm{re}} = 3$ for this Si-60. The results are shown in the fifth column of Table 8.2: σ_a decreases significantly if A_{SR} is used instead of $A_{\mathrm{N_2}}$. The second technique is the use of a calibration curve for σ and m. The closest we have at the moment is (8.34) (Fig. 8.11), which worked for Larsen's data. By using it, we obtain the data in the sixth column, which are similar to those in the fifth column. Thus these two columns corroborate each other.

8.7 Conclusion

The examples collected in this review seem to indicate that the promises of fractal geometry have, at least in part, been fulfilled. Fractal geometry has provided the proper language and the necessary vocabulary to reformulate some classical problems in heterogeneous chemistry, and the crucial importance of being able to do so for scientific progress cannot be underestimated. And, if nothing else, fractal analysis of these problems has provided at least an indication of the trend of change of reaction parameters with changes in the morphology and structure of the chemically reactive environments. As detailed in Sect. 8.1, the geometry problem in chemistry had been in the past unapproachable. This, we believe, has been changed for good.

Acknowledgments. Scientific activity in this area is currently supported by the US - Israel Bi-National Science Foundation. D.A. is a member of the F.Haber Research Center for Molecular Dynamics and of the Farkas Center for Light Energy Conversion. R.G. thanks M. Sheintuch for supervising the study described in Sect. 8.5. D.F. thanks the Valatzi - Pikovski Foundation for support.

References

8.1   T.R. Thomas, ed.: *Rough Surfaces* (Longman, Harlow 1992)

8.2   B. Mandelbrot: *The Fractal Geometry of Nature* (Freeman, San Francisco 1982)

8.3   H.-O. Peitgen, P.H. Richter: *The Beauty of Fractals* (Springer, Berlin, Heidelberg 1986)

8.4   C.K.R. Varma: Phil. Mag. **1**, 959 (1967), Fig. 8.1.

8.5   D. Avnir, ed.: *The Fractal Approach to Heterogeneous Chemistry: Surfaces, Colloids, Polymers* (Wiley, Chichester 1989, 3rd corrected reprinting 1992)

8.6   P.A. Burrough: Nature **294**, 240 (1981)

8.7   L.Z. Fong: Modern Phys. Lett. **1**, 601 (1986)

8.8   M. Sato et al: Chem. Phys. Lett. **181**, 526 (1991)

8.9   A. Hurd, D.W. Schaefer, D.M. Smith, S.B. Ross, A. Le-Mehaute, S. Spooner: Phys. Rev. B **39**, 9742 (1989)

8.10   D. Avnir, D. Farin, P. Pfeifer: New J. Chem. **16**, 439 (1992)

8.11   D. Farin, D. Avnir: J. Phys. Chem. **91**, 5517 (1987)

8.12   M. Yaworski, J.G. Byrne: Metal. Trans A **19**, 1371 (1988)

8.13   B.H. Kaye: *A Random Walk Through Fractal Dimensions* (VCH, Weinheim 1989)

8.14 A. Seri-Levy, D. Avnir: Surf. Sci. **248**, 258 (1991)

8.15 D. Avnir, D. Farin, P. Pfeifer: Nature **308**, 261 (1984)

8.16 D. Farin, D. Avnir: Proceedings of the IUPAC Symposium on Characterization of Porous Solids, Germany, 1987; K.K. Unger, D. Behrens, H. Kral, eds. (Elsevier, Amsterdam 1987)

8.17 D. Avnir, D. Farin, P. Pfeifer: J. Colloid Interface Sci. **103**, 112 (1985)

8.18 S.H. Ng, C. Fairbridge, B.H. Kaye: Langmuir **3**, 340 (1987)

8.19 This is true for particles which are not mass fractals, and this is usually the case for mechanically stable particles: M. Ben-Ohoud, F. Obrecht, L. Gatineau, P. Levitz, H. Van-Damme: J. Colloid Interface Sci. **124**, 156 (1988) H. Van-Damme, P. Levitz, L. Gatineau, J.F. Alcover, J.J. Fripiat: J. Colloid Interface Sci. **122**, 1 (1988)

8.20 D. Farin, A. Volpert, D. Avnir: J. Am. Chem. Soc. **107**, 3368, 5319 (1985)

8.21 D. Farin, D. Avnir, in [8.5], Sect. 4.1.2

8.22 C.F. Wenzel: *Lehre von der Verwandschaft der Körper* **8** (Dresden 1777); R. Winderlich: J. Chem. Educ. **27**, 56 (1950)

8.23 A. Coniglio, H.E. Stanley: Phys. Rev. Lett. **52**, 1068 (1984)

8.24 M. Silverberg, D. Farin, A. Ben-Shaul, D. Avnir: Ann. Isr. Phys. Soc. **8**, 451 (1986)

8.25 P. Meakin: CRC Crit. Rev. Solid State Mater. Sci. **13**, 143 (1987)

8.26 G.E. Blonder: Phys. Rev. Condens. Mater. **33**, 6157 (1986)

8.27 M. Matsushita, K. Honda, H. Toyoki, Y. Hayakawa, H. Kondo: J. Phys. Soc. Jpn **55**, 2618 (1986)

8.28 C.-L. Yang, Z.-Y. Chen, M.A. El-Sayed: J. Phys. Chem. **91**, 3002 (1987)

8.29 H.E. Stanley, N. Ostrowsky, eds.: *On Growth and Form* (Nijhof, Dordrecht 1986); NATO ASI Ser.

8.30 K.S. Love, C.W. Whittaker: J. Agric. Food Chem. **2**, 1268 (1954)

8.31 G.R. Holdren, P.M. Speyer: Geochim. Cosmochim. Acta **49**, 675 (1985)

8.32 I. Bergman: J. Appl. Chem. **12**, 336 (1962); **13**, 356 (1963)

8.33 D. Farin, D. Avnir: J. Pharm. Sci. **81**, 54 (1992)

8.34 A.A. Noyes, W.R. Whitney: J. Am. Chem. Soc. **19**, 930 (1897)

8.35 A.W. Hixon, J.H. Crowell: Ind. Eng. Chem. **23**, 923 (1931)

8.36 R.J. Hintz, K.C. Johnson: Int. J. Pharm. **51**, 9 (1989)

8.37 D. Farin, J. Kiwi, D. Avnir: J. Phys. Chem. **93**, 5851 (1989)

8.38 B.L. Yang, F. Hong, H.H. Hung: J. Phys. Chem. **88**, 2531 (1984)

8.39 M. Boudart: Adv. Catal. **20**, 153 (1969)

8.40 D. Avnir, J.J. Carberry, O. Citri, D. Farin, M. Gratzel, A.J. McEvoy: Chaos **1**, 397 (1991)

8.41 D. Farin, D. Avnir: J. Am. Chem. Soc. **110**, 2039 (1988)

8.42 D. Avnir, A. Seri-Levy, D. Farin, O. Citri, in: Proceedings of the 23rd Biannual Conference of the Royal Spanish Society of Chemistry, Salamanca, Spain, 1990; ed. by A. San Feliciano, M. Grande, J. Casado, pp. 309-319

8.43 D. Farin, D. Avnir, in: Proceedings of the 9th Int. Congress on Catalysis, ed. by M.J. Phillips and M. Ternan, Vol. 3, Chem. Inst. Ottawa, Canada, 1988; pp. 998-1005

8.44 S.D. Jackson: React. Kinet. Catal. Lett. **39**, 223 (1989)

8.45 A. Seri-Levy, J. Samuel, D. Farin, D. Avnir: Stud. Surf. Sci. Catal. **47**, 353 (1989)

8.46 D. Avnir, O. Citri, D. Farin, M. Ottolenghi, J. Samuel, A. Seri- Levy, in: *Optimal Structures in Heterogeneous Chemistry* ed. by P. Plath, Springer Ser. Syn. (Springer, Berlin, Heidelberg 1989), pp. 65 - 81

8.47 R. Mutin, J.M. Basset, M. Prette: C.R. Acad. Sc. Paris **273C**, 1704 (1971)

8.48 L. Guczi, B.S. Gudkov: React. Kinet. Catal. Lett. **9**, 343 (1978)

8.49 P. Bak: Physics Today, Dec. 1986, p. 39

8.50 T. Kleiser, M. Bocek: Z. Metalkde **77**, 582 (1986)

8.51 R. Gutfraind, M. Sheintuch, D. Avnir: Chem. Phys. Lett. **147**, 8 (1990)

8.52 P. Meakin, A. Coniglio, H.E. Stanley, T.A. Witten: Phys. Rev. A **34**, 3325 (1986)

8.53 H.E. Stanley, P. Meakin: Nature **335**, 405 (1988)

8.54 J. Lee, H.E. Stanley: Phys. Rev. Lett. **61**, 2945 (1988)

8.55 R. Blumenfeld, A. Aharony: Phys. Rev. Lett. **62**, 2977 (1989)

8.56 S. Havlin, B.L. Trus: J. Phys. A **21**, L731 (1988)

8.57 T.C. Halsey, M.H. Jensen, L.P. Kadanoff, I. Procaccia, B.I. Shraiman: Phys. Rev. A **33**, 1141 (1986)

8.58 R. Gutfraind, M. Sheintuch, D. Avnir: J. Chem. Phys. **95**, 6100 (1991); for general discussions see the chapters by H.E. Stanley, A. Aharony, and B. Mandelbrot and C. Evertsz in: *Fractals and Disordered Systems*, ed. by A. Bunde and S. Havlin (Springer, Berlin Heidelberg 1991)

8.59 D. Avnir, D. Farin, P. Pfeifer: J. Chem. Phys. **79**, 3566 (1983)

8.60 D. Avnir, D. Farin: New J. Chem. **14**, 197 (1990)

8.61 D. Farin, D. Avnir: J. Chromatogr. **406**, 317 (1987)

8.62 G. Guo, Y. Chen, Y. Tang, X. Cai, S. Lin: Preprints of the IUPAC Symp. on Characterization of Porous Solids, Germany, 1987, p. 77

8.63 H. Frank, H. Zwanziger, T. Welsch: Fres. Z. Anal. Chem. **326**, 153 (1987)

8.64 S.V. Christensen, H. Topsoe: Private Communication (1987)

8.65 H. Spindler, P. Szargan, M. Kraft: Z. Chem. **27**, 230 (1987)

8.66 W.I. Friesen, R.J. Mikula: J. Colloid Interface Sci. **120**, 263 (1987)

8.67 H. Spindler, G. Vojta: Z. Chem. **28**, 421 (1988)

8.68 C. Fairbridge: Catal. Lett. **2**, 191 (1989)

8.69 D.H. Everett: Private Communication; D.H. Everett: Ph.D. Thesis (Oxford 1942)

8.70 E. Ignatzek, P.J. Plath, U. Hundorf: Z. Phys. Chem (Leipzig) **268**, 859 (1987)

8.71 D.L. Stermer, D.M. Smith, A.J. Hurd: J. Colloid Interface Sci. **131**, 592 (1989)

8.72 K. Nakanishi, N. Soga: J. Non-Cryst. Solids **100**, 399 (1988)

8.73 J.W. Larsen, P. Wernett: Energy Fuels **2**, 719 (1988)

8.74 S. Ozeki: Langmuir **5**, 186 (1989)

8.75 A.D. Palmer, M. Cheng, J.C. Goulet, E. Furimsky: Fuel **69**, 183 (1990)

8.76 G. Cheng, C. Tao, S. Pang: Proceedings Natl Sci. Congr. on Theory and Application of Fractals, Sichuan Univ. (Chengdu, China 1989) p. 35

8.77 L. Piscitelle, R. Segars, R. Kramer, R.L. Bagalawis, in: *Fractal Aspects of Materials*, MRS Extended Abstracts EA-20, ed. by J.H. Kaufman, J.E. Martin, P.W. Schmidt (MRS Pittsburgh 1989), pp. 115 - 118

8.78 R.R. Mather: Pigments and Dyes **14**, 49 (1990)

8.79 F. Riedo, M. Czencz, O. Liardon, E. Sz. Kovats: Helv. Chim. Acta **61**, 1912 (1978)

8.80 G. Korosi, E. Sz. Kovats: Colloids Surf. **2**, 315 (1981)

8.81 N. Le Ha, J. Ungvarai, E. Sz. Kovats: Anal. Chem. **54**, 2410 (1982)

8.82 J. Gobet, E. Sz. Kovats: Ads. Sci. Technol. **1**, 111 (1984)

8.83 K.K. Unger: *Porous Silica* (Elsevier, Amsterdam 1979)

8.84 R.K. Iler: *The Chemistry of Silica* (Wiley, New York 1979)

8.85 A.Y. Meyer, D. Farin, D. Avnir: J. Am. Chem. Soc. **108**, 7897 (1986)

8.86 D. Avnir: J. Am. Chem. Soc. **109**, 2931 (1987)

8.87 P. Larsen, O. Schou: Chromatographia **16**, 204 (1982)

8.88 Y.E. Kutsovskii, E.A. Paukshtis, Y. Aristov: React. Kinet. Catal. Lett. **46**, 57 (1992)

8.89 S. Shioji, K. Tokami, H. Yamamoto: Bull. Chem. Soc. Jpn **65**, 728 (1992)

8.90 [8.83], pp. 116 - 117.

8.91 C.H. Lochmuller, D.R. Wilder: J. Chromatogr. Sci. **17**, 574 (1979)

8.92 I. Bergman, S. Paterson: J. Appl. Chem. **11**, 369 (1961)

8.93 L.M. Walter, J.W. Morse: J. Sediment. Petrol. **54**, 1081 (1984)

8.94 J. Gross: U.S. Bur. Mines Bull. **1**, 402 (1938)

8.95 N. Kaneniwa, N. Watari: Chem. Pharm. Bull. **22**, 1699 (1974)

8.96 N. Watari, N. Kaneniwa: Chem. Pharm. Bull. **24**, 2577 (1976)

8.97 C.W. Curtis, J.A. Guin, B.L. Kamajian, T.E. Mody: Fuel Process Technol. **12**, 111 (1986)

8.98 Z.W. Wilchinsky, R.W. Looney, E.G. Tornqvist: J. Catal. **28**, 351 (1973)

8.99 G.R. Taylor, R. Hughs: J. Chem. Technol. Biotechnol. **29**, 8 (1979)

8.100 A.V. Artemov, E.F. Vainshtein: Zh. Prikl. Khim. **60**, 1578 (1987)

8.101 O. Rodrigues, K. Setinek: J. Catal. **39**, 449 (1975)

8.102 B.J. Hwang, T.-C. Chou: I&EC Res. **26**, 1132 (1987)

8.103 S. Tsuruya, H. Kuwahara, M. Masai: J. Catal. **108**, 369 (1987)

8.104 J. Kiwi, M. Gratzel: J. Am. Chem. Soc. **101**, 7214 (1979)

8.105 J. Barbier, A. Morales, R. Maurel: Bull. Soc. Chim. Fr. **1-2**, I-31 (1978)

8.106 J. Barbier, P. Marecot, R. Maurel: Nouv. J. Chim. **4**, 385 (1980)

8.107 J. Barbier, P. Marecot, A. Morales, R. Maurel: Bull. Soc. Chim. Fr. **7-8**, I-309 (1978)

8.108 S. Fuentes, F. Figueras, R. Gomez: J. Catal. **68**, 419 (1981)

8.109 Y. Takai, A. Ueno, Y. Kotera: Bull. Chem. Soc. Jpn **56**, 2941 (1983)

8.110 A. Benedetti, G. Cocco, S. Enzo, F. Pinna: React. Kinet. Catal. Lett. **13**, 291 (1980)
8.111 T.A. Dorling, R.A. Moss: J. Catal. **5**, 111 (1966)
8.112 J.C. Wu, P. Harriott: J. Catal. **39**, 395 (1975)
8.113 M. Jarjoui, P.C. Gravelle, S.J. Teichner: J. Chim. Phys. **75**, 1069 (1978)
8.114 H. Topsoe, N. Topsoe, H. Bohlbro, J.A. Dumesic: Stud. Surf. Sci. Catal. **7**, 247 (1981)
8.115 S. Bhatia, N.N. Bakhshi, J.F. Mathews: Can. J. Chem. Eng. **56**, 575 (1978)
8.116 J. Barbier, A. Morales, R. Maurel: Nouv. J. Chim. **4**, 223 (1980)

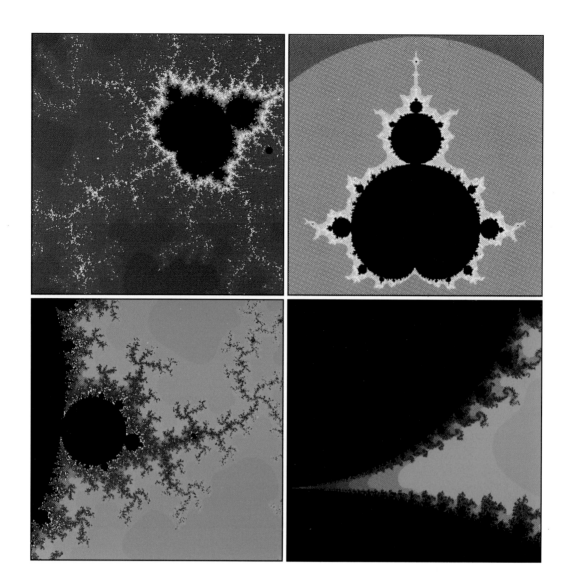

9 Computer Exploration of Fractals, Chaos, and Cooperativity

Dennis C. Rapaport and Martin Meyer

9.1 Introduction

One of many important lessons to have emerged from modern science is that complex behavior can be exhibited by simple systems. It is almost paradoxical that complex output demands neither complex input nor a complex autonomous mechanism; the three categories of behavior enumerated in the title are all examples of such complexity. This chapter describes the set of software demonstrations to be found on the diskette accompanying this volume; this software has been designed to permit the reader to undertake a personal exploration of a range of phenomena that fall into one or more of these categories. Each serves as an example of the surprising coexistence of simple formulation and complex behavior.

Fractals owe much of their prominence to the images they create [9.1,2]. Landscapes (coastlines, mountains, rivers), lifeforms (animal, vegetable), a diversity of other natural phenomena (snowflakes, dielectric breakdown, fluid fingers, clouds), human invention (mechanical structures, abstract art, stock prices) and many other classes of entities can be subjected to fractal analysis, although the degree of applicability of this description varies considerably. The fact that such a purely mathematical entity can embody enough by way of shape, texture, and other detail to convince the human eye-brain complex of the reality of a scene is truly remarkable, and it is in this context that the term fractal forgery was coined. Limitations notwithstanding, there is no denying the visual fascination attached to anything that is even remotely fractal, so great is the contrast with the more classical mathematical forms typified by simple geometrical shapes and perfect solids. Pictures convey some of this feeling, but nothing can match the experience of growing one's own fractal object – either

---

◄ **Fig. 9.0.** Mandelbrot set: midpoint (m), region size (d), and cycle count (n) for each image are as follows: (a) $m = 1.30 + 0i$, $d = 2.60$, $n = 150$; (b) $m = -0.805 + 0.137i$, $d = 0.246$, $n = 150$; (c) $m = 0.2746 - 0.5899i$, $d = 0.0403$, $n = 300$; (d) $m = 0.27318 - 0.59506i$, $d = 0.00295$, $n = 550$

as a mathematical exercise, or as part of a simulated physical process – and it is a goal of the demonstrations in this collection to provide this capability

Chaos is a subject that has attracted widespread attention [9.3-5]. The notion that a simple mechanical or mathematical problem can exhibit complex – indeed utterly unpredictable – behavior has profound consequences, both practical and philosophical; fractals are often seen in chaotic systems, sometimes simply by looking at the behavior, and sometimes via a deeper analysis of the trajectories in phase space. The software collection includes examples of systems displaying chaotic behavior, designed in a manner that encourages user exploration.

Statistical mechanics provides the theoretical basis for studying the cooperative behavior of interacting many-body systems, such as the Ising model, at equilibrium. The elucidation of the nature of the critical point has proved to be one of the successes in this field [9.6], and the scale invariance that is characteristic of critical behavior suggests the role fractals play in this field. Away from equilibrium, the absence of theoretical guidance remains a disappointment, but even at equilibrium there are a great many questions that theory cannot answer. As a consequence, much of the progress in this field has relied heavily on computation in general, and simulation in particular [9.7-9]. Examples of such simulations, reduced in scope for real-time demonstration, are to be found in this collection.

The emphasis of the software is on visualizing qualitative features; most quantitative details are concealed and the viewer is exposed only to graphic renderings of the phenomena. The user interfaces for the programs are simple and intuitive, but only very specific aspects of each problem have been addressed. To provide a deeper treatment of each problem and the tools for more thorough exploration would have required a much greater expenditure of effort; this may be a project for the future if interest so warrants.

9.2 The Software Collection

The demonstrations form a somewhat eclectic collection of material related to the title of this chapter. The visual aspects are stressed, and in the majority of cases the programs aim to encourage exploration. An important consideration governing the selection is the speed with which images appear and change. Many of the more interesting problems associated with fractal growth, chaotic dynamics and cooperative behavior at phase transitions call for complex and time-consuming computations. Such calculations are for research scientists with powerful computers and boundless patience, the typical reader cannot be expected to be so patient. Thus the effects demonstrated here call for relatively simple computations, and results emerge reasonably quickly (how quickly is a function of computer performance); it should be kept in mind that some of the demonstrations based on real-time simulation are indeed reproducing frontier research of not so very long ago.

The collection includes demonstrations involving fractals (deterministic shapes, stochastic landscapes), cellular automata (evolving cell arrays), cluster growth (diffusion-limited aggregation, invasion percolation), cooperative phenomena (Ising model, percolation), self-organizing many-body systems (soft-disk fluid), chaos (logistic map, double pendulum), other kinds of system exhibiting collective behavior (polymers, sandpiles), and lastly, iterative processes (affine mappings, Mandelbrot set). This classification is arbitrary to a considerable degree since the categories overlap; in part the division reflects those features emphasized in the demonstrations.

Another factor governing this particular selection of material is the authors' background and preference; however it is hoped that this has not introduced too strong a bias, and that the average reader will find something of interest. The fact that many of the readers have associations with statistical mechanics has also guided the choice and presentation of much of the material.

An issue that had to be addressed was the type of computer to be used for the project. Since graphics, interactivity and fast response are all key elements in this software, there is little opportunity for producing anything truly portable over a range of hardware platforms and operating systems. Therefore the decision was made to focus on the Apple Macintosh (see, e.g., [9.10] for details) and MS-DOS personal computers, and practically all the demonstrations were designed to run on the most basic models presently available, in order that the material be as accessible as possible. An earlier collection of computer demonstrations designed for a different computer, which overlaps the present work to some extent, was described previously [9.11].

The description of each program in this chapter begins with a general background to help set the scene. Computational techniques and other algorithmic details are then mentioned. This discussion has been kept to a minimum by overall space constraints; little is lost however, because most of the demonstrations involve only relatively simple calculations, and in most cases examples of such programs, or at least descriptions of the computational techniques, are to be found elsewhere. Those programs which make use of random numbers employ a simple shift-register generator [9.12].

Much of the effort that went into preparing the software was devoted to the graphics and the user interface; while the principles involved are quite general, the implementation is very specific to the computing platform, and this is not the appropriate place to discuss the subject. Features included in each demonstration program are summarized in the descriptions, but the complete (and concise) operating instructions will be found in the help window that is a part of each program. A selection of pictures taken from the screen of the Macintosh versions of the programs (at screen resolution) during the runs is included. Sources of further information and background material are listed in the bibliography.

9.3 Fractals

The guided tour begins with two demonstrations of shape generation by fractal processes. The first is purely deterministic and produces such well-known figures as the Koch curve, the second is an interactive tool for creating primitive landscapes.

9.3.1 Deterministic Fractals

While much of the application of fractals [9.13,14] is to random processes, some of the most interesting fractal objects – and here they are indeed fractals in the truest mathematical sense – are those produced by deterministic processes. In such objects, self-similarity over a range of length scales is guaranteed. This demonstration (see e.g. Fig. 9.1) produces images of well-known fractal objects formed by recursive subdivision of elementary geometrical shapes [9.1,15].

Fig. 9.1. The Sierpinski sponge – a deterministic fractal

The demonstration offers a selection of a few of the most familiar fractals in this class (see Sect. 1.2), with the user being able to choose the number of subdivision cycles that generate the finer features, up to a limit where the size of the details approaches the resolution limit of the image. Simple iterative (or recursive) geometric operations on line segments and polygonal areas is all that is required mathematically in these computations.

9.3.2 Stochastic Landscapes

The intriguing patterns described by a random walk led to the idea that surfaces might also be produced by random processes [9.16]. Totally random surfaces, in which heights are uncorrelated, are uninteresting as they bear little visual relevance to natural landscapes, but if the altitude changes imposed on an initially horizontal grid follow certain rules, surprisingly realistic pseudo-landscapes emerge – especially when artistically augmented with color and texture. The rules for generating surfaces (explained in detail in Sect. 1.5) are fractally based: the variance σ of the random perturbation applied to regions of given size L (i.e., the roughness) scales as $\sigma \propto L^H$ (so-called fractional Brownian motion), and perturbations are applied successively over a range of length scales [9.16,17]. The Hurst exponent H is identical to α defined in Chap. 4.

The demonstration (see Fig. 9.2) attempts to provide a limited overview of the capabilities of the approach and produces mountainous landscapes and related scenery by successive subdivision and random displacement of a two-dimensional square grid. Transparent grid views of surfaces characterized by adjustable parameters can be quickly examined from different orientations, and views of the same surface with hidden portions removed can be generated. The computational requirements depend on the degree of detail (displaying hidden surface images can involve considerable calculational effort [9.18]), and the latter is kept fairly low in this demonstration to ensure acceptable response on the average computer.

The grid generation follows methods described in [9.15-17] and in Sect. 1.5. An initial peak of user-determined height is used to start the computation on the lowest (3×3) resolution lattice. Successive midpoint displacements are then made using a normal distribution with diminishing variance, with the rate at which the variance drops as the scale decreases also being a user-regulated parameter (greater realism could be achieved by giving more thought to the choice of distribution [9.17]). Before drawing the surface, negative grid values are set to zero, thereby introducing an effective sea-level; the grid boundaries are also fixed at zero elevation for convenience. There are a number of ways of hiding those portions of the image that are not visible to the observer; this is an old but central question in computer graphics [9.18]. Here a simple approach is used in which the quadrilaterals that tile the grid surface are simply drawn as filled projected polygons (the fact that in three-dimensional space the polygons are not planar does not present a problem); the only algorithmic effort required is to decide on the order in which the grid is scanned to ensure that the polygons are rendered so that nearer polygons can obscure (by drawing them later) those further away.

Whether the image is reminiscent of islands, fjordal coastlines, rolling hills, or rugged alpine territory, depends on the choice of parameters, and the degree of subdivision selected, not to mention the throws of the stochastic dice that perturb altitude. As with other fractal imagery, one sees a forgery, since there is no physical mechanism underlying the picture; the more refined the mechanism controlling the subdivision process the less apparent the forgery.

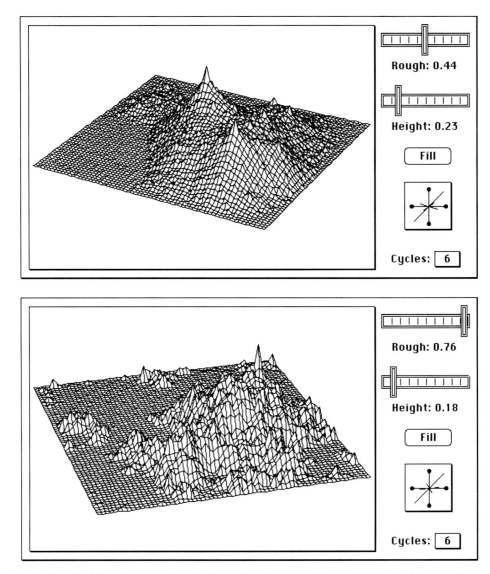

Fig. 9.2. Random landscapes: simple forgeries of mountains and islands (the small set of axes is used to rotate the view, and the sliders change the surface parameters)

9.4 Cellular Automata

This section describes one of the most familiar applications of the cellular automaton idea, namely its use as a method for growing cell patterns according to certain classes of deterministic rules, as typified by the game of Life.

9.4.1 Cell Arrays

There exist an enormous variety of cellular automata [9.19,20] (see also Chap. 9 (by D. Stauffer) in [9.21]) that have been designed for various purposes; applications are to be found in almost all branches of science, including physics, chemistry, biology, and the earth sciences. What is so attractive about the approach is the ease with which models can be formulated and solved. Where the alternative is the solution of partial differential equations – if there is an alternative – the cellular automata approach is often more appealing. In addition to those automata designed with specific problems in mind, a further motivation for exploring the problem is the complexity inherent in these simple models (a recurring theme throughout this chapter), and systematic studies of cellular automata families have been carried out to connect the degree of pattern complexity with the nature of the rules governing growth.

The automata used in the demonstration are based on two-state models – lattice sites are occupied or vacant, or, to lend a more biological flavor, the cell occupying each site is 'alive' or 'dead' – that obey totalistic rules involving nearest neighbors [9.22]. The meaning of the term 'totalistic' is that one simply counts the number of living neighbors of a given cell and then applies a transition rule based on this number and the state of the cell; the rule either changes the state of the cell or leaves it the same.

The game of Life [9.23,24] is one example of this family of cellular automata – by far the most famous – whose development predates interest in the more general problem; it is also the default case with which the demonstration begins. The cells themselves occupy the sites of a square lattice. The rules of Life are simple in the extreme: a living cell dies either from loneliness if it has less than two living neighbors, or from overcrowding if four or more of the eight (nearest and next nearest) neighboring cells are alive. Neither is the system averse to parthenogenesis, and a cell springs to life if exactly three neighbors are already alive. Other combinations leave the state unchanged. It does not require much effort to observe the rich variety of lifeforms which can develop, and in the heyday of the subject an entire 'zoology' was described.

In the program, the number of neighbors considered can be either four (for nearest neighbors only) or eight (including both nearest and second-nearest neighbors), and the transition rule can be varied. It is convenient to use short-hand notations for these rules. Perhaps the most straightforward notation is simply to state the numbers of neighbors that bring a dead (−) cell to life or keep a living cell (+) alive. For example, the notation [−123456,+0236] indicates that a dead cell (−) springs to life if there are between 1 and 6 living neighbors, and a live cell (+) survives if it has 0, 2, 3, or 6 living neighbors (see Fig. 9.3). The notation for the game of Life is [−3,+23].

In the commonly used more compact notation the transition rule is described by only one number, with each bit of the binary representation of this number used to determine whether a cell will be alive or dead during the next cycle, if it currently has the corresponding number of living neighbors. Thus the binary representation of 13814 indicates that a live cell survives if it has 0, 2, 3, or 6 living neighbors, and a dead cell springs to life if there are between 1 and 6 living neighbors; the Life game corresponds to rule 224.

All the computational algorithm must do is scan the cell grid and apply the appropriate rule to each cell in turn, based on its current environment; the only point that should be noted is that the new state of a cell only takes effect when

Fig. 9.3. Growth processes – cellular automata: (left) rule [−123456,+0236] (rule 13814 in the compact notation) started from single site; (right) rule [−4678,+35678] (rule 976, central site not treated specially, in the compact notation) started from random 50% occupation

the scan is complete (here, unlike the Ising model Monte Carlo study discussed later, the updating process is simultaneous rather than sequential). The user can alter certain aspects of the problem: the neighborhood size, whether the cell state is used – as in Life – to pick the rule set, the initial state (even to the extent of being able to draw and edit one's own), and the total grid size (the boundary conditions are chosen so that cells outside the grid remain dead).

Nothing could be less sophisticated, but even prescriptions of this kind lead to a fascinating range of patterns – stationary, periodic, reproductive, random, fractal-like, and more. In this particular implementation, the totalistic rules and square lattice introduce an eightfold symmetry into the patterns if the initial state is just a single site (or some other configuration with this symmetry). With over a quarter of a million different sets of rules and neighborhoods, multiplied by an enormous number of essentially different starting states, this is close to being the ultimate computer game.

9.5 Cluster Growth

The demonstrations in this section show two approaches to generating clusters on lattices. One is diffusion-limited aggregation, where particles adhere to a growing cluster; the other is invasion percolation, where the cluster grows by forcing itself through a random porous medium.

9.5.1 Diffusion-Limited Aggregation

Diffusion-limited aggregation – DLA – [9.9,14] (see also Chap. 1 (by H. E. Stanley) in [9.21]) is a deceptively simple process for growing clusters by capturing randomly diffusing particles. The lattice-based version of the problem was one of the earliest growth models, and though originally thought to produce fractal clusters and now known to yield forms that are less readily characterized, it remains popular as a vehicle for both the study and visualization of growth phenomena.

The underlying dynamics are as follows: An initial seed particle (or row of particles) is placed on the lattice. Additional particles are then injected, one at a time, at remote random locations, and allowed to wander randomly over the lattice bonds until a site adjacent to one already occupied is reached. When this happens, the particle joins the growing cluster. To accelerate growth rate without biasing cluster shape, particles can either be allowed to take larger steps when far away from the cluster, or the injection points can be brought closer to the current cluster boundary.

Variations of the basic procedure have been proposed that embellish the simple model in a variety of ways in order to alter the properties of the generated clusters. Different lattices (or even the computationally more demanding continuum) and spatial dimensions can be tried, the rules for walking, adhering to the cluster, and particle injection can all be varied, with clearly visible consequences on the growth process.

The present version of the simulation (see Fig. 9.4) employs a square lattice. The sticking rule can be adjusted to require that a site be visited a specified number of times before adhesion occurs (the degree of stickiness increases with each visit). New particles are injected from a boundary that gradually recedes from the growing cluster, and particles attempting to cross this temporary boundary are forced back into the interior. A further increase in overall perfor-

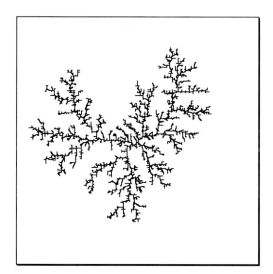
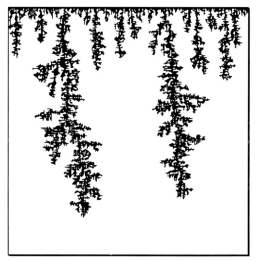

Fig. 9.4. Diffusion-limited aggregation: (left) single seed particle, cluster of 4000 particles; (right) row of seeds at top, 5 visits required before sticking, 10000 particles

mance results from not showing all particle moves; the default is to show very few of them, thereby allowing the cluster to form as quickly as possible.

The computation employs a matrix whose elements represent the state of the lattice sites, with different values denoting sites that are empty or occupied, sites at the boundary, and sites adjacent to those already in the cluster (if multiple visits are included then other site values record the number of visits). In this way there is never any need for a particle to inquire where it is in relation to either the cluster or the boundary, and all decisions as to legal moves and terminating the motion are made on the basis of information associated with a single lattice site. The neighboring sites must of course be updated whenever the cluster changes, but this occurs on average only once in several thousand diffusion steps.

9.5.2 Invasion Percolation

The invasion percolation approach to modeling cluster formation [9.25,26] (see also Chap. 2, and Chap. 4 (by A. Aharony) in [9.21]) differs from DLA in that growth is now from within, rather than a process of aggregation from without. The cluster is typical of that formed when one liquid displaces another in a porous medium. The displaced fluid is assumed to be infinitely compressible; a more computationally demanding version of the model accounts for incompressibility.

The porous medium is represented by a lattice whose sites have randomly assigned values corresponding to local pore diameter. The expanding fluid emerges at a constant rate from either a point or line source and, at each step, penetrates the narrowest pore on the cluster perimeter. The additional inclusion of incompressibility would require recognizing the appearance of closed loops in the growing cluster and prohibiting further growth into the voids so formed; the simpler version of the model demonstrated here will eventually fill all lattice sites.

9.6 Cooperative Phenomena

Two examples of systems that exhibit cooperativity, though in very different ways, are discussed in this section. The first is the familiar Ising model of ferromagnetism, the second is the percolation problem, and both are treated by appropriate Monte Carlo methods. Cluster formation plays a prominent role in both problems; percolation clusters represent tangible objects generated by the process itself, while Ising clusters are transient islands of ordered spins whose average properties and fluctuations determine the macroscopic behavior.

9.6.1 Ising Model

The Ising model is the all-time classic of statistical mechanical modeling, as well as one of the best known models in theoretical physics. Originally proposed as a simple model to explain ferromagnetism, in which the spin variable associated with each crystal lattice site corresponds to the alignment direction ('up' or 'down') of the magnetic dipole at that site, it has over the years been adapted and applied to the study of a variety of phenomena, including lattice gases, binary alloys, spin glasses, and protein folding. It has remained the subject of intensive study from the earliest days of computer simulation, and despite the fact that exact analytic solutions are confined to two dimensions or less, the broad range of numerical information available leaves little doubt as to the details of its behavior in general. The two-dimensional model treated here does have analytic results available for some properties, although the particular feature stressed in the demonstration – cluster formation – is one where theory provides only limited help.

The positive exchange energy J (see Appendix 9.B.1) favors a parallel alignment of neighboring spins and therefore the formation of ordered spin clusters. The emphasis here is on the way these spin clusters develop within the framework of standard Metropolis Monte Carlo simulation [9.7,9,27]; different colors are used to denote the two spin directions. At high temperature the spin clusters are small and evanescent, but with reduced temperature cluster sizes and lifetimes increase. The critical temperature T_c of the infinite system (where spontaneous magnetization first appears as temperature is reduced) is marked by a diverging correlation length, which is associated with spin clusters whose mean size also diverges. The small system simulated does not have a singular critical point (in common with all finite-size systems), but over a range of temperature near T_c ($= 2.269...$, expressed in units of the spin interaction energy) the clusters are of substantial size relative to the lattice and subject to large fluctuations. Below T_c the clusters take on a certain permanence, corresponding to the onset of long-range order, an effect which becomes more pronounced as the temperature drops even further. Though quantitative results are not shown in this simulation, the method can produce a highly accurate picture of the equilibrium behavior of the Ising model.

The simulation shows a lattice of 4096 spins that are coupled ferromagnetically and obey periodic boundary conditions. The updating follows the standard procedure [9.27] (see Appendix 9.B.1) in which random numbers are used to decide whether to flip spins based on the consequent energy change; the lattice is swept in an ordered manner, and the environment of each spin at any instant is actually a mixture of neighbors in pre- and post-update states (sequential rather than parallel updating is used). The display is refreshed after each complete sweep through the lattice. The user regulates the temperature, and this can be altered by any amount at any time.

9.6.2 Percolation

The problem of lattice percolation is another of the widely studied models that capture the essence of certain kinds of spatially random processes; see Chap. 1, Chaps. 2 and 3 in [9.21], and [9.28]. Percolation is a rather special class of problem because it falls into the category of systems, such as the Ising model, for which the powerful methodology developed for critical phenomena can be brought to bear.

One of the methods widely used for investigating percolating systems is Monte Carlo. The approach differs from that of the Ising model – instead of generating a succession of evolving configurations each state is produced independently. For the site percolation problem on a lattice, the Monte Carlo procedure consists simply of randomly filling the sites with a given probability; subsequent work entails identifying and measuring the properties of connected site clusters produced by this process.

Theoretical scaling predictions exist for the distribution of cluster sizes and related properties, and testing these predictions unambiguously requires the study of large clusters, which, in turn, implies a need for large lattices. While the largest such studies to date have focused on lattices with over 10^{11} sites [9.29], the present demonstration is many orders of magnitude smaller. This does not prevent the user from discerning the essential difference between the kinds of cluster present below and above the critical concentration p_c (see Chap. 1), whose value for the (infinite) square lattice is close to 0.593.

The demonstration allows the user to generate configurations of sites occupied with any desired probability and observe the clusters that form (see Fig. 9.5). Different lattice sizes can be chosen, and the type of clusters displayed can also be selected (such as all clusters, or a few of the largest clusters). The

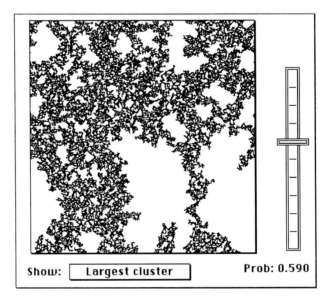

Fig. 9.5. Percolation: the largest cluster present in a particular configuration of occupied sites (256 × 256 lattice)

lattice itself is subject to periodic boundary conditions, so that clusters are often observed to wrap around the system (large clusters almost always do).

Of the algorithms underlying the demonstrations in this collection, the one used for assigning occupied sites to clusters is one of the most complex. It (the Hoshen-Kopelman algorithm) also happens to be a widely documented technique, so that little need be said about it here [9.9,30,31] (see also Chap. 2 in [9.21]). The naive approach would be to scan the lattice and, each time an occupied site is encountered, identify the cluster to which it belongs and note any changes by tracing back along a chain of previously analyzed sites in the same cluster. As clusters become larger, and mergers become increasingly frequent, much superfluous data manipulation occurs as sites change cluster membership. The efficient approach is to make use of pointers that, by referencing clusters indirectly, allow deferring the task of establishing whether two apparently different clusters are each a peninsula of a single cluster.

9.7 Many-Body Systems

An example of an interacting many-body system – the soft-disk fluid – that exhibits cooperative behavior is explored here by means of molecular dynamics simulation. Larger versions of simulations of this kind can be used to study the solid-liquid phase transition in great detail, but the real-time constraint restricts the present demonstration to systems so small that only a mere hint of such behavior is apparent.

9.7.1 Soft-Disk Fluid

Molecular dynamics simulation embraces a family of techniques based on generating the detailed phase-space trajectories of interacting, classical, many-body systems. It is widely used in studying equilibrium and transport properties of matter in all its states [9.8,32,33]. One of the earliest achievements of this approach was to demonstrate the existence of a solid-liquid phase transition in a hard-disk fluid, and one of the first images produced by these simulations [9.34] shows coexisting solid and liquid regions that are clearly delineated by the highly localized particle trajectories of the former and the wide-ranging trajectories of the latter.

While it is not the molecular trajectories themselves (whether in phase or configuration space) that are normally of concern, but rather those averages of physical properties derived from them that correspond to experimentally measurable quantities, the trajectories can provide a useful means for understanding aspects of the detailed dynamical processes that are at work. Molecular movement in an equilibrium fluid is diffusive in nature (after the passage of several mean collision times) and resembles the random walk of Brownian motion; the fractal dimension of these self-similar trajectories can actually be measured [9.35].

Fig. 9.6. Soft-disk fluid: (left) disks at high density and their (localized) trajectories; (right) trajectories only (over a short time interval) at lower density

This demonstration supplies a soft-disk fluid whose parameters (pressure and temperature) can be varied to generate both the fluid (there is no clear distinction between liquid and gas in this system) and solid phases. Melting and crystallization phenomena can be studied. The trajectories of the particles (see Fig. 9.6) can also be followed and the behavior characteristic of the solid and fluid states examined.

The simulation is a standard constant-temperature treatment of a fluid with a soft repulsive potential [9.8] (also see Appendix 9.B.2). Normal molecular dynamics conserves the total energy of the system (subject to the accuracy of the numerical integration technique used), and corresponds to the microcanonical ensemble; the constant-temperature variant forgoes energy conservation in order to generate results corresponding to the canonical ensemble. Constant temperature is maintained by the rather crude method of rescaling velocities at each timestep to ensure constant kinetic energy; for these purely qualitative demonstrations such an approach is adequate, although more refined methods are available for quantitative studies. The widely-used cell-list technique is employed to improve the efficiency of the computation, and the boundaries are periodic. Note that the sizes of system treated here interactively are typical of those which, in the early days of molecular dynamics, required long and laborious computation to produce results – such is the measure of progress (on the other hand, current simulations can involve 10^5–10^6 particles).

The temperature is altered in response to user requests by rescaling the velocities by small amounts. Pressure is controlled indirectly by varying the density: this is done by dragging the corner of the (periodic) container to produce the desired size and shape, and the system then gradually adjusts itself to the new configuration. The number of disks in the simulation is determined by the initial size of the container. Either or both disks and trajectories can be watched as the simulation proceeds.

9.8 Chaos

One does not have to look very far to find chaos. Two very different examples of systems exhibiting chaotic behavior under suitable conditions form the subject of this section. The first is a simple mathematical exercise, namely the logistic map; the second is a mechanical device – a pendulum made from two masses coupled together.

9.8.1 Logistic Map

This ostensibly simple one-dimensional quadratic map [9.36,37] embodies much of the complexity of chaotic systems, although at first acquaintance it appears remarkably benign (for more details see also Chap. 1). After all, what could be simpler than choosing a value of x between 0 and 1 and applying the transformation $x_{n+1} = 4rx_n(1 - x_n)$ to it, over and over again, where $0 < r < 1$? The answer turns out to depend on the value of r (r corresponds to $\lambda/4$ in Chap. 1). For r not too large there is indeed an r-dependent fixed point to which the iterations converge rapidly, but this is followed by a range of r for which the limit oscillates between 2 values, then 4, 8, and so on, until an apparently random (but reproducible) sequence of x values is obtained. Tucked away inside the range of r for which the sequence appears random are further regions with limit cycles of period 3, 6, ..., 5, 10, ..., and a great deal more. Even random sequences may contain almost periodic subsequences. As with some of the other examples in this collection, there is plenty of scope for roaming through unfamiliar territory.

The demonstration (see Fig. 9.7) permits the user to study the problem at any level of resolution by zooming into smaller and smaller subranges of the parameter r. The chosen range can be swept automatically, or individual points can be selected for study. For each value of r successive sets of points of the iteration sequence are shown, and these eventually converge to the limiting value or set of values (which may or may not be small in size); the main display allows the formation of a composite picture of the behavior over the selected parameter range (where the role of fractals may be suggested). There is also a scrolled view of the most recent values in the sequence, as well as an automatically updated display of the power spectrum computed from the latest points which can aid in extracting periodicity from a noisy sequence.

With a certain amount of exploration the user will discover periodic sequences of various lengths, the location of the bifurcations where period doubling occurs, chaos, intermittency, and more. There is little that need be said about the computations, except for the fact that a fast Fourier transform is used to generate the power spectrum, and that the raw data sequences used in this analysis are first passed through a triangular windowing filter to remove discontinuities at the endpoints [9.38].

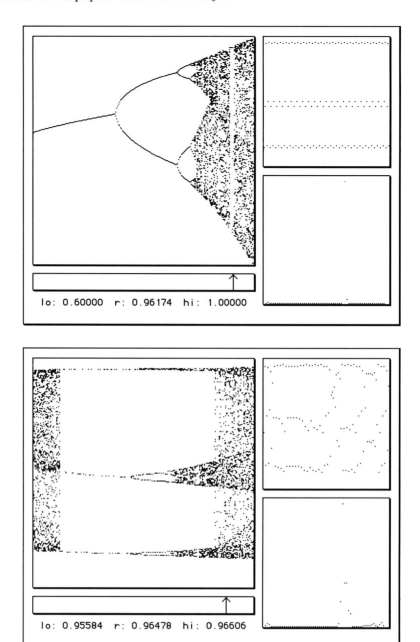

Fig. 9.7. Logistic map: (top) scan of the interesting parameter range; (bottom) narrow range with 3-cycles and chaos; the sequence and power spectrum for the current parameter setting (indicated by the arrow) are also shown

9.8.2 Double Pendulum

The simple (single mass) pendulum encountered as an elementary exercise in mechanics exhibits purely harmonic motion, and has a reputation for making a highly reliable timepiece. Even if the oscillations are enlarged, the problem, though a little more complicated, produces no surprises. But the situation changes drastically when a second mass is suspended from the first, and this demonstration shows the unpredictable behavior that can ensue because of extreme sensitivity to the initial state.

The user is allowed to change the pendulum arm lengths and mass sizes, and set the initial state. Motion is displayed directly on the screen, as well as by means of graphs showing the time dependence of the angular coordinates. The computer will also oblige by playing a sound whose frequency changes with the angular velocity of the arm suspending the smaller mass (the tune is devoid of musical merit). The underlying numerics employ a fourth-order Runge-Kutta integration method [9.38] to solve the coupled quartet of first-order differential equations derived by elementary Lagrangian methods; derivation of these equations of motion is a relatively simple exercise in classical mechanics (see Appendix 9.B.3).

9.9 More Collectivity

Two further highly idealized physical models are discussed under this heading. Both are treated by stochastic methods and display forms of cooperative behavior, although these small demonstrations may not reveal a great deal of it. The first is a collection of randomly moving polymer chains, the second shows sandpile development.

9.9.1 Polymers

Reptation [9.39] is a mechanism proposed for describing the dynamics of polymer chains that move among one another in a melt or concentrated solution. The analogy is with the motion of a snake – the body follows the path traced out by the head. In the reptation picture it is assumed that this kind of motion dominates, with each chain moving in a tube created by its neighbors; the lifetime of a tube must of course be long enough for the chain to slide a significant distance along its own length in order for the model to be a useful one. Tests using molecular dynamics simulation tend to support the mechanism [9.40]. Reptation is also useful in Monte Carlo studies of equilibrium chains because it provides a simple means for generating sequences of states. (A more detailed discussion of the subject appears in Chap. 6.)

This demonstration (see Fig. 9.8) shows what happens when a collection of self-avoiding chains are placed on a two-dimensional lattice with hard boundaries and allowed to evolve by random reptational motion. At each step the head of each chain advances to a randomly selected neighbor site; if the choice is unavailable further attempts are made, but eventually the chain simply reverses direction (as would a two-headed snake). Observing the behavior may sometimes lead to the impression that the chains know what they are doing, but at other moments they appear extremely unintelligent, which indeed they are. In a truly dynamic situation, where state-dependent forces drive the motion, there is obviously a higher degree of organization at work.

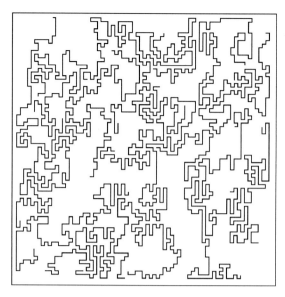

Fig. 9.8. Polymers: a set of randomly reptating chains

9.9.2 Sandpiles

Sandpile models are examples of (stochastic) dynamical systems that exhibit a form of self-organization [9.41,42]. Grains fall randomly onto a pile growing in the vertical plane, and whenever the local slope exceeds a prescribed threshold, reorganization of the pile occurs via lateral motion. A succession of grain descents by lateral movement may be regarded as an avalanche, and questions related to the distributions of avalanche sizes have attracted attention. The problem is simplified by restricting grain positions to the sites of a lattice.

This rather limited demonstration shows how different kinds of sandpile evolve. The pile forms on a base which may have a wall on one side, or on a base with holes. Sand flows out through these holes, as it does across the open ends of the base. The height difference (or drop) between adjacent columns of grains that can be sustained before lateral motion begins is also adjustable. A scrolling display shows the number of grains engaged in lateral motion at any moment – a measure of the tendency for avalanches to occur.

9.10 Iterative Processes

The description of this software collection concludes with two further demonstrations of iterative mathematical processes, in this case processes that produce beautiful abstract (and sometimes not so abstract) imagery. The first is based on linear affine transformations, the second on a quadratic mapping in the complex plane that generates the hypnotic Mandelbrot set. Both programs offer the user the opportunity for almost unlimited exploration. Fractal-like behavior and chaos (the extreme sensitivity to initial conditions) are evident in both examples.

9.10.1 Affine Mappings

Affine transformations [9.15,43] combine rotations, scale changes and translations; though they do not share the nonlinearity of some of the other mappings that have attracted attention, they are nevertheless also capable of exhibiting rich visual behavior. Since relatively few numbers are required to specify a set of transformations capable of recreating the details of complex images, the approach finds application in highly efficient image compression schemes.

The demonstration program shows what happens when a series of affine transformations is applied to a single initial point x_0 in the plane. An affine transformation is defined by $x_{n+1} = Ax_n + b$, where A is a 2×2-matrix and b is a 2-component vector. A set of m transformations $\{A_k, b_k\}$, $k = 1, ..., m$, initially chosen by the user are applied over and over again in random order (for this reason the method is sometimes called the Chaos Game) with probabilities proportional to the determinant of A_k, and the images consist of a series of points x_n (the first few hundred points are omitted from the picture). Although the procedure is obviously probabilistic, the image is fully specified by the chosen set transformations, and is independent of the order in which they are applied, the choice of x_0, and the probabilities themselves; however, the speed (i.e., the number of iterations necessary to show the image with a given level of detail) does depend on the probabilities, and the criterion used here is known to work well. Several transformation sets corresponding to well-known images [9.15] are included as examples that can be studied and manipulated.

The number m of transformations, as well as each of their parameters (six per transformation) can be changed – visual feedback is practically instantaneous. The user can also investigate the plane in which the images are displayed, zooming in and out of regions of particular interest, although enlarged images will develop more slowly if many of the computed points lie outside the visible domain.

Some of the images generated (see Fig. 9.9) appear fractal in nature (even resembling deterministic fractals), others are of geometric patterns or familiar everyday objects, and yet others are more amorphous in form, which, like Rorschach spots, leave a great deal to the imagination. It comes as a surprise to realize that precisely the same mathematical technique is responsible for all the images.

9.10.2 Mandelbrot Set

Again the familiar rhetorical question is posed: What could be simpler than the mapping $z_{n+1} = z_n^2 + c$, where c is point in the complex plane, and the iterations begin with $z = 0$? The answer is that there are probably few mathematical creations that are more complicated. Discovered almost by accident long after it should have been (and even then only thanks to the computer) this mapping has generated some of the most beautiful images to have emerged from the scientific enterprise [9.1-3].

The images are constructed by counting the iterations until the mapping diverges (it is enough to reach a value of z for which $|z| > 2$). Each pixel on the

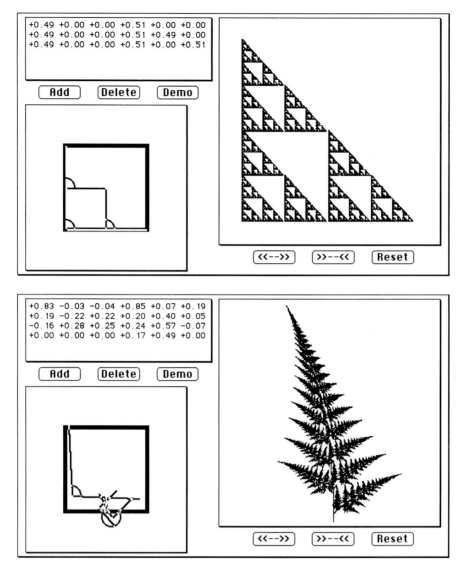

Fig. 9.9. Affine mappings: a stochastically generated gasket, and a minor variation on a fern leaf; the transformation parameters and the graphical device used for changing them are also shown

screen corresponds to a point c in the complex plane. Pixel color is determined by the iteration count, if known, with black being used if divergence has not, or will not occur; properly used color coding can lead to attractive images (see Fig. 9.10). No matter how much one magnifies a region that includes the boundary of the set (the set is defined to contain all points for which divergence does not occur), the intricate details and familiar signature patterns continue to appear.

The demonstration provides a basic set of tools for the explorer. The size of the displayed image can be altered (the smaller the faster), and almost unlimited zooming is provided (a linear magnification of close to 10^5 is achievable before loss of numerical precision becomes apparent). To provide for optimal color choice the available colors can be cycled; saturated colors are used, and it is the hue that is a function of iteration count. The computations are simple, as befits the problem, but to enhance performance on the 'average' computer floating point arithmetic is not used; instead, scaled integers are used to represent fixed point values with similar precision, but with a dynamic range just adequate for the calculation.

9.11 Summary

In this chapter we have described the set of computer demonstrations on the diskette packaged with this book. The figures illustrate some of the features that can be observed (though not the menus and dialog boxes used to control the programs), but it is the capability for personal exploration that makes the programs far more valuable than static images and this companion text. Detailed operating instructions are incorporated into each program.

The authors hope that these examples will both entice the uninitiated as well as allow the more experienced to enjoy the fun of numerical experimentation. The next step for the novice reader is personal software development, rather than merely being content with what is available; then there really are no limits as to where exploration can lead. It goes without saying that the authors will greatly appreciate all manner of feedback in order to enhance both the form and content of future versions of this collection.

9.A Appendix: Alphabetical Program List

The programs included on the Macintosh and DOS versions of the diskette are listed below (names are limited to 8 characters because of DOS restrictions), together with the numbers of the sections in which they are discussed. The Readme program provides general information and some notes on usage.

Ca2d: cellular automata patterns (9.4.1)
DLA: diffusion-limited aggregation (9.5.1)
FractPic: deterministic fractal pictures (9.3.1)
Invasion: invasion percolation (9.5.2)
Ising: Ising model Monte Carlo simulation (9.6.1)
Itfun: iterated affine maps (9.10.1)
Landscap: fractal landscapes (9.3.2)
Mandel: Mandelbrot set (9.10.2)
MolDyn: molecular dynamics of soft-disk fluid (9.7.1)
Pendulum: dynamics of double pendulum (9.8.2)
Percol: site percolation cluster formation (9.6.2)
Polymer: lattice polymer reptation (9.9.1)
Qdmap: logistic map (9.8.1)
Sandpile: one-dimensional sandpiles (9.9.2)

9.B Appendix: Mathematical Details

While most of the computer demonstrations are closely related to material covered in other chapters of these volumes, a few are not discussed elsewhere. In the interest of completeness further details of several of these problems are included here.

9.B.1 Ising model: The energy in zero magnetic field is

$$E = -J \sum_{n.n.} s_i s_j - H \sum_i s_i$$

where the first sum is over nearest neighbor pairs, J is the exchange energy, H the applied field, and $s_i = \pm 1$. The simulation is carried out at temperature T. The Monte Carlo procedure considers each spin s_i in turn, and applies the following operations. Replace s_i by $-s_i$ and compute the energy change ΔE. Now choose a uniformly distributed random number r in the interval $(0,1)$. If $r < 1/[1 + \exp(\Delta E/T)]$ this new spin value replaces the original, otherwise the previous value is restored.

9.B.2 Soft-disk fluid: The interaction between particles separated by distance r is

$$V(r) = 4\epsilon \left[(\sigma/r)^{12} - (\sigma/r)^6 \right]$$

for $r < r_c = 2^{1/6}\sigma$, and zero otherwise; here σ and ϵ are units of distance and energy respectively, and r_c is the cutoff distance. This truncated form of the Lennard-Jones potential produces a repulsive interaction between particles which approach too closely. The equations of motion are then (assuming a system of units in which the particles have unit mass, and σ and ϵ are both unity)

$$\ddot{\mathbf{r}}_i = 48 \sum_{j(\neq i)} \left(r_{ij}^{-14} - \tfrac{1}{2} r_{ij}^{-8} \right) \mathbf{r}_{ij}, \qquad i = 1, \dots, N.$$

9.B.3 Double pendulum: Start with the Lagrangian

$$L = \tfrac{1}{2} m_1 l_1^2 \omega_1^2 + \tfrac{1}{2} m_2 [l_1^2 \omega_1^2 + l_2^2 \omega_2^2 + 2 l_1 l_2 \cos(\Delta\theta) \omega_1 \omega_2]$$
$$+ m_1 g l_1 \cos\theta_1 + m_2 g (l_1 \cos\theta_1 + l_2 \cos\theta_2)$$

where $\omega_i = \dot{\theta}_i$ are the angular velocities of the pendulum arms, $\Delta\theta = \theta_1 - \theta_2$, and m_i and l_i are the masses and arm lengths. The Lagrange equations are then

$$\dot{\omega}_1 + \frac{l_2 m_2}{l_1(m_1 + m_2)} \left[\dot{\omega}_2 \cos(\Delta\theta) + \omega_2^2 \sin(\Delta\theta) \right] + \frac{g}{l_1} \sin\theta_1 = 0$$

$$\dot{\omega}_2 + \frac{l_1}{l_2} \left[\dot{\omega}_1 \cos(\Delta\theta) - \omega_1^2 \sin(\Delta\theta) \right] + \frac{g}{l_2} \sin\theta_2 = 0.$$

These can be rearranged, and the result is a set of four first-order equations in which the left-hand sides are $\dot{\theta}_1$, $\dot{\theta}_2$, $\dot{\omega}_1$, and $\dot{\omega}_2$.

References

9.1   B.B. Mandelbrot: *The Fractal Geometry of Nature* (Freeman, New York 1982)

9.2   H-O. Peitgen, D. Saupe, eds.: *The Science of Fractal Images* (Springer-Verlag, New York 1988)

9.3   J. Gleick: *Chaos: Making a New Science* (Penguin, New York 1987)

9.4   I. Stewart: *Does God Play Dice?* (Blackwell, Cambridge MA 1989)

9.5   P. Cvitanovic, ed., *Universality in Chaos* (Adam Hilger, Bristol 1984)

9.6   C. Domb, M.S. Green, eds.: *Phase Transitions and Critical Phenomena*, vols. 1-6; C. Domb, J. L. Lebowitz, eds.: *Phase Transitions and Critical Phenomena*, vols. 7- (Academic Press, New York 1972-)

9.7   K. Binder, ed.: *Applications of the Monte Carlo Method in Statistical Physics* (Springer-Verlag, Berlin 1984); K. Binder, ed.: *The Monte Carlo Method in Condensed Matter Physics* (Springer-Verlag, Berlin 1992)

9.8   M.P. Allen, D.J. Tildesley: *Computer Simulation of Liquids* (Clarendon Press, Oxford 1987)

9.9   H. Gould, J. Tobochnik: *An Introduction to Computer Simulation Methods* (Addison-Wesley, Reading MA 1988); also see the monthly column in Computers in Physics

9.10 *Inside Macintosh*, vols. 1-6 (Addison-Wesley, Reading MA 1985-91)

9.11 D.C. Rapaport: Computers in Physics **3**:5, 18 (1989); J. Stat. Phys. **58**, 775 (1990)

9.12 S. Kirkpatrick, E. Stoll: J. Comput. Phys. **40**, 517 (1981)

9.13 J. Feder: *Fractals* (Plenum, New York 1988)

9.14 T. Vicsek: *Fractal Growth Phenomena* (World Scientific, Singapore 1989)

9.15 H-O. Peitgen, H. Jürgens, D. Saupe: *Fractals for the Classroom* (Springer-Verlag, New York 1992)

9.16 A. Fournier, D. Fussell, L. Carpenter: Comm. ACM **25**, 371 (1982); reprinted in [9.18], p. 114

9.17 R.F. Voss: in [9.2], p. 21

9.18 K.I. Joy, et al., eds.: *Computer Graphics: Image Synthesis* (Computer Society Press, Washington DC 1988)

9.19 S. Wolfram: *Theory and Applications of Cellular Automata* (World Scientific, Singapore 1986)

9.20 T. Toffoli, N. Margolus: *Cellular Automata Machines* (MIT Press, Cambridge MA 1987)

9.21 A. Bunde, S. Havlin, eds.: *Fractals and Disordered Systems* (Springer-Verlag, Berlin 1991)

9.22 N.H. Packard, S. Wolfram: J. Stat. Phys. **38**, 901 (1985)

9.23 E.R. Berlekamp, J. H. Conway, A.K. Guy: *Winning Ways*, vol. 2 (Academic Press, New York 1982), p. 817

9.24 M. Gardner: *Wheels, Life and Other Mathematical Amusements* (Freeman, New York 1983)

9.25 R. Chandler, J. Koplik, K. Lerman, J.F. Willemsen: J. Fluid Mech. **119**, 249 (1982)

9.26 D. Wilkinson, J.F. Willemsen: J. Phys. A **16**, 3365 (1983)

9.27 K. Binder, D.W. Heermann: *Monte Carlo Simulation in Statistical Physics* (Springer-Verlag, Berlin 1988)

9.28 D. Stauffer, A. Aharony: *Introduction to Percolation Theory*, 2nd ed. (Taylor and Francis, London 1992)

9.29 D.C. Rapaport: J. Stat. Phys. **66**, 679 (1992)

9.30 J. Hoshen, R. Kopelman: Phys. Rev. B **14**, 3438 (1976)

9.31 D.C. Rapaport: Comput. Phys. Repts. **5**, 265 (1987)

9.32 G. Ciccotti, W.G. Hoover, eds.: *Molecular Dynamics Simulation of Statistical Mechanical Systems* (North-Holland, Amsterdam 1986)

9.33 C.R. Catlow, S.C. Parker, M.P. Allen, eds.: *Computer Modeling of Fluids Polymers and Solids* (Kluwer, Dordrecht 1990)

9.34 B.J. Alder, T.E. Wainwright: Phys. Rev. **127**, 359 (1962)

9.35 D.C. Rapaport: J. Stat. Phys. **40**, 751 (1986)

9.36 R.M. May: Nature **261**, 459 (1976)

9.37 M.J. Feigenbaum: Los Alamos Science **1**, 4 (1980); reprinted in [9.5], p. 49

9.38 W.H. Press, et. al.: *Numerical Recipes* (Cambridge University Press, Cambridge 1986)

9.39 M. Doi, S.F. Edwards: *The Theory of Polymer Dynamics* (Clarendon Press, Oxford 1986)

9.40 K. Kremer, G.S. Grest: J. Chem. Phys. **92**, 5057 (1990)

9.41 P. Bak, C. Tang, K. Wiesenfield: Phys. Rev. A **38**, 364 (1988)

9.42 L.P. Kadanoff, S.R. Nagel, L. Wu, S. Zhou: Phys. Rev. A **39**, 6524 (1989)

9.43 M.F. Barnsley, in [9.2], p. 219

Index

A. Bunde, S. Havlin (Eds.)

Fractals and Disordered Systems

1991. XIV, 350 pp. 163 figs. 10 tabs. ISBN 3-540-54070-9

Fractals and disordered systems have recently become the focus of intense interest in research. This book discusses in great detail the effects of disorder on mesoscopic scales (fractures, aggregates, colloids, surfaces and interfaces, glasses, and polymers) and presents tools to describe them in mathematical language. A substantial part is devoted to the development of scaling theories based on fractal concepts. In 10 chapters written by leading experts in the field, including E. Stanley and B. Mandelbrot, the reader is introduced to basic concepts and techniques in disordered systems and is led to the forefront of current research. In each chapter the connection between theory and experiment is emphasized, and a special chapter entitled "Fractals and Experiments" presents experimental studies of fractal systems in the laboratory.

The book is written pedagogically. It can be used as a textbook for graduate students, by university teachers to prepare courses and seminars, and by active scientists who want to become familiar with a fascinating new field.

Tm.B94.3.1

Springer-Verlag
and the Environment

We at Springer-Verlag firmly believe that an international science publisher has a special obligation to the environment, and our corporate policies consistently reflect this conviction.

We also expect our business partners – paper mills, printers, packaging manufacturers, etc. – to commit themselves to using environmentally friendly materials and production processes.

The paper in this book is made from low- or no-chlorine pulp and is acid free, in conformance with international standards for paper permanency.